Complete Manual of

INDUSTRIAL SAFETY

S. Z. Mansdorf

PRENTICE HALL
Englewood Cliffs, New Jersey 07632

Prentice-Hall International (UK) Limited, London
Prentice-Hall of Australia Pty. Limited, Sydney
Prentice-Hall Canada, Inc., Toronto
Prentice-Hall Hispanoamericana, S.A., Mexico
Prentice-Hall of India Private Limited, New Delhi
Prentice-Hall of Japan, Inc., Tokyo
Simon & Schuster Asia Pte. Ltd., Singapore
Editora Prentice-Hall do Brasil, Ltda., Rio de Janerio

10 9 8 7 6 5 4 3 2 1

Library of Congress Cataloging-in Publication Data

Mansdorf, S.Z. (Seymour Zach)
    Complete Manual of Industrial Safety / S.Z. Mansdorf
       p. cm.
    Includes index.
    ISBN 0-13-159633-0
    1. Industrial safety—Handbooks, manuals, etc. 2. Industrial
safety—United States—Handbooks, manuals, etc. I. Title
T55.M352 1993
670' .28'9—dc20   93–4986
                                          CIP

ISBN 0-13-159633-0

PRENTICE HALL
Career & Personal Development
Englewood Cliffs, NJ 07632

Simon & Schuster, A Paramount Communications Company

Printed in the United States of America

HOW THIS BOOK WILL HELP YOU

The *Complete Manual of Industrial Safety* will help you develop or improve upon your programs for safety and health. This will result in better employee relations and morale, improved production and efficiency, reduced workers compensation costs, better product quality, and an improved bottom line.

This book has been written by seasoned professionals for ease of use by business managers. It emphasizes the practical application of proven techniques that can be applied equally well by the manager with part-time responsibilities, full-time safety professionals, or a harried plant manager. It will also serve as a valuable reference for other managers or supervisors who want or need to know more about safety and health including those in finance, production, engineering, maintenance, human relations, and others.

The *Manual* has been written with an emphasis on making sure that it is easy to understand. It is full of easy to read tables, useful lists, practical advice, and ready to use forms. It also contains sources of more information including free literature of value. It is intended to take the mystery out of what OSHA requires and provide usable advice on how to establish and manage an effective safety program.

You will find this book to be a timely, informative, and an immediately useful guide to health and safety in manufacturing and related industries. In The *Manual,* you will:

❖ Learn how and why OSHA conducts inspections, issues citations and penalties, and how to handle an OSHA inspection or appeal.

❖ Learn how to develop a safety program including how to write a safety and health policy, prepare a safety and health manual, staff the program and promote its effectiveness to both management and labor.

❖ Learn what OSHA requires for employee training, what other training is important, and how to conduct and evaluate the effectiveness of the training.

❖ Find out what records of occupational injuries or illness are required by OSHA, learn how to complete the OSHA forms, and learn how to use these records as a management tool to improve your performance and reduce costs to improve profits.

❖ Find out what reports, records of inspections, and other recordkeeping is required by the OSHA regulations.

❖ Learn how to plan for the safe layout and design of new plant construction or renovation and the code restrictions that apply.

❖ Find practical ways to reduce acccidents and their costs.

❖ Learn why it is important to conduct accident investigations, find a recommended step by step procedure and forms for conducting investigations, and how to use the report findings to prevent further accidents.

❖ Learn how to analyze jobs for safety hazards, find the forms that can be used on a step by step basis for the analysis, and learn how the results of this work will help you.

❖ Learn about injuries that can result from poorly designed work stations or work practices and how to eliminate them and improve work efficiency and production.

❖ Learn how to develop a program to find and fix problems before they cause equipment damage, accidents, and injuries. Find extensive checklists for conducting inspections that can be used by workers, supervisors, and managers.

❖ Learn what protective equipment should be used, how to select it, and how to maintain it. Learn what OSHA requires you do when the protection is used, how to train your employees, and find example programs that can be used in your plant.

❖ Find out what planning you should do to handle emergencies from fire, weather, chemicals, or other dangerous events. Learn how to develop a plan and what to do after the emergency is over.

❖ Learn what OSHA and others require in the design and use of working areas including floors, stairs, runways, aisles, exits, and ladders.

❖ Find out how to identify and fix materials handling problems including reducing back injuries from improper lifting.

❖ Learn how to effectively guard machinery and comply with OSHA's rules on lockout/tagout.

❖ Learn how to avoid injuries from hand and portable power tools and find checklists that can be used for inspecting for damaged or dangerous equipment.

❖ Learn about what precautions are needed when handling hazardous materials and learn about the principles of hazard control.

❖ Learn about the safe storage and use of flammable materials, how to protect against a fire, and what OSHA requires for worker training.

❖ Learn what type of powered industrial truck is best for your application, how to establish an inspection and training program for your drivers, and about effective maintenance.

❖ Find out how the National Electrical Code impacts manufacturing and what lockout/tagout procedures apply for maintenance activities. Learn about electrical hazards and training required for employees who work with electrically powered equipment.

❖ Learn about the hazards of welding and cutting and how to control them as well as what training your welders should receive.

❖ Learn about the hazards of pressure vessels such as boilers, reactors, pressure tanks, and others. Find out about code requirements from ASME and others. Learn about design considerations, inspections, maintenance, and recordkeeping requirements.

❖ Find out what OSHA requires when workers are exposed to high levels of noise. Learn how to develop a hearing conservation program and worker training, what protective equipment is needed, and what audiometrical testing will tell you.

❖ Learn about the selection and use of hoisting devices such as chains, hoists, and slings. Find out what inspections should be performed and what OSHA requires.

❖ Find out when the rules for construction safety apply to manufacturing and what needs to be done to comply with them.

❖ Find out where to go to get expert help and assistance including free services and information.

INTRODUCTION

OVERVIEW AND PURPOSE OF THE MANUAL

The National Safety Council estimated that workplace accidents cost this country 63.3 billion dollars in 1991. More importantly, it estimated that almost 10,000 workers were killed on the job and that more than six million were injured or made ill. For business, this meant over sixty million lost workdays. That is equivalent to the loss of a workforce of over 30,000 workers for a year. These statistics dramatize the importance of safety.

In today's business environment, safety is recognized as an obligation of the business owner or manager from a moral, ethical, legal, and financial standpoint. The *Complete Manual of Industrial Safety* has been written to assist the manager in meeting these obligations. It has been written for those without any prior specialized training or credentials in safety or industrial hygiene. It is aimed at manufacturing businesses of a small to medium size (1,000 employees or less) although others may find it useful as well.

All managers, if asked, would state that they are in favor of plant safety. However, the prevention of accidents and illnesses is a demanding responsibility. On the shop floor, it is often perceived to compete with other job responsibilities, to compromise efficiency and to act as a drain on business resources. The intent of the guide is therefore twofold. First, the intent is to establish the value of a comprehensive safety program beyond the ethical and moral obligations associated with management's role, i.e., to demonstrate the financial advantages of establishing safety as a regular business system. Second, it is intended to help the already overtaxed manager establish, promote, and manage a safety and health program in as simple a manner as possible using easily understood and pragmatic guidance.

The *Complete Manual of Industrial Safety* has been organized into three parts containing 26 Chapters and an Appendix. Part I (Chapters 1 through 5) deals with the establishment and management of a plant safety program. Part II (Chapters 6 through 13) describes programs to reduce the potential for accidents. Part III covers specific manufacturing hazards. Finally, the Appendix provides additional sources of safety-related help.

SAFETY FROM THE HISTORICAL PERSPECTIVE

At the beginning of this century, accidents were thought to be unavoidable and a matter of "bad luck." Most managers believed accident prevention was not possible as accidents were thought to be an inherent by-product of production. Everyone knew that accidents occurred nearly everyday in the workplace. Or as might have been said in that era, accidents were simply and inevitably a cost of doing business.

How serious was the problem of industrial accidents or was it a problem at all? Historically, there were no requirements to maintain accident statistics to substantiate the scope of the problem. In 1906 the Russell Sage Foundation commissioned a study to determine the seriousness of what was perceived to be a problem. Its investigation covered the period from July 1906 through June 1907. Steel mills, coal mines, railroad yards, and other industrialized businesses were visited. Workers, members of management, and people from the general area were interviewed. Records from local hospitals, doctor's offices, and the county coroner were reviewed for information about industrial injuries and fatalities. The study was completed and published as "The Pittsburgh Survey." Although the study was confined to one specific industrial area of the country the results were so dramatic they shocked the entire nation.

A chart in the study became known as "The Death Calendar of Allegheny County." The chart revealed that 528 fatalities occurred during the 12-month period of the study, or an average of 10 fatalities a week for just one county of one state of the country. The study revealed that in addition to the 528 fatalities there were amputations of 45 legs, 30 arms, 20 hands and 60 partial hands. There were 70 eyes lost and 45 arms injured so severely they were unusable.

Even Theodore Roosevelt in his Presidential Message of 1908 said: "The number of accidents which result in the death or crippling of wage earners... is simply appalling. In a very few years it runs up a total far in excess of the aggregate of the dead in any major war."

The Pittsburgh Survey proved the seriousness of the problem and is often believed to have fostered the safety and health movement in the United States. Shortly after the Survey was completed, the first effective workers' compensation law was passed in 1910, the National Safety Council was started in 1913, and The American Society of Safety Engineers was organized in 1915. Since the Pittsburgh Survey, the safety and health movement has continued to evolve into a sophisticated discipline that stresses a systems approach for accident prevention, hazard evaluation, and risk analysis. The philosophy has changed from one that considered accidents as a matter of luck and a cost of doing business to a philosophy that all accidents are preventable. Secondly, that accident causes can be generally traced to unsafe acts or unsafe conditions, or to a combination of the two.

Our legislative process has supported the safety and health movement and continues to support that movement. In 1970 the most comprehensive and far-reaching safety and health law up to that time was passed: Public Law 91-596, The Occupational Safety and Health Act, more commonly referred to as the OSH Act. The OSH Act contains a clause, known as the General Duty Clause, which requires the employer to provide a workplace free of recognized hazards. In addition, the OSH Act also contains specific safety and health standards with which the employer must comply.

Recently, OSHA has been under increasing pressure to become more stringent in its enforcement of existing rules and to speed the development of new rules and regulations. One consequence of this public pressure has been a seven-fold increase in the monetary value of fines and encouragement to press for criminal sanctions where appropriate. This pressure has also resulted in newspaper headlines touting monetary fines against companies in excess of several million dollars under their egregious (flagrant) penalty policy. Further, it is very likely that OSHA will undergo legislative reform within the next few years resulting in an enforcement agency with even more clout and a greater mandate to pursue companies which have excessive worker injuries or illnesses. While the adverse effects of OSHA citations will encourage many to meet the minimum requirements, there are more reasons to develop, establish, promote, and maintain safety than simply compliance.

A BUSINESS APPROACH TO SAFETY

Aside from the penalties for noncompliance, it should be evident that providing a safe and healthful workplace makes good business sense. Paying a worker's salary while off work does not make good business sense nor does paying overtime or a new employee's salary to cover for the injured person. Other costs incurred include administrative costs to process the injury claim, direct medical costs, and the costs for lost production that may result when an accident requires modification or repairs to the equipment involved in the incident. There is also the negative impact on public relations with the community and customers. Accidents and illnesses and injuries are expensive.

What does an accident cost? Statistics compiled by safety and health professionals generally show that every dollar of direct costs, such as medical bills, results in four to ten dollars of indirect and sometimes hidden costs.

For the majority of companies, their most valuable asset is their employees. With this perspective, it is just common sense to protect them. In the era of World War II, the national slogan was to "conserve the fighting strength" by keeping the industrial worker healthy so that we could build more tanks, planes, and other weapons and supplies than our enemies. This concept of keeping the worker healthy to win the battle (now the profit battle) has the same value today as it did during World War II. The responsibility for improving the safety and health of the workforce belongs to every member of the organization from the CEO to the first line supervisor. The failure of anyone in the organization to make safety a daily part of their lives can result in accidents. Accident prevention and safety must be an integral part of our work, it's a moral obligation, it's the law, and it's good business!

S.Z. Mansdorf

ACKNOWLEDGEMENTS

This was a group effort of many safety and health professionals at Mansdorf & Associates, Inc. I served as the principal author and editor. The Introduction was written by me as was Chapter 1, "Complying with Safety Legislation and Regulations." Dr. J. G. Collins, CIH, also contributed to that chapter. Chapter 2, "Establishing a Plant Safety Program" was written by me with Jim Sesic, CSP and Jay Brown, CIH, as contributors. Chapter 3, "Managing Plant Safety" was written by me with the assistance of Jim Stone, IHIT. Ms. Jenni Ticer, CIH, was my coauthor on Chapter 4, "Safety and Health Training." Chapter 5, "Recordkeeping" was written by Jim Sesic, CSP and me. Chapter 6, "Plant Design and Layout for Safety" and Chapter 7, "Reducing Accidents and Illness" were written by me with Dr. J. G. Collins, CIH contributing. I was the author for Chapter 8, "Accident Investigations" while Bill Youngblut, CIH was the author of Chapter 9, "Job Analysis." Chapter 10, "Ergonomics" was written by me and Jim Sesic, CSP. Chapter 11, "Plant Inspections" was written by Tom Knupp, CIH. Chapter 12, "Personal Protective Equipment" was written by Jay Brown, CIH and myself. Chapter 13, "Emergency and Disaster Planning" was done by me. Dr. Tony Lott, CIH served as the author of Chapter 14, "Walking and Working Surfaces" while I wrote Chapter 15, "Materials Handling and Storage." Chapter 16, "Machine Guarding" was written by Jenni Ticer, CIH. Tom Knupp, CIH was the author of Chapter 17, "Hand and Portable Power Tools" with contributions by me. Chapter 18, "Hazardous Materials" was written by Jim Stone, IHIT with contributions by Dr. J. G. Collins, CIH and myself. Mr. Bill Nichols was my coauthor on Chapter 19, "Fire Safety." Chapter 20, "Powered Industrial Trucks" was written by Jim Sesic, CSP with me as a contributor. Chapter 21, "Electrical Safety" was written by Dr. Tony Lott, CIH. Jim Stone, IHIT and I were coauthors of Chapter 22, "Welding and Cutting." Dr. Tony Lott, CIH wrote Chapter 23, "Pressure Vessels". Jenni Ticer, CIH and I wrote Chapter 24, "Noise and Vibration." The coauthors of Chapter 25, "Ropes, Chains, and Slings" were me and Bill Youngblut, CIH. Chapter 26, "Construction Safety" was written by Jim Sesic, CSP and myself. Also contributing to the overall effort through excellent editing were Bob Modrell and Paul Foster. Finally, Ms. Debby Catanese provided the thousands of pages of drafts and redrafts—always with a smile.

S. Z. MANSDORF
STOW, OHIO

ABOUT THE CONTRIBUTORS

S. Z. Mansdorf, Ph.D., CIH, CSP

Dr. Mansdorf has an international reputation and over twenty years of broad experience in occupational safety, health and environmental control. He is nationally certified in both the comprehensive practice of industrial hygiene and safety. He earned Masters degrees in environmental health sciences from the University of Michigan and in industrial safety from Central Missouri State University. His Doctoral degree is in environmental engineering from the School of Engineering at the University of Kansas.

He has managed and conducted hundreds of projects and studies, both nationally and internationally, for industrial, academic, and governmental clients.

Dr. Mansdorf has published or presented over forty papers and is the author of three books and a number of manuals. He has been a Senior Editor for the American Industrial Hygiene Association Journal and an active member of the AIHA. He is also a contributing editor for Occupational Hazards magazine. He is a full or professional member of the American Society of Safety Engineers, the Air and Waste Management Association, the National Environmental Health Association, the American Academy of Industrial Hygiene, the American Academy of Sanitarians, the American Society for Testing and Materials and others.

Dr. Mansdorf has been involved in a large number of teaching activities for industry, private interest groups, and academia. He is an Associate Professor of Environmental Engineering (adjunct) at Cleveland State University, where he teaches environmental engineering courses in their Graduate School. In addition, he is an Environmental Science Officer in the U.S. Army Reserves (Medical Service Corps).

Antone L. Lott, Ph.D., CIH

Dr. Lott has over sixteen years of experience in occupational and environmental health in the petroleum and petrochemical industry, the health care field, and in consulting. He also has experience in the electrical power generating industry. He is nationally certified in the comprehensive practice of industrial hygiene by the American Board of Industrial Hygiene. He earned Masters degrees in organic chemistry from the University of Michigan, and in public health from the University of Oklahoma. His Doctoral degree is in inorganic chemistry from the University of Michigan.

Dr. Lott has managed and conducted a large number of health and safety projects at various locations including Europe, Africa, and the Far East.

Dr. Lott is a special consultant for health and safety to the American College of Pathologists and has published or presented over thirty technical papers. He is a full member of the American

Industrial Hygiene Association and the American Academy of Industrial Hygiene. Dr. Lott has broad teaching experience in academia and industry in the areas of chemistry, occupational and environmental health and safety, asbestos, and hazardous waste operations.

J. G. Collins, Ph.D., CIH

Dr. Collins has over fourteen years of experience in occupational and environmental health. He is nationally certified in the comprehensive practice of industrial hygiene by the American Board of Industrial Hygiene. He earned an undergraduate degree in chemistry from the University of Manitoba, and a Doctoral degree in physical chemistry from the University of Alberta. Dr. Collins also conducted post doctoral research at Rice University. He has eighteen years of experience in the chemical, petrochemical, plastics, rubber, and textile industries and over three years of experience in consulting.

Dr. Collins has conducted or managed a variety of sampling surveys in the industrial sector. He has been auditor or project manager on dozens of occupational and environmental health and safety audits and studies at various industrial, manufacturing, and academic institutions.

Dr. Collins has published or presented over seven technical papers. He is a full member of the American Industrial Hygiene Association, has served on several technical committees, and was a reviewer for the American Industrial Hygiene Association Journal. He is a diplomate of the American Academy of Industrial Hygiene and a member of the American Chemical Society. He is a professional member of the American Society of Safety Engineers.

James R. Sesic, CSP

Mr. Sesic has over thirteen years of broad experience in safety and environmental engineering. He is nationally certified in the comprehensive practice of safety by the Board of Certified Safety Professionals. His undergraduate degree is in business administration with post baccalaureate studies in chemistry from the University of Akron. Mr. Sesic has had experience in rubber and chemical manufacturing operations, in research and development and pilot plant operations, and in consulting.

Mr. Sesic is a professional member of the American Society of Safety Engineers and other technical organizations such as the American Industrial Hygiene Association.

Thomas W. Knupp, CIH

Mr. Knupp has over sixteen years of experience in occupational and environmental health. He is nationally certified in the comprehensive practice of industrial hygiene by the American Board of Industrial Hygiene. He earned an undergraduate degree in biology from Ohio State University, and has completed over fifty graduate hours in physiology at Kent State University. Mr. Knupp has had experience in the rubber and chemical industries, and in consulting.

He has conducted hundreds of industrial hygiene and safety surveys—both nationally and abroad. He has been industrial hygienist or project manager on many occupational health studies and audits for a variety of clients in the industrial and governmental areas.

Mr. Knupp is a full member of the American Industrial Hygiene Association. He is also a diplomate of the American Academy of Industrial Hygiene. He is an active member of the Akron Council of Engineering and Scientific Societies.

William R. Youngblut, CIH

Mr. Youngblut has an undergraduate degree in safety engineering and management from Indiana University of Pennsylvania and a Master of Science in industrial hygiene from Central Missouri State University. Furthermore, he has over fourteen years experience in environmental compliance and health, industrial hygiene and safety engineering in the alloy, ceramics, foundry and gas transmission industries. He is nationally certified in the comprehensive practice of industrial hygiene by the American Board of Industrial Hygiene.

He has been involved with a variety of industrial hygiene sampling surveys. He has conducted several occupational and environmental health and safety audits and studies both nationally and internationally.

Mr. Youngblut is a full member of the National and local chapters of the American Industrial Hygiene Association , a professional member of the American Society of Safety Engineers, and a full member of the Air & Waste Management Association.

Jay L. Brown, CIH

Mr. Brown has an undergraduate degree in industrial hygiene from Ohio University. He has over three years experience in environmental health, industrial hygiene, and consulting. He is certified in the comprehensive practice of industrial hygiene by the American Board of Industrial Hygiene.

Mr. Brown is a member of the American Society of Safety Engineers, and is also a member of the American Industrial Hygiene Association.

Jennifer M. Ticer, CIH, CHMM

Ms. Ticer has an undergraduate degree in industrial hygiene from Ohio University. She has over three years experience in environmental health, industrial hygiene and consulting.

Ms. Ticer is certified in the comprehensive practice of industrial hygiene by the American Board of Industrial Hygiene. She has also been certified by the Institute of Hazardous Materials Management as a Certified Hazardous Materials Manager. She is a member of the American Industrial Hygiene Association.

James A. Stone, Jr., M.S.P.H., IHIT

Mr. Stone has an undergraduate degree in industrial hygiene and environmental toxicology from Clarkson University and a Masters degree in Environmental Health Sciences from Tulane University. He has experience in the petrochemical industry and in industrial hygiene consulting.

Mr. Stone has successfully completed the first part of the certification process for industrial hygiene and is an IHIT.

William W. Nichols

Mr. Nichols has an undergraduate degree in Environmental Health Sciences from Ohio University. He has experience with an international petroleum company where projects have included air sampling for a variety of gas, vapor, and particulate materials, compilation of an ergonomics reference manual for refinery managers, development of a basic industrial hygiene training manual for supervisors, and the preparation of guidelines for handling lead based paint.

Mr. Nichols is a member of the American Industrial Hygiene Association.

TABLE OF CONTENTS

Part Three
Control of Hazards in Manufacturing Plants

PART ONE

MANAGING
THE PLANT
SAFETY PROGRAM

CHAPTER
1

COMPLYING WITH SAFETY LEGISLATION AND REGULATIONS

This chapter briefly introduces the roles and responsibilities of the three key federal agencies: The Occupational Safety and Health Administration (OSHA), the National Institute for Occupational Safety and Health (NIOSH), and the Environmental Protection Agency (EPA). This is followed by a more detailed discussion on the Occupational Safety and Health Act, regulatory standards, OSHA inspections and priorities, how inspections are conducted, information on citations and penalties, the need for developing a plan of action and standard procedures for a governmental inspection, and a summary of the organization of the OSHA General Industry Standards by subpart in the Code of Federal Regulations (CFR).

The key federal agency responsible for regulation of safety and health in the workplace is the Occupational Safety and Health Administration (OSHA). Its purpose is to assure, so far as possible, every working man and woman in the nation safe and healthful working conditions. The agency was established by the Occupational Safety and Health Act of 1970 as part of the Department of Labor and is responsible for establishing and enforcing workplace safety and health standards. As an enforcement agency, OSHA has the authority to levy and collect fines of up to $70,000 per violation.

The National Institute for Occupational Safety and Health (NIOSH) was also established by the Occupational Safety and Health Act of 1970 and is currently part of the Department of Health and Human Services. NIOSH is responsible for research on the safety and health effects of exposures in the workplace and for providing training and information to workers and employers. NIOSH has no enforcement powers (cannot assess penalties or levy fines); however, it does have a right of entry to businesses as part of its research role.

The Environmental Protection Agency (EPA), also founded in 1970, is the largest noncabinet agency within the executive branch and has been recently proposed for cabinet status. The EPA is responsible for the protection and enhancement of the environment. Its role

1

is to abate and control pollution in air, water, and land. There are some areas in which the regulations overlap between OSHA and EPA, e.g., the regulation of hazardous wastes, emergency response to chemical spills, asbestos abatement and control, hazard communication, and several other areas. As a result of this overlap, there is a cooperative agreement between EPA and OSHA for joint activities (including inspections and referrals).

THE OCCUPATIONAL SAFETY AND HEALTH ACT

As stated earlier, the purpose of the Occupational Safety and Health Act of 1970 is "...to assure so far as possible every working man and woman in the Nation safe and healthful working conditions and to preserve our human resources." The Act provides several means of attaining its purpose, including:

❖ authorizing enforcement of safety and health standards developed under the Act

❖ assisting and encouraging the states in their efforts to assure safe and healthful working conditions

❖ providing for research, information, education and training in the field of occupational safety and health

Who Is Covered

The Act applies to all employment performed in any commercial business in any of the 50 states, the District of Columbia, Puerto Rico, the Virgin Islands, American Samoa, Guam, the Outer Continental Shelf Lands, Johnston Island, and the Canal Zone. Presently, the Act excludes other federal, state, and local governmental agencies. However, there is an Executive Order which requires federal agencies to conform to OSHA standards where feasible. Also, some states follow the OSHA regulations for their public employees. The Act covers more than 60 million employees in approximately 5 million workplaces. The Act defines an employer as "a person engaged in a business affecting commerce who has employees." This essentially includes everyone who is in business.

General Duty Clause

Section 5 of the Act is one of its shortest yet most important sections. It is commonly known as the "General Duty Clause" and has been used as a catch-all regulation. It is intended to cover those situations where there are no specific written regulations or prohibitions yet the situation presents a recognized hazard. The employer's duty under Section 5 of the Act is stated as follows:

1. shall furnish to each of his employees employment and a place of employment which are free from recognized hazards that are causing or are likely to cause death or serious physical harm to his employees;

2. shall comply with occupational safety and health standards promulgated under this Act.

A recognized hazard is a condition that is:

of common knowledge or general recognition in the particular industry in which it occurs and is detectable by means of the senses, or of such wide, general recognition as a hazard in the industry that there are generally known and accepted tests for its existence which should make its presence known to the employer.

Employees also have some duties under this section:

Each employee shall comply with occupational safety and health standards and all rules, regulations, and orders issued pursuant to this Act which are applicable to his own actions and conduct.

Nevertheless, the courts have held that the *employer* is responsible for ensuring that the employees comply with the OSHA standards. It is a rare circumstance when the employer can prove that the employee could not be made to comply with the rules (short of dismissal).

The Authority to Issue Standards

The Act gives the Secretary of Labor the authority to issue, revise, modify and revoke health and safety standards. An occupational safety and health standard is a regulation that designates working conditions or practices necessary to prevent injury or illness. These may be either consensus standards (e.g., national electrical code) or OSHA developed standards (e.g., hazard communication). These standards are published for public comment and require advanced notice of rule-making, public hearings, and other measures. Therefore, they are rarely a surprise. The regulations and standards are all available through local public libraries and through subscription services.

Enforcement

The Secretary of Labor and authorized representatives may enter any work area without delay and at reasonable times. They are authorized to inspect, investigate accidents, review records of occupational injuries and illnesses, and privately question any employer, owner, operator, agent, or employee. When an inspection reveals an apparent violation or hazard, OSHA's Area Directors may issue a citation, propose a penalty for the alleged violation, and set a date by which the employer must abate the hazard.

The Act also provides a means for employers and employees to appeal cases to the OSHA Review Commission and the U.S. Circuit Court of Appeals. It makes clear that an employee who files a complaint about conditions in his workplace or who exercises any of their other rights under the Act may not be discriminated against. The Act also provides for enforcement procedures in dealing with safety and health conditions so serious that they require immediate action.

State Programs

The Act encourages states to develop plans for operating and enforcing their own programs. The state programs must be "at least as effective as" the federal program. Standards issued under state programs must also remain "as effective as" improved or expanded federal standards. They need not cover all the areas federal laws cover, however, and need not go beyond the limits of state law. Federal laws and enforcement cover areas or issues that state programs do not. For example, Connecticut's state program only covers public employees; therefore, federal OSHA covers all others. A listing of OSHA approved state plans is shown in Table 1-1.

Table 1-1
States with Approved Plans

Alaska	Minnesota
Arizona	New York
Connecticut	South Carolina
Hawaii	Tennessee
Indiana	Utah
Iowa	Virginia
Kentucky	Virgin Islands
Maryland	Wyoming

Small Business Administration Assistance

The Act also provides for employer assistance by authorizing the Small Business Administration (SBA) to make or guarantee a loan to a small business if the changes the business must make to comply with the Act threaten it with substantial economic injury.

OSHA INSPECTIONS

Authority to Inspect

To enforce its standards, OSHA is authorized under the Act to conduct workplace inspections. Every establishment covered by the Act is subject to inspection by OSHA compliance officers. These are usually technical persons (safety and/or industrial hygiene professionals) who have been trained in the requirements of the OSHA standards and in the recognition of safety and health hazards. Similarly, states having their own occupational safety and health programs conduct inspections using their own state compliance officers.

Under the Act, "upon presenting appropriate credentials to the owner, operator or agent in charge," an OSHA compliance officer is authorized to:

Enter without delay and at reasonable times any factory, plant, establishment, construction site or other areas, workplace, or environment where work is performed by an employee of an employer; and to

Inspect and investigate during regular working hours, and at other reasonable times, and within reasonable limits and in a reasonable manner, any such place of employment and all pertinent conditions, structures, machines, apparatus, devices, equipment and materials therein, and to question privately any such employer, owner, operator, agent or employee.

With very few exceptions, inspections are conducted without advance notice. In fact, alerting an employer in advance of an OSHA inspection can bring a fine and/or a criminal penalty.

There are, however, special circumstances under which OSHA may indeed give notice to the employer, but even then, such notice will usually be less than 24 hours. These special circumstances include:

Imminent danger situations which require correction as soon as possible;

Inspections which must take place after regular business hours, or which require special preparation;

Cases where notice is required to assure that the employer and employee representative or other personnel will be present; and/or

Situations in which the OSHA Area Director determines that advance notice would produce a more thorough or effective inspection.

Employers receiving advance notice of an inspection must inform their employees' representative or arrange for OSHA to do so. If an employer refuses to admit an OSHA compliance officer, or if an employer attempts to interfere with the inspection, the Act permits appropriate legal action.

Based on a 1978 Supreme Court ruling (*Marshall v. Barlow's Inc.*), OSHA may not conduct warrantless inspections without an employer's consent. OSHA may, however, inspect after acquiring a judicially authorized search warrant based upon administrative probable cause or upon evidence of a violation.

Inspection Priorities

OSHA does not have the staff or resources to inspect all 5 million workplaces covered by the Act. Therefore, OSHA has established a system of inspection priorities based on a "worst-first" approach.

Imminent Danger. Imminent danger situations are given top priority. An imminent danger is any condition where there is reasonable certainty that a danger exists that can be expected to cause death or serious physical harm imminently (or before the danger can be eliminated through normal enforcement procedures).

Serious physical harm is any type of harm that could cause permanent or prolonged damage to the body or which, while not damaging the body on a prolonged basis, could cause such temporary disability such as in-patient hospital treatment. OSHA considers that "permanent or prolonged" damage has occurred when for example, a part of the body is crushed or severed, a leg or finger is amputated, or sight in one or both eyes is lost. This kind of damage also includes that which renders a part of the body either functionally useless or substantially reduced in efficiency on or off the job, for example, bones in a limb shattered so severely that mobility or dexterity will be permanently reduced.

Health hazards may constitute imminent danger situations when they present a serious and immediate threat to life or health. This is sometimes referred to as "immediately dangerous to life and health" or an IDLH condition. For a health hazard to be considered an imminent danger, there must be a reasonable expectation that (1) toxic substances such as dangerous fumes, dusts or gases are present and (2) exposure to them will cause immediate and irreversible harm to such a degree as to shorten life or cause reduction in physical or mental efficiency, even though the resulting harm is not immediately apparent.

The OSHA Area Director is required to review the information and to determine immediately whether there is a reasonable basis for the allegation. If it is decided the case has merit, the Area Director is required to alert the OSHA Regional Administrator and the Regional Solicitor, and to assign a compliance officer to conduct an immediate inspection of the workplace.

Upon inspection, if an imminent danger situation is found, the compliance officer will ask the employer to voluntarily abate the hazard and to remove endangered employees from exposure. Should the employer fail to do this, OSHA, through the Regional Solicitor, may apply to the nearest Federal

District Court for appropriate legal action to correct the situation. Before the OSHA inspector leaves the workplace, he or she will advise all affected employees of the hazard. Should OSHA "arbitrarily or capriciously" decline to bring court action, the affected employees may sue the Secretary of Labor to compel the Secretary to do so. Such action can produce a temporary restraining order (immediate shutdown) against the operation or section of the workplace where the imminent danger exists.

Catastrophes and Fatal Accidents Second priority is given to investigation of fatalities and catastrophes resulting in hospitalization of five or more employees. Such situations must be reported to OSHA by the employer within forty-eight hours. Investigations are made to determine if any OSHA standards were violated.

Employee Complaints Third priority is given to employee complaints of alleged violation of standards or of unsafe or unhealthful working conditions. The Act gives each employee the right to request an OSHA inspection when the employee feels he or she is in imminent danger from a hazard or when he or she feels that there is a violation of an OSHA standard which threatens physical harm. OSHA will maintain confidentiality if requested, will inform the employee of any action it takes regarding the complaint and, if requested, will hold an informal review of any decision not to inspect. Just as in situations of imminent danger, the employee's name will be withheld from the employer, if the employee so requests.

Programmed High-Hazard Inspections Next in priority are programs of inspection aimed at specific high-hazard industries, occupations and chemicals. Industries are selected for inspection on the basis of such factors as the, injury and illness incidence rates and employee exposures to toxic substances. Special emphasis may be regional or national in scope, depending on the distribution of the workplaces involved. Comprehensive safety inspections in manufacturing will be conducted only in those establishments with lost work-day injury rates at or above the most recently published Bureau of Labor Standard (BLS) national rate for manufacturing. States with their own occupational safety and health programs may use somewhat different systems to identify high-hazard industries for inspection.

Follow-Up Inspections

A follow-up inspection is normally conducted to determine if the previously cited violations have been corrected. If an employer has failed to abate a violation, the compliance officer informs the employer by a "Notification of Failure to Abate" with additional daily penalties while the alleged violation continues.

OSHA INSPECTION PROCESS

Prior to an inspection, the compliance officer usually tries to become familiar with as many relevant facts as possible about the workplace, taking into account such things as the history of the establishment, the nature of the business and the particular standards likely to apply.

Inspector's Credentials

An inspection begins when the OSHA compliance officer arrives at the establishment. He or she or she displays official credentials and asks to meet an appropriate employer repre-

sentative. Employers should always insist upon seeing the compliance officer's credentials. An OSHA compliance officer carries U.S. Department of Labor credentials bearing a recent photograph and a serial number that can be verified by phoning the nearest OSHA office. Anyone who tries to collect a penalty at the time of inspection, or promotes the sale of a product or service at any time, is not an OSHA compliance officer.

Opening Conference

In the opening conference the compliance officer (CSHO) explains why the establishment was selected and determines if it should be subject to a comprehensive safety inspection based on its lost workday injury case rate. (This rate is the number of lost work-day injuries per 100 full-time employees.) The CSHO also will ascertain whether an OSHA-funded consultation program is in progress or whether the facility is pursuing or has received an inspection exemption.

The CSHO then explains the purpose of the visit, the scope of the inspection, and the standards that apply. The employer will be given copies of applicable safety and health standards, as well as a copy of any employee complaint that may be involved. If the employee has so requested, his or her name will not be revealed.

The employer is asked to select an employer representative to accompany the CSHO during the inspection. An authorized employee representative also is given the opportunity to attend the opening conference and to accompany the CSHO during the inspection. If the employees are represented by a recognized bargaining representative, the union ordinarily will designate the employee representative to accompany the CSHO. Similarly, if there is a plant safety committee, the employee members of the committee will designate the employee representative (in the absence of a recognized bargaining representative). If neither employee group exists, the employee representative may be selected by the employees themselves, or the CSHO will determine if any employee suitably represents the interest of other employees. The employer cannot select the employee representative for the walk-around. The employer can, however, select a representative of management to accompany the others. The Act does not require that there be an employee representative for each inspection. However, if there is no authorized employee representative, the CSHO must consult with a reasonable number of employees concerning safety and health matters in the workplace.

Inspection Tour

After the opening conference, the CSHO and accompanying representatives proceed through the establishment, inspecting work areas for compliance with OSHA standards. The route and duration of the inspection are determined by the CSHO. While talking with employees, the CSHO is required to minimize any work interruptions. The CSHO observes conditions, consults with employees, may take photos (for record purposes), takes instrument readings, and examines records.

Trade secrets observed by the CSHO are required to be kept confidential. A CSHO who releases confidential information without authorization is subject to a $1,000 fine and/or one year in jail. The employer can require that the employee representative have clearance for any areas that are considered sensitive or containing special trade secrets.

Employees are consulted during the inspection tour. The CSHO may stop and question workers, in private if necessary, about safety and health conditions and practices in their workplaces. Each employee is protected, under the Act, from discrimination for exercising their safety and health rights.

Posting and recordkeeping are checked. The CSHO will inspect records of deaths, injuries and illnesses which the employer is required to keep. He or she will check to see that a copy of the totals from the last page of the OSHA No. 200 form has been prominently displayed. If records of employee exposure to toxic substances and harmful physical agents are required to be maintained, they are also examined.

During the course of the inspection the CSHO will point out to the employer any unsafe or unhealthful working conditions observed. At the same time the CSHO will discuss possible corrective action if the employer so desires. Some apparent violations detected by the CSHO can be corrected immediately. When they are corrected on the spot, the CSHO records such corrections to help in judging the employer's good faith in compliance. Even though corrected, the apparent violations may still serve as the basis for a citation and notice of proposed penalty.

An inspection tour may cover part or all of an establishment, even if the inspection resulted from a specific complaint, fatality or catastrophe. The inspection can also be conducted by a "team" of CSHO's, depending on the size of the facility and scope of work.

Closing Conference

After the inspection tour, a closing conference is held between the CSHO and the employer or the employer representative. The CSHO discusses all unsafe or unhealthy conditions observed on the inspection and indicates all apparent violations for which citations may be issued or recommended. The employer is also informed of appeal rights by the CSHO. The CSHO will not indicate any proposed penalties. Only the OSHA Area Director has the authority to propose penalties after having received a full report from the CSHO.

During the closing conference, the employer may wish to produce records to show compliance efforts and to provide information which can help OSHA determine how much time may be needed to abate an alleged violation. When appropriate, more than one closing conference may be held. This is usually necessary when health hazards are being evaluated or when laboratory reports are required.

A closing discussion will be held with the employees, or their representative if requested, to discuss matters of direct interest to employees. The employees' representative may be present at the closing conference.

CITATIONS AND PENALTIES

Citations Issued by the Area Director

After evaluating the CSHO's findings, the Area Director determines what citations, if any, will be issued and what penalties, if any, will be proposed. Citations inform the employer and employees of the regulations and standards alleged to have been violated, and of the proposed length of time set for their abatement. The employer will receive citations and notices of proposed penalties by certified mail. The employer must post a copy of each citation at or near the place a violation occurred, for three days or until the violation is abated, whichever is longer.

Types of Violations

Listed below are the general classes of violations which may be cited:

❖ Other Than Serious Violation—A violation that has a direct relationship to job safety and health, but probably would not cause death or serious physical harm.

❖ Serious Violation—A violation where there is substantial probability that death or serious physical harm could result, and where the employer knew, or should have known, of the hazard.

❖ Willful Violation—A violation that the employer intentionally and knowingly commits. The employer either knows that what he or she or she is doing constitutes a violation, or is aware that a hazardous condition exists and has made no reasonable effort to eliminate it. This is the most severe category for an initial violation.

❖ Repeated Violation—A violation of any standard, regulation, rule or order where, upon reinspection, a substantially similar violation is found.

Penalties

OSHA applied a new penalty policy as of March 1, 1991 which significantly increased the maximum monetary penalties for certain violations (from $10,000 to $70,000). OSHA's position as stated in their Field Operations Manual is as follows:

1. ...penalties are not designed as punishment for violations nor as a source of income for the Agency. The Congress has made clear its intent, however, that penalty amounts should be sufficient to serve as an effective deterrent to violations.

2. Large proposed penalties, as Congress has clearly recognized, serve the public purpose intended under the Act...

The following section, also extracted from the OSHA Field Operation Manual, summarizes the new penalty structure.

Type of Violation as a Factor. In proposing civil penalties for violations, a distinction is made between serious violations and other violations. There is no statutory requirement that a penalty be proposed when the violation is not serious; but a penalty must be proposed when the violation is serious. The maximum penalty that may be proposed for a serious or an other-than-serious violation is $7,000. In the case of willful or repeated violations, a civil penalty of up to $70,000 may be proposed; but the penalty may not be less than $5,000 for a willful violation. For other specific violations of the Act, civil penalties of up to $7,000 may be proposed. Penalties for failure to correct a violation may be up to $7,000 for each calendar day that the violation continues beyond the final abatement date.

Statutory Authority. Section 17 provides the Secretary with the statutory authority to assess civil penalties for violations of the Act.

a. Section 17(b) of the Act provides that any employer who has received a citation for an alleged violation of the Act which is determined to be of a serious nature shall be assessed a civil penalty of up to $7,000 for each violation.

b. Section 17(c) provides that, when the violation is specifically determined not to be of a serious nature, a proposed civil penalty of up to $7,000 may be assessed for each violation.

c. Section 17(i) provides that, when a violation of a posting requirement is cited, a civil penalty of up to $7,000 shall be assessed.

Minimum Penalties. The following guidelines apply:

a.   The proposed penalty for any willful violation shall not be less than $5,000. This is a statutory minimum and not subject to administrative discretion.

b.   When the adjusted proposed penalty for an other-than-serious violation (citation item) would amount to less than $100, no penalty shall be proposed for that violation.

c.   When, however, there is a citation item for a posting violation, this minimum penalty amount does not apply with respect to that item since penalties for such items are mandatory under the Act.

Penalty Factors. Section 17(j) of the Act provides that penalties shall be assessed on the basis of four factors:

a.   The gravity of the violation,

b.   The size of the business,

c.   The good faith of the employer, and

d.   The employer's history of previous violations.

These four factors can significantly reduce the prescribed fines. However, there are very detailed rules on the application of these factors. Additional information is available through your local OSHA office. Finally, it is also important to note that citation and penalty policies can be different in state administered programs.

APPEALS PROCESS

Appeals by Employees

If an inspection were initiated due to an employee complaint, the employee or authorized employee representative may request an informal review of any decision not to issue a citation. Employees may not contest citations, amendments to citations, penalties or lack of penalties. They may contest the time proposed in the citation for abatement of a hazardous condition. They also may contest an employer's "Petition for Modification of Abatement" (PMA) which requests an extension of the abatement period. Employees must contest the PMA within 10 working days of its posting or within 10 working days after an authorized employee representative has received a copy.

Within 15 working days of the employer's receipt of the citation, the employee may submit a written objection to OSHA. The OSHA Area Director forwards the objection to the Occupational Safety and Health Review Commission, which operates independently of OSHA. Employees may request an informal conference with OSHA to discuss any issues raised by an inspection, citation, notice of proposed penalty or employer's notice of intention to contest.

Appeals by Employers

When issued a citation or notice of a proposed penalty, an employer may request an informal meeting with OSHA's Area Director to discuss the case. This informal conference is used very effectively by most knowledgeable companies to "work out" an equitable agreement on compliance issues. The Area Director is authorized to enter into settlement agreements that revise citations and penalties to avoid prolonged legal disputes.

Petition for Modification of Abatement

Upon receiving a citation, the employer must correct the cited hazard by the prescribed date unless it contests the citation or abatement date. However, factors beyond the employer's reasonable control may prevent the completion of corrections by that date. In such a situation, the employer who has made a good faith effort to comply may file for a PMA date.

The written petition should specify all steps taken to achieve compliance, the additional time needed to achieve complete compliance, the reasons such additional time is needed, all temporary steps being taken to safeguard employees against the cited hazard during the intervening period, that a copy of the PMA was posted in a conspicuous place at or near each place where a violation occurred, and that the employee representative (if there is one) received a copy of the petition.

Notice of Contest

If the employer decides to contest either the citation, the time set for abatement, or the proposed penalty, the employee has fifteen working days from the time the citation and proposed penalty are received in which to notify the OSHA Area Director in writing. An orally expressed disagreement will not suffice. This written notification is called a "Notice of Contest."

There is no specific format for the Notice of Contest. However, it must clearly identify the employer's basis for filing—the citation, notice of proposed penalty, abatement period or notification of failure to correct violations. A copy of the Notice of Contest must be given to the employees' authorized representative. If any affected employees are not represented by a recognized bargaining agent, a copy of the notice must be posted in a prominent location in the workplace, or else served personally upon each unrepresented employee.

Review Procedure

If the written Notice of Contest has been filed within the required 15 working days, the OSHA Area Director forwards the case to the Occupational Safety and Health Review Commission (OSHRC). The commission is an independent agency not associated with OSHA or the Department of Labor. The Commission assigns the case to an Administrative Law Judge (ALJ).

The ALJ may investigate and disallow the contest if it is found to be legally invalid, or a hearing may be scheduled for a public place near the employer's workplace. The employer and the employees have the right to participate in the hearing; the OSHRC does not require that they be represented by attorneys (however they usually are represented by counsel). Once the Administrative Law Judge has ruled, any party to the case may request a further review by OSHRC. Any of the three OSHRC commissioners may also, at his or her own motion, bring a case before the Commission for review. Commission rulings may be appealed to the appropriate U.S. Court of Appeals.

Appeals in State Plans

States with their own occupational safety and health programs have a state system for review and appeal of citations, penalties and abatement periods. The procedures are generally similar to federal OSHA's, but cases are heard by a state review board or equivalent authority.

In conclusion, these rights and responsibilities for employers are based on published policies, instructions, or regulations that are available through your state plan office or through your local, area, or regional OSHA office. They are listed in the phone book under the U.S. Department of Labor.

GUIDELINES FOR AN INSPECTION PLAN

A written plan should be in place to provide guidance on how to handle OSHA inspections, as well as for any other governmental inspection. The following policy questions should be answered and be a part of the plan.

1. Will you require a warrant to be obtained or will you allow entry? If you are going to require warrants there should be a valid reason for doing so, an antagonistic relationship with a government inspector is of no benefit.

2. Who is to be notified when an inspector arrives at the gate? The guard or receptionist should have a list of people to notify. It is not recommended that the inspector be kept waiting for very long. The notification list should include:

❖ the person responsible for safety and health
❖ the manager in charge of the facility
❖ legal counsel (if available)
❖ labor representation.

If the inspector does not offer credentials, ask for them. If there is any question as to the legitimacy of the inspector, call the OSHA Area Director for verification.

Learn the nature of the visit. Is it a complaint, general inspection or follow up inspection? If everything is in order continue with the inspection. The inspector should be taken by the most direct route to the area of interest. <u>Remember: the inspector is obligated to cite the employer for violations observed while at the site. Therefore, do not involve plant areas not specifically included.</u>

Keep meticulous notes on what is said by the inspector and what is said to the inspector. All members of management should be instructed to cooperate with the inspector and to answer all questions. It is recommended that you do not volunteer information beyond what is requested, do not embellish and do not lie.

If the inspector is to monitor the air and if you have the capabilities to duplicate the monitoring then do so. If pictures are requested, the employer has the right to take duplicate pictures, to approve the pictures or to have copies of all pictures taken. Photographing should be conducted in a way which shows no more background than is absolutely necessary. If a violation is seen in the picture, a citation can be issued.

The employer has a right to a closing conference. If a closing conference is not offered, request that one be given.

Additional information on OSHA inspection is available free of charge from your local, area, or regional office of OSHA.

SUMMARY OF THE ORGANIZATION OF OSHA STANDARDS

Requirements (standards) for health and safety for general industry can be found within the Code of Federal Regulations, Title 29, Part 1910. Part 1910 is further divided into subparts. They are organized as follows:

OSHA Subpart A
This section covers general regulatory information such as purpose and scope of the OSHA regulations for general industry.

OSHA Subpart B
This subpart covers adoption and extension of established federal standards.

OSHA Subpart C
Access to employee exposure and medical records requirements are found in this subpart.

OSHA Subpart D
The standards in this subpart cover the requirements for walking and working surfaces, fixed and portable ladders, stairs, guarding of floor and wall openings, and scaffolding.

OSHA Subpart E
This subpart covers emergency evacuation plans and standards for means of egress.

OSHA Subpart F
This subpart covers powered platforms, manlifts, and vehicle-mounted work platforms.

OSHA Subpart G
This subpart covers the standards regulating occupational noise, radiation, and ventilation.

OSHA Subpart H
This subpart covers hazardous materials such as compressed gases as well as other materials (e.g., explosives) and emergency response.

OSHA Subpart I
This subpart covers personal protective equipment requirements as well as the respiratory protection standard.

OSHA Subpart J
Included in this subpart are the standards for sanitation, color coding, lockout/tagout, and warning signs.

OSHA Subpart K
This subpart contains the requirements for medical and first aid.

OSHA Subpart L
This subpart addresses fire protection requirements.

OSHA Subpart M
This subpart contains the requirements for compressed air equipment.

OSHA Subpart N
The standards found in this subpart include the requirements for general materials handling; powered industrial trucks; overhead gantry cranes and slings.

OSHA Subpart O
General requirements for machine guarding, mechanical power presses, and forging machine standards are found in this subpart.

OSHA Subpart P
The general standards for hand and portable power tools and for guarding of hand and portable power tools are included in this subpart.

OSHA Subpart Q
The entire subpart is devoted to standards regulating welding, cutting, and brazing operations.

OSHA Subpart R
This subpart addresses regulations affecting special industries, such as grain handling and agricultural operations.

OSHA Subpart S
This subpart covers electrical safety standards.

OSHA Subpart T
This subpart covers commercial diving regulations.

OSHA Subpart U-Y
These subparts are reserved for future use.

OSHA Subpart Z
This subpart covers allowable exposures to air contaminants. The hazard communication standard is also found in this subpart.

CHAPTER
2

ESTABLISHING A PLANT SAFETY PROGRAM

This chapter describes the basic elements needed to establish a proactive safety program. This includes the four major aspects identified by OSHA as being essential, as well as how to organize a safety manual and to promote safety. Like other business systems, an effective safety program requires goals, staffing, and recordkeeping to allow for an evaluation of performance.

ESSENTIAL ELEMENTS OF A SAFETY AND HEALTH PROGRAM

OSHA published a safety and health guidance document listing its suggested minimal requirements for an effective safety and health program. The title of the document is, "Managing Worker Safety and Health." It was developed by the Office of Cooperative Programs and was published in draft form on October 1992. Since it is several hundred pages in length and available from OSHA it will not be repeated here. Nevertheless, the four major aspects of its program approach deserve mention. These major program elements are:

- ❖ Management Commitment and Employee Involvement
- ❖ Worksite Analysis of Hazards
- ❖ Hazard Prevention and Control
- ❖ Safety and Health Training.

OSHA feels that these four key elements are critical to developing an effective safety and health program. The essence of each major element is summarized briefly below.

Management Commitment and Employee Involvement

Experience has clearly shown and common sense suggests that no program will be successful without the true support of management. This means a clearly communicated policy fully supported by management. It also means a program instilled with goals and performance measures so that attainment of these goals results in some type of reward. Finally, it means a commitment by management to assign the responsibility, to give the authority, and to provide the resources necessary to accomplish these goals. OSHA also clearly supports the concept of joint labor and management safety committees. This issue is discussed later in this chapter.

Worksite Analysis of Hazards

A program of worksite analysis of hazards is the second of the four essential program aspects recommended by OSHA. This program element should include regular workplace inspections, analysis of proposed or planned process changes in the workplace, job hazard analysis, comprehensive health and safety surveys, investigation of accidents and near misses, and a mechanism for employees to notify management of potential or perceived safety and health problems (such as the formation of joint labor/management health and safety committees).

Hazard Prevention and Control

Hazard prevention and control is the third of the four essential elements. It consists of the use of engineering techniques to reduce or eliminate hazards in the design stage. For hazards that cannot be eliminated in the design stage, it is necessary to establish work practices, administrative procedures, or personal protective equipment controls to reduce the hazard. This aspect of an overall safety and health program should also include the maintenance of equipment and control systems, as well as planning and preparing for emergencies.

Safety and Health Training

The last of the four essential elements is safety and health training. Like other aspects of business, such as production, training is an essential component to the safe conduct of work. It is also important to note that safety training should be accomplished for all levels of employees (e.g., worker, supervisor, middle management, top management) so that everyone knows the nature and extent of the hazards they work with and their roles and responsibilities for establishing and maintaining safe work conditions. Training is an especially important aspect of safety and health programs since one of the most common causes of unsafe acts is a lack of knowledge or information.

Each one of these essential program elements is discussed in the chapters that follow this one. The next section of this chapter discusses how to develop and implement a policy statement on safety and health to establish management's commitment. This is followed by suggestions for developing a safety manual to organize and house the safety policy, safety rules, procedures, and other aspects of the overall safety and health program. Once the manual has been developed, it is necessary to promote the safety program; this is covered in the next chapter. The remaining sections of the chapter deal with staffing and implementing safety as a business system, hiring and motivating safety staff, dealing with consultants and unions, and finally, establishing safety and health committees which help promote employee involvement.

SAFETY POLICY STATEMENT

The safety policy statement is the foundation of the safety program. Therefore, devote considerable thought to its development and be prepared to commit company resources to meet the requirements of the policy. The safety policy statement can be a driving force or a destructive force for the entire safety program.

It is imperative that the policy statement be approved and issued over the signature of the top official in the organization. Without such a commitment the safety program will be continually compromised by opposing viewpoints of line managers. The safety policy statement must be perceived as *COMPANY POLICY*, not the safety person's policy, not the human relation's department policy, or any individual's policy.

In developing a safety policy, first decide on a definition of "safety." Safety can have many different meanings to people, depending on their job, experiences, and work circumstances. Webster's Dictionary simply defines safety as: "1. free from injury or risk. 2. secure from liability to injury or risk. 3. involving little or no risk." However, these dictionary definitions do not satisfy the complex and far-reaching meaning of safety in today's industrial environment.

There probably is no universal definition of safety applicable to every situation. However, a short definition of safety based on a proactive approach is as follows:

> Safety is an attitude that influences the behavior of individuals in a positive manner in their relationships with others, in doing routine tasks, and in reactions to situations that may occur.

What is also important to a policy on safety is the following concept:

> For a safety-minded person or company, safety is an integrated part of daily work that receives a priority equal to or greater than other priorities and when properly implemented reduces risk of injury and property loss.

Understanding the meaning of safety permits progression to the next step: the creation of a safety policy statement. The following questions should be addressed in the policy statement.

1. What does management intend to accomplish through the safety policy?

2. What are the overall goals of the policy?

3. What is the intended scope of the policy? Will it include:

    ❖ on-the-job-safety

    ❖ home safety

    ❖ recreational safety

    ❖ highway safety

    ❖ fire and property damage?

4. Who are the individuals responsible for the functional aspects of the safety program?

5. What are each individual's responsibilities?

6. Who will be granted authority to carry out the program functions and to what extent will the authority be granted?

7. How will individuals be held accountable for meeting the goals of the policy and what will be their respective responsibilities?

Figure 2-1 provides an example of a safety policy statement that can be modified for your use.

Clauses to Avoid

The following list of excerpts from safety policy statements illustrates clauses to avoid, why they should be avoided, and suggested alternatives.

(1) "Provide a workplace that is free of known hazards."

The only workplace that is free of known hazards is one that is not being used. Hazards are inherent with nearly everything we do. It should be our goal to reduce the risks of recognized hazards.

Suggested approach: It is our goal to control the risks associated with workplace hazards to prevent events that may cause injury or illness to our employees.

(2) "Safety is our first priority," or, "Safety is of equal priority with increasing production and reducing costs."

There is nothing wrong with these statements, but you must be dedicated to such a commitment. The first time a supervisor or an employee circumvents a safety provision, the policy statement is worthless and the entire system is in jeopardy of being lost.

(3) "All accidents must be prevented."

A utopian idea but hardly practical. Employees know that such an idealistic goal is not attainable or if attainable, only for a brief period of time. Accidents happen but injuries can be prevented.

Suggested approach: The only acceptable level of safety performance is one that prevents injuries.

(4) "All personnel should work in a safe manner."

"Should" is an arbitrary directive and enables personnel to make value judgements. The worst case is where management thinks it is a good idea to work safely, but does not make it a requirement.

Suggested approach: Safety is a responsibility that must be shared equally and without exception.

Solicit Advice from Key Managers

When you have finished the policy statement, distribute it to the key members of your organization that will be involved with the day-to-day enforcement of the policy and

Figure 2-1
Safety Policy Statement

The management of *(company name)* recognizes the importance of safety and health in making life more rewarding, and is committed to providing a workplace for our employees in which recognized hazards are controlled.

The philosophy and objectives behind this commitment are:

1. The safety and health of all *(company name)* employees is our first priority.

2. The only acceptable level of safety and health performance is one that prevents injury and accidents.

3. Safety and health are an integral part of production and all other business functions that cannot be separated or by-passed.

4. Safety and health are a responsibility that must be shared equally and without exception by everyone within the organization, and;

5. Supervision and management will be held accountable for the safety and health of the personnel for whom they are responsible. All employees will be required to make their safety and the safety of their fellow workers a first priority. As a condition of employment, each individual within the organization will be expected to conduct their daily tasks in a manner that is consistent with the philosophy and objectives in this policy.

SIGNATURE _____

TITLE _____

DATE _____

ask for their comments. When revising the policy, do not dilute the overall thrust of the statement simply because someone thinks it will reduce production output, increase down-time or "just take more of my time from my real job." This type of response must be dealt with immediately and with authority to assure the integrity of the entire safety program. When the policy has been finalized it should be signed and readied for communication to all employees.

An example of a very well-stated policy which is simple but direct and incorporates product and environmental safety as well is that of the DuPont Company. Their policy states:

"We will not make, handle, use, sell, transport, or dispose of a product unless we can do so safely and in an environmentally sound manner."

You may wish to incorporate other company philosophies into your statement. Finally, consider presenting the policy in person by scheduling plant-wide meetings to emphasize your commitment to the policy and to answer questions concerning the policy. Simply posting the policy on the bulletin board reduces the impact and the importance of the statement.

SAFETY AND HEALTH MANUALS

A safety and health manual is a repository for safety policies, procedures, rules, forms, and programs. This does not mean that you must include all governmental standards and regulations in the manual. The safety manual is intended to provide management's approach to safety and health issues. A suggested outline for a model safety and health manual is provided in Figure 2-2.

Safety and health manuals are the key to effective administration of the safety programs. Manuals must be accessible to employees at all times. Well organized safety manuals placed throughout the facility are an excellent means of ensuring employee cooperation and commitment.

SAFETY RULES

A safety program is guided by standards generated by the safety policy statement, governmental regulations, good safety methods, and specific facility safety and health rules.

Earlier we also discussed the procedure for developing a safety policy statement. The facility must develop and issue specific safety and health rules, regulations, and policies that are the standards for driving the safety program. Obviously, many of these will be the result of government mandated requirements. Safety rules should have the following common characteristics:

1. Purpose—A clearly defined purpose for the rule establishes credibility for its issuance and can be an important promotional tool for obtaining employee acceptance and voluntary compliance.

2. Scope—Who is covered by the rule? What area(s) of the facility will be covered? When is the rule in effect?

3. Requirement(s)—Each requirement of the rule must be written clearly, and using words and terms that are easily understood. Arbitrary, ambiguous, and confusing words must be avoided.

4. Dates—The date the rule was issued and the dates of subsequent revisions are important to ensure that the rule is current.

5. Signature—The signature of the person authorizing the rule adds credibility and strength to the rule for enforcement.

Safety rules must be accessible to all employees, must be kept up-to-date, and must be reviewed periodically to assure they are accurate and meet current standards and needs. Once the safety rules are developed, they can be organized into the safety and health manual.

**Figure 2-2
Safety and Health Manual
Section Outline**

Section 1

This section should contain an introduction to the manual, the reason for its publication, an executive overview supporting the manual, and any special acknowledgements that may be appropriate.

Section 2

Section 2 could include an introductory overview of the manual contents, identification of essential elements for your safety and health system, and supportive information for reducing injuries and illnesses.

Section 3

Section 3 could be devoted to providing the procedures for administration of your safety system, including but not limited to:

❖ reporting and recording injuries and illnesses

❖ safety committees and their charters

❖ safety policy statement

❖ accident investigation procedures

❖ procedures for handling governmental inspections

Section 4

This section could contain the specific policies, procedures, rules, and regulations for driving the safety system, such as:

❖ lockout/tagout procedures

❖ confined space program

❖ hazard communication program

❖ emergency response program

Section 5

This section could contain resource information:

❖ safety and health reference publications

❖ governmental publications

❖ training materials

❖ safety and health professional organizations

SAFETY PROMOTION

If safety is not continually promoted, the level of safety performance will decline. Early warning signs may be an increase of first aid injuries, a general increase in all or specific types of accidents, an increase in near miss incidents, or even housekeeping problems. These indicators will tell you that the system needs attention or promotion. When you attempt to discover the cause(s) for an unfavorable safety performance trend and do not find specific factors, such as, training deficiencies, process changes, increased production schedules, or an influx of new personnel, you may find that the program needs additional promotion. Safety promotion can be achieved through a direct or indirect approach or through a combination of the two.

Direct Promotional Approach

Direct safety promotion involves direct intervention into the system, such as:

❖ Modification of safety training programs, including retraining.

❖ Providing short safety talks either at the start of each shift, following lunch, or during coffee breaks.

❖ Holding safety meetings led by the line manager or the top executive officer in the facility explaining management's concerns over declining safety performance.

❖ Development of a safety information program. Such programs often include the posting of accident-free work hours. This method utilizes bulletin boards, recording the number of hours worked since the last reportable accident and/or the number of hours worked since the last lost-time accident. Display safety posters pertinent to plant safety problems or place warning signs that an accident took place and the date of the accident. If you use posters as a regular promotional tool or on special occasions to address a specific problem, remember these two points:

   (1) Do not display the posters for too long a period of time. A poster that has been "hanging around" in the same place with the same message soon becomes invisible to the workers. Posters covered with grease, obliterated by graffiti, or torn can do more harm than good in promoting safety.

   (2) Do not use posters just for the sake of using posters. Safety posters warning about the hazards of cotton dust in a ball bearing factory do not serve a purpose. If safety messages are meaningless, or perceived as meaningless by your employees, then soon employees will not take the time to read any of the messages.

❖ Evaluating safety performance on a regular basis. This can be accomplished by supervisors and line managers making routine evaluations of job performance. This type of promotion requires instant feedback on how well each individual is performing their assigned tasks. If they are doing everything correctly they should be complimented by the evaluator at the time of the observation or informed of what they are doing incorrectly.

A positive approach should be used for whatever means of direct safety promotion is selected. If you must point out that an individual is not performing a task safely you may want to take a tactful approach illustrated in the following example:

Supervisor: "Paul, I couldn't help noticing you were lifting those 100 pound spools of wire from the floor to the rewind table. I know that we need to get this run out by the end of the shift, but I am more concerned about the possibility of you hurting yourself."

Paul: "I know, but if I have to wait for the fork truck, we'll never get this job done."

Supervisor: "Paul, I appreciate your interest and your hard work. But, I want you to wait for the fork truck or I will get someone to help you move those spools. Just let me know when you need help. We can't afford to lose a good worker like you because of a back injury."

If you use safety meetings to promote the improvement of safety performance, approach the subject by referring to a specific time when safety performance was at an acceptable level. Do not set an improvement goal that is unrealistically attainable in a reasonable period of time. Once you have reversed the unfavorable performance trend and have reached the first goal, you may want to set higher performance goals. Remember, you are trying to improve performance, not reach a zero accident level in one step.

Indirect Promotional Approach

An indirect promotional approach is one in which you use a secondary means to influence the behavior of the primary concern. The use of a contest (secondary means) to improve safety performance (the primary concern) is an example. The safety contest has become a widely used means of promoting safety throughout industry. There are pros and cons concerning the use of safety contests to promote safety. Many companies have used a variety of safety contests as part of their on-going safety programs. We will not discuss the particulars of each type of contest but a list of commonly used safety contests is provided in Table 2-1.

The premise for using a contest to stimulate safe work performance is that employees will enjoy competing for the recognition and/or prize and that the competitive drive will favorably influence safety performance.

Safety contests have proven to be very successful in many major corporations and have been part of their ongoing safety systems for years. Other companies have used safety contests as a kickoff for a new safety effort or commitment, while some have been successful in using safety contests as a quick-fix for turning around unfavorable safety performance levels. However, safety contests are indirect promotional tools and should not be used in place of a well organized safety program backed by a strong management commitment.

The argument for not using safety contests (specifically incentive contests) is that you are providing additional compensation for employees to perform their daily tasks for which they were hired. That is, why should you reward someone for doing what is expected? Modern management theory is that you should both encourage and reward positive or desired behavior. Most companies view safety contests this way.

Managers who do not use safety contests often support the use of spontaneous awards when safety performance exceeds expectations or is judged to be above peer group performance levels. An example would be providing an employee dinner or an award for outstanding safety

Table 2-1
Commonly Used Safety Contests

1. SAFETY BINGO—Numbers are drawn as in regular bingo, but only at the end of a day or week for which there have been no accidents.

2. STEAKS or FRANKS and BEANS—Usually a contest between departments where the winners are served steaks by the losers who must eat franks and beans during the dinner.

3. PERFORMANCE INCENTIVE—Prizes are awarded to employees who complete a specified work period without an accident. Prizes are commensurate with the accident-free time period, such as:

   ❖ 1 week—coffee and donuts

   ❖ 1 month—free lunch and/or small prize

   ❖ 3 months—free dinner and/or moderate prize

   ❖ 1 year—Family picnic/dinner and/or large prize, such as a car awarded by a lottery drawing of eligible employees.

4. PROMOTIONAL ITEMS—Employees are given a common use item, such as a shirt or jacket, for achieving a safety goal. These items usually identify the award (e.g., "2000 hours accident free") and if worn, reinforce the safety message.

performance, such as completing one work year without a lost-time accident. The dinner is in appreciation for outstanding performance and not a carrot; and perhaps just as important, you are not establishing a condition that may become a precedent or expected employment benefit.

You may find that once you initiate a safety contest you have created a "tail that is wagging the dog." It is not unusual to hear employees complain: "Oh the same old coffee and donuts again" or "Why should I work so hard being safe when all I get is the same beverage mug, pen, or key ring...." or "I can't believe it, Bill won the car in the safety contest and he never wears his safety glasses or gloves." This would indicate that it is time to change your approach.

Circumstances may warrant the use of both direct and indirect promotion tools. You may have a safety meeting to express concern over the poor housekeeping in several areas of the plant and how it has led to several accidents. You may first review housekeeping procedures with the personnel in the areas and organize a steak or frank and beans contest with the department showing the most improvement getting the steaks. Table 2-2 provides a list of advantages for using contests and Table 2-3 lists possible disadvantages. In conclusion, safety contests along with an effective safety program can help promote safety. However, they must have an objective and achieve that objective to be of benefit.

Table 2-2
Advantages for Using Safety Contests

❖ Generally, you can achieve positive results quickly.

❖ Safety contests do not require large financial commitments. Effective contests are self-supporting or money saving when contest expenses are compared to reductions in workers' compensation costs.

❖ All employees are involved.

Table 2-3
Possible Disadvantages in Using Safety Contests

❖ Interest often drops off quickly.

❖ Winners in drawings for a big prize are not always the best examples of safety-minded employees.

❖ Employees may demand larger prizes or become disinterested.

HOW TO STAFF AND IMPLEMENT A SAFETY PROGRAM

Safety programs in large corporations require a well-defined and qualified staff for implementation and management. Smaller companies cannot afford the nonproduction staffing that is usually found in the larger companies. Nevertheless, the larger companies can serve as models. In a large company, the safety staff will generally be comprised of employees with full time safety and health responsibilities, employees with safety and health responsibilities as part of their other job assignments, and consultants hired on an as-needed basis to supplement the safety program. For the purpose of this discussion, four specialty divisions will be considered: safety, industrial hygiene, fire protection, and environmental engineering. Figure 2-3 shows a hypothetical organizational chart for a large health and safety department in a major corporation which would include all of the specialty divisions.

In smaller businesses (representing the majority of companies) the safety organization may not have full-time personnel. A line manager, often the human resource manager, may handle the safety function along with his other duties, such as equal employment opportunity. Small organizations often have to rely on consultants from the state on-site consultation services, insurance carrier, trade organization, or independent contractors to assist them in managing their safety systems.

Line management and particularly supervision must be actively involved in the implementation of the safety system. Safety programs that segregate the implementation and management of the safety function from the responsibilities of the line managers and supervisors and that rely solely on the safety staff will not perform satisfactorily. Stated another way, implementation of the safety program should be the responsibility of all management in the organization. When the implementation of the safety program is the exclusive and separate responsibility of the safety department, the safety program will not reach its full potential. Safety must be integrated into all daily functions.

Figure 2-3
Example Organization Chart

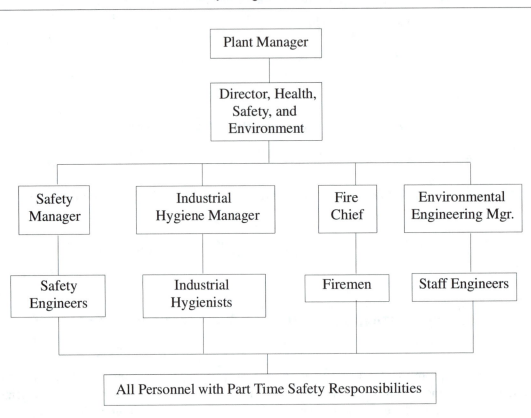

HIRING, TRAINING, AND MOTIVATING A SAFETY STAFF

Not so many years ago, a typical safety department for some large companies consisted of one person whose duties usually involved keeping track of accident statistics, holding occasional safety meetings, and perhaps placing safety posters on plant bulletin boards. Further, it was common for this person to have been a good worker who was injured and could no longer handle the assigned work tasks. The fact that he or she had been injured was considered an advantage since it showed others the consequences of poor safety habits. In some cases, it might have been a manager or supervisor who was no longer capable of performing his or her job satisfactorily. These safety people were motivated directly or indirectly by the knowledge that if they "didn't rock the boat" they would have a job until retirement. Today's highly technical safety standards and government regulations require more capability than simply placing posters on walls.

Qualifications (hence salaries) will vary among potential candidates for your safety staff. These include certified safety professionals (CSP) or certified industrial hygienists (CIH), those with college degrees in safety or industrial hygiene, individuals currently practicing safety or industrial hygiene, and, lastly, employees with little or no safety or industrial hygiene education or experience.

❖ Certified Professionals (CSP and CIH)—These individuals have successfully met the academic and experience requirements as well as having successfully completed a certification examination required by the Board of Certified Safety Professionals and/or the American Board of Certified Hygiene.

❖ Safety & Industrial Hygiene Degrees—Individuals with college degrees specifically in safety or industrial hygiene may be difficult to find because of the limited number of programs at an undergraduate level. Individuals with industrial hygiene or safety degrees have the essential background to meet the technical demands for most safety and industrial hygiene positions, but recent graduates usually lack practical field experience. Students whose programs included internships may have some experience. The amount of industrial experience can be of paramount importance when immediate results are expected or when operations are complex or unique.

❖ Experienced Personnel—Individuals with satisfactory work experience can sometimes provide the necessary expertise; however, it can be difficult to determine the candidate's actual qualifications. It may be necessary to supplement the candidate's expertise by providing special training through professional development courses such as those at the OSHA Training Institute in Chicago or Associations such as the American Society of Safety Engineers.

❖ Little or No Experience—There may be an employee that would like to become involved in the safety or industrial hygiene functions of your safety program. An employee who is intimately familiar with work tasks and processes can provide unique insight into safety and health related risks. With additional training in safety and industrial hygiene, the company may be well rewarded for its patience and the additional cost involved in developing the employee's expertise.

The selection of employees for fire protection and environmental engineering are similar in that the employee pool will contain a mixture of academically qualified, experienced, and inexperienced persons.

CONSULTANTS

Why should you use consultants? When should you use consultants? Consider exactly what a consultant can provide:

1. Supplemental assistance to expedite projects that are too large for the in-house staff or projects that the staff cannot complete in a reasonable or desired time frame.

2. Expertise that may be beyond that of in-house staff.

3. Independent thinking and opinions. In some cases it is advantageous to involve an outside party to avoid allegations of conflict of interest.

4. Validation of safety system performance and safety staff performance through auditing. Safety audits are structured similarly to financial audits. The safety audit inspects the safety system to determine if the system is being operated as it was designed and if there are errors or omissions. It provides an overall evaluation of actual system performance compared to perceived performance. Objective safety audits by a third party can reveal conditions that may be overlooked because of familiarity with the system. Consultants are not disadvantaged by either familiarity or conflict of interests.

Consultants can be cost effective when used properly. Just as qualifications vary among staff members, they also vary among consultants. When selecting consultants, be sure that they have the qualifications to achieve the work desired. It is advisable to have periodic, independent audits of your safety system by outside consultants to validate its functioning, just as it is advisable to have independent audits of your financial system.

UNIONS

The implementation and enforcement of safety systems in facilities with organized labor agreements is not too different than in facilities without such agreements. Implementation and enforcement of systems are usually covered by the same considerations as in any other work rule. It is common for safety and health issues to be addressed separately from other issues in today's agreements. The following guidelines will be helpful in implementing and enforcing a successful safety system:

1. Compliance with safety and health rules and regulations must be demonstrated by management before hourly employees can be expected to comply.

2. Enforcement must be uniform.

3. Rules, regulations, and policies must be presented to the employees in an understandable manner prior to compliance.

4. Involvement of hourly employees in developing the safety system is beneficial but not absolutely required unless it was previously stipulated in a negotiated agreement.

This chapter has provided the information for developing a basic framework for your facility safety system. The system is dynamic and will evolve through various changes before it fits comfortably within your overall business plan. However, the effort to obtain this fit will provide real bottom-line benefits.

CHAPTER
3
MANAGING PLANT SAFETY

Low accident rates are the product of effective safety programs. If you manage your safety program efficiently and effectively the savings in accident costs will be reflected by company profits. Hold the workers and management accountable for performance and you will improve the system. Also, use dollar measures to evaluate and report on safety performance, just as in any other business system. These similarities to other traditional business measures will allow you to manage your safety program as a profit-making entity rather than simply as a cost of doing business.

THE HIGH COSTS OF ACCIDENTS

Accident costs will adversely affect your plant's bottom line. This alone justifies the need for continuous safety improvement. You and your management should accept safety as a good business practice, as well as the ethical thing to do. There are two places your safety dollar can be invested: in the costs associated with preventing accidents (your safety program) and in the costs of accidents after they occur.

When accidents are not prevented you incur indirect costs as well as the more readily observed direct costs of the accident itself. These indirect costs can exceed the direct costs by as much as *four times*. The following list illustrates some common cost elements associated with accidents.

Direct Costs:

1. Damage to machinery, tools, and materials

2. Wage losses

3. Insurance premiums

4. Medical payments

5. Workers' Compensation

Indirect Costs:

1. Fines from regulatory agencies, e.g., OSHA

2. Profit loss due to overhead without production and time lost due to damaged equipment, failure to fill orders promptly

3. Post-accident time for supervisory, production, clerical, and management staff (i.e., investigations, paperwork)

4. Overtime work

5. (Re)Training expenses

6. Fringe benefits which are administered even though employee is not working

7. Intangible costs, i.e., good will, business reputation, employee relations

8. Efficiency costs due to the loss of experienced workers

9. Suffering and expenses incurred by the worker, his family, and society

In general, indirect costs are "uninsured costs." Direct costs or "insured costs" represent payments under special labor agreements, workers' compensation laws and other costs usually covered by insurance or special funds. The direct costs associated with an accident may seem somewhat remote to you as a manager since the majority of them are typically borne by the insurance company. However, the amount your company pays for insurance depends on what you cost the insurance company. Insurance rates are based on three factors: (1) average industrial accident rates, (2) the company's past accident rate record, and, (3) occasionally, the company's current record. What this means to you is that preventing accidents will reduce the cost of your insurance. Simply stated, the less the insurance company needs to pay, the less you will pay the insurance company and the better your bottom line will be.

INFLUENCING SAFETY PERFORMANCE WITH FINANCIAL FACTORS

Parties who control safety performance must be held accountable for that performance and be rewarded or penalized accordingly. To illustrate and emphasize the cost of accidents and to motivate your employees you can employ the following financial "incentives":

1. Prorate insurance premiums: A department or division can be charged a specified amount of insurance premium based on its share of the total accident costs. This makes it difficult for an area with a poor safety record to show a profit.

2. Charge the cost of accidents to the operating department. Subtract the cost of accidents from working budgets to emphasize the importance of safety performance.

3. Make the safety record a part of the supervisor's personnel record. This adds a degree of personal accountability and should be used for the supervisor's performance reviews.

4. Include safety performance in compensation and performance evaluations for all personnel. Give financial rewards for good safety performance, thereby generating a significant interest in safety. This can be applied from the lowest level employee to upper levels of management.

5. Outline safety responsibilities in everyone's job description. This makes working safely a condition of employment.

ORGANIZING THE SAFETY SYSTEM

In Chapter 2, organization and staffing of a safety program were discussed. In order for the safety program to be effective, each employee must know his or her duties and be held accountable for the execution of them. The worker must be aware of how to perform work tasks in a safe manner and then to actually perform them that way. Management, on the other hand, must keep the safety program running and adapt it as necessary. It is important to recognize that management has always been held responsible for the safe conduct of company employees. This is a legal framework which has been repeatedly upheld by the courts even though the worker may act improperly. Management, therefore, has the greatest degree of responsibility and accountability for the safety program.

Management Accountability

As discussed in the previous section, the responsibility for safety and health is vested in management. Each level of management should be accountable to its superior level for safety and health performance. One responsibility, however, is shared by management at all levels: the promotion of safety and health *by example*. This requires uniform enforcement of safety and health policies and procedures and makes safety and health equal to or above all other priorities.

In the following outline of management responsibilities, the management titles are selected on the basis of common usage and reflect a large organization. These responsibilities are common to most safety programs, although actual assignments may be different in any particular plant.

Executive Officer   The executive officer, usually the vice president or general manager, responsible for all plant operations of a division, has overall responsibility for the facility safety and health program. This role is mostly administrative. This person should serve as the example. This person must follow all safety and health rules without exception, must show an interest in programs, and must demonstrate his or her approval of exceptional performance (e.g., letters of commendations and performance rewards). This person must understand fully that he or she can and will be held accountable by federal agencies, state agencies, and the courts for serious violations of regulatory standards whose penalties can include fines and even criminal prosecution.

Plant Manager This individual is usually responsible for all aspects of an operation or plant. Responsibility for program performance is vested here. The plant manager should be held accountable by the executive officer for safety performance. Specific safety program duties may include:

Administrative Functions

❖ Approves, adopts, and enforces all safety rules and regulations and single objective safety and health or accident prevention programs.

❖ Regularly reviews the performance of the plant safety and health program and addresses problems, as needed.

❖ Assigns and oversees managers who have demonstrated ability in effective administration of the safety and health programs, and provides motivation and assistance to meet minimum program requirements.

❖ Requires a safety and health inspection by production, engineering, and safety and health personnel before new equipment is placed in operation.

Control Functions

❖ Makes necessary plant audits with the safety and health officer to appraise the program's effectiveness.

❖ Reviews the plant safety and health program annually to determine its effectiveness and need for changes.

❖ Serves as chairman of a committee investigating all fatal accidents and major catastrophes.

Foreman This is the individual responsible for an assigned area within a department or operation who supervises and coordinates the activities of the supervisors. The foreman is accountable to the plant manager or department head and is responsible for health and safety within his or her assigned area.

Administrative Functions

❖ Reviews all injury cases and other safety concerns for assigned work area, taking any necessary corrective actions.

❖ Ensures that supervisors meet safety program requirements via the performance review process.

❖ Manages safety program elements, such as housekeeping, in assigned areas.

❖ Assigns specific pieces of equipment to the supervisors for weekly inspection for unsafe conditions, housekeeping, and general operation and reviews their reports.

❖ Schedules employees for medical evaluations as required, ensures that supervisors make such employees available as scheduled, and keeps supervisors advised of employee limitations as reported by the medical department.

Control Functions

❖ Routinely conducts spot checks, as well as formal safety inspections of unsafe acts or conditions.

❖ Reports all injuries or illnesses to the manager or department head as soon as practical, reviews supervisors' accident reports with them, and in case of injury, investigates and submits reports to the manager or department head describing corrective action.

Training Functions

❖ Ensures that all personnel are properly trained.

❖ Provides instruction to supervisory personnel for specific safety and health responsibilities.

❖ Attends randomly selected training sessions to evaluate effectiveness of supervisory personnel's training skills.

Supervisor   The front-line manager maintains daily contact with production and maintenance workers. Employee safety and health and safe working conditions for a given area are vested in this person.

Administrative Functions

❖ Makes at least one individual safety and health contact per month with each employee.

❖ Reviews reported accident causes and unsafe acts and directs corrective action.

❖ Makes daily inspections of work areas and immediately corrects unsafe or unsatisfactory conditions. Reports to the foreman any conditions which cannot be immediately corrected.

❖ Requires employees to inspect items such as tools, wire ropes, chains, clevises, pins, spreaders, etc., and powered material handling equipment before each use, makes daily spot checks of conditions, conducts monthly inspections using a check list, and submits these lists to the foreman noting the type of defect and action taken.

❖ Maintains utility identification in his areas.

❖ Enforces the medical department's recommendations with respect to employee's physical limitations, reports apparent physical limitations to the foreman and requests a medical opinion if there is any question about an employee's physical condition.

❖ Enforces the use of personal protective equipment by making spot checks to verify usage and condition.

❖ During accidents or near-misses, assures that injured employees receive prompt medical attention, isolates areas or shuts down equipment, as necessary, and immediately reports to the foreman the facts of the illnesses and corrective actions taken.

❖ Conducts thorough investigations of all accidents, near-miss incidents, and injuries.

Training Functions

❖ Attends all scheduled and assigned safety and health training meetings.

❖ Provides periodic instruction to employees in health and safety rules and regulations.

❖ Provides on-the-job training and instruction to employees for safely carrying out their job assignment.

❖ Instructs each new employee on all relevant safe working procedures.

❖ Keeps records of all training activities.

Safety and Health Professional This professional is responsible for staff coordination of the local safety and health programs and for staff services within the organization. This person's authority is delegated by the plant manager, to whom he or she acts as an advisor and counselor. Together they seek to promote, coordinate, and evaluate the safety and health program. For smaller plants, this position may well be a part-time function of the personnel manager or others.

Administrative Functions

❖ Assesses plant-wide safety performance, with the plant manager.

❖ Reviews routinely, with the plant manager, the accident experience and formulates recommendations and corrective actions on a program basis.

❖ Provides management with necessary safety and health information and advice on new regulatory requirements.

❖ Provides direction, recommendations and general support in emergency situations.

❖ Participates in the review of all near-miss incidents and lost time accidents, and leads investigations for fatal accidents.

❖ Develops and maintains an effective safety and health promotional program.

❖ Consults with the plant engineering department on original plans and sees that all plans and specifications for new or proposed changes in processes, equipment, or methods are reviewed for compliance with safety and health standards before acceptance or use.

❖ Investigates potential health hazards in collaboration with the medical department.

❖ Selects the proper personal protective equipment (PPE), establishes a program for testing and evaluation of PPE, and approves initial purchases.

❖ Maintains active membership in professional safety and health organizations.

Control Functions

❖ Reports all disabling injuries to division and corporate safety and health offices within a prescribed (e.g., twenty-four hours) period from occurrence.

❖ Formulates and distributes statistical reports of safety and health data to all plant management.

Training Functions

❖ Develops and coordinates the training activities.

❖ Personally trains safety and health trainers and inspectors.

MONITORING SAFETY PERFORMANCE

Monitoring of the safety program is necessary to assure effectiveness and to improve it. The best approach to measuring safety performance is to study the quality of the system rather than the quality of the results. In other words, determine the effectiveness of the program rather than specific incidents. This reduces the likelihood of repeating an incident.

Inspections and Audits

Inspections and audits have always been the principal method of evaluating safety. The details of how to conduct plant inspections are discussed in Chapter 11.

The safety inspection must examine two things to be effective: unsafe acts and unsafe conditions. Regular inspections should be conducted by the supervisor and should be routinely audited by a safety professional or by that person's supervisor, for example. The supervisor must be held accountable for his or her department's safety performance. The focus should be on examining why a guard is missing or not installed, or why a hazard has not been addressed rather than the specific hazard or deficiency. Inspections and audits, therefore, should look for flaws in the system that allow unsafe conditions to arise. It is important that these inspections and audits become a permanent part of management's role.

Accident Statistics

Another means of evaluating safety performance is with accident and injury records. You can compare your plant's records to industry standards, such as those issued by OSHA, or to past plant performance. Before using statistics as a basis for evaluation, however, you must consider several things.

When numbers are used as safety objectives, you set quotas—a practice deemed bad by many managers. Within any organization, accident rates can vary randomly. By setting a quota or by reacting to normal and random variation, a frustrating situation results—you "chase" events over which you have no control. Statistical tests should be used to determine what is actually outside this normal variation and requires specific attention. For example, in a department of 500 employees with one accident, a second accident is a 100 percent increase but still a very low frequency rate.

Another problem that arises is that the employee reporting the numbers is often evaluated on the basis of those numbers. A conflict is created if this employee must choose between what is best for himself and what is best for the company. This is the major problem with quotas which can result in not all injuries being reported (generally only the serious ones) thereby making the data statistically biased.

In addition, numbers do not always accurately reflect reality. Luck is often the only difference between a serious accident and a minor one, therefore all accidents and "near-misses" must be investigated and recorded. For example, an employee slips and falls in your plant. An older employee will probably experience greater injury than a younger one. If reporting is based on injury severity, the incident would not be reported in the younger employee's case and the flaw in the safety system, e.g., a housekeeping problem, may go uncorrected. If all accidents or near misses are reported, you can address the circumstances surrounding the fall without regard to the age of the "victim" (you attack frequency, not severity). There is a more detailed discussion of this concept in Chapter 8, Accident Investigations.

The bottom line is that any numerically based system of performance must address the statistical significance of the data without singling out individuals or departments. Numbers must be regarded not as a means for attaining an accident-free environment, but rather as a means for gathering information on what is wrong with the safety system itself.

Accident and Injury Investigations

Accident investigations can be one of the tools to help monitor the safety system. Accident and injury investigations have traditionally been delegated to the supervisor. Forms are generally used, however, they do not usually address the underlying cause of the accident or the flaw in the system that may exist. Require the inspector to propose several possible causes for the accident. As a follow-up, more than one step should be taken to prevent a recurrence. To reinforce accountability, require supervisors to investigate all accidents, not just lost-time accidents. Ensure that accident reports circulate up the management line so that everyone is aware of problems and can contribute to solutions (see Chapter 8 for additional detail on accident investigations).

Recording Injury Experience

There are a number of methods for handling the data generated during the monitoring of a safety system. The American National Standards Institute's (ANSI) methods of recording and measuring work injuries have been the standard since the late 1930s (ANSI Z16.1, "Method of Recording and Measuring Work Injury Experience"). These methods use ratios comparing unwanted events to man-hours of work exposure. The standard defines three measures of injury experience.

1. Disabling Frequency Rate:

$$F = \frac{\text{Number of Disabling Injuries} \times 1,000,000}{\text{Employee Hours of Exposure}}$$

2. Disabling Injury Severity Rate

$$S = \frac{\text{Total Days Charged} \times 1,000,000}{\text{Employee Hours of Exposure}}$$

3. Average Days Charged Per Disabling Injury $= F/S$

These measures can also be calculated using dollar figures rather than injury totals. Because of the considerable lag-time between occurrence and receipt of medical bills, loss appraisals, and claim settlements, data may not be available in a timely manner. This leads to the use of estimates which may affect the accuracy of the final figures. OSHA has slightly different rules regarding measurement of safety performance. These rules are covered in detail in Chapter 5, Recordkeeping.

INFORMATION MANAGEMENT SYSTEMS

The large quantity of information generated by a safety program can make computerization a real advantage. The computer program or system used must be capable of two basic activities: recordkeeping and reporting. Some examples of possible system capabilities include:

1. Accident, incident, and occupational illness recordkeeping

2. Accident reporting and recordkeeping

3. Safety inspection reports and recordkeeping

4. Safety audit reports

5. Corrective action records

6. Data summaries and comparisons

7. Internal reports for management

8. Workers' compensation cost reports and recordkeeping

9. OSHA log data reports and recordkeeping

The question will inevitably arise as to whether it is better to buy a commercially available system or to write your own. Very few companies have the resources to develop their own software in-house. This is generally the more expensive option, however, you are guaranteed that you will get exactly what you want. There is a wide variety of health and safety related software now available that can be readily customized to meet your individual needs. In general, assess your particular system needs and evaluate commercially available systems. If these can provide at least half of what you require from a system it is probably cheaper to go that route and to have it customized, or to use it even with the existing limitations.

Evaluating Information Management Systems

1. Define and prioritize what you need from a system.

2. Evaluate the vendor's experience and reputation. Points to consider include:

    ❖ Are they stable financially and managerially?

    ❖ Can they provide upgrade support?

    ❖ What other software and/or services do they have to offer?

    ❖ What do other customers and/or user groups say about the system and vendor?

3. Ensure that the system is compatible with your needs. Points to consider are:

    ❖ Is your hardware compatible (PCs, minis, mainframe, communications)?

    ❖ Can they alter the system to your specifications for the amount you want to spend?

    ❖ Is the system easy and inexpensive to upgrade?

    ❖ Are agency approved formats used or available for reports?

4. Make sure that the system can be customized and expanded to fit your requirements.

5. The system must be easy to use. Find out whether or not the vendor will train your staff in its use.

6. Does the vendor offer technical assistance and support? A 24-hour 800 number is ideal.

Safety and Health System Software

The following list of companies in Table 3-1 currently offer health, safety, and/or environmental oriented software. This list is by no means complete and does not constitute an endorsement. It is simply meant to serve as a "jumping-off point."

Table 3-1
Currently Available Software: Injury/Illness Records

C. Alexander & Associates
460 Vista Roma
Newport Beach, CA 92660
(714) 644-5829

A. V. Systems, Inc.
924 Woodlawn Street
Ann Arbor, MI 48104
(313) 662-0355

The Chemtox System
Division of Resource Consultants, Inc.
7121 Cross Roads Blvd.
P.O. Box 1848
Brentwood, TN 37024
(615) 373-5040
Fax: (615) 370-4339

Control Software Group, Inc.
Safety Management Resources, Inc.
5106 N.W. 8th Avenue
Gainesville, FL 32605
(800) 535-7107
Fax: (904)372-8676

Donley Technology
P.O. Box 335
Garrisonville, VA 22463
(703) 659-1954

EcoAnalysis, Inc.
221 E. Matilija Street
Suite A
Ojai, CA 93023

(805) 646-1461
Fax: (805) 646-4141

ENSR Consulting & Engineering
35 Nagog Park
Acton, MA 01720
(800) 722-2440
Fax: (508) 635-9180

Excel Data Systems, Inc.
FiSerWare Division
47 Joseph Lane
Glendale Hts., IL 60139
(708) 690-2780

General Research Corp.
Flow Gemini Information Systems
1900 Gallows Road
Vienna, VA 22182
(703) 506-5166
Fax: (703) 356-4289

The Hawkwa Group, Inc.
P.O. Box 321
Mundelein, IL 60060
(708) 949-8488

HazMat Control Systems, Inc.
3409 Lakewood Blvd.
Suite 2C
Long Beach, CA 90808
(310) 429-9055

Health & Hygiene Inc.
420 Gallimore Dairy Road
Greensboro, NC 27407
(919) 665-1818

Industrial Training Systems Corp.
9 E. Stow Road
Marlton, NJ 08053
(609) 983-7300
Fax: (609) 983-4311

Institute of Health Management
P.O. Box 9957
St. Louis, MO 63122
(314) 965-3383

International Loss Control Ins.
4546 Atlanta Hwy.
P.O. Box 1898
Loganville, GA 30249
(404) 466-2208

Management & Communication Consultants (MC-2)
3016 Raintree Road
Oklahoma City, OK 73120
(405) 751-0555

Pro-Am Safety
Software Division
4432 Route 910
P.O. Box 1290
Gibsonia, PA 15044
(412) 443-0410
Fax: (412) 443-5655

Safety Management Systems
4664 Jamestown Ave. #103
Baton Rouge, LA 70808
(504) 928-4661

Safety Sciences Inc.
7586 Trade St.
San Diego, CA 92121
(619) 578-8400

Safety Software, Inc.
2030 Spottswood Road

Suite 200
P.O. Box 5225
Charlottesville, VA 22905-5225
(804) 296-8789
Fax: (804) 296-1660

Software 2000
One Park Center
Drawer 6000
Hyannis, MA 02601
(508) 778-2000

SPECTRUM Human Resource Systems
Corp.
1625 Broadway
Suite 2800
Denver, CO 80202
(303) 534-8813
Fax: (303) 595-9970

Stewart-Todd Associates, Inc.
700 American Ave.
King of Prussia, PA 19406
(215) 962-0166
Fax: (215) 962-0124

TRACOM
Training Communication Corp.
443-B Carlisle Dr.
Herndon, VA 22070
(800) 872-2660

Training Communications Corp.
443 Carlisle Dr.
Herndon, VA 22070
(703) 478-9600

U.S. Occupational Health Inc.
205 W. Randolph St.
Suite 720
Chicago, IL 60606
(312) 641-1449
Fax: (312) 372-0330

Web-Wolf Data Systems, Inc.
7306 Hooking Rd.
McLean, VA 22101
(703) 790-8842

CHAPTER
4

SAFETY AND HEALTH TRAINING

Many standards promulgated by OSHA explicitly require the employer to train employees in the safety and health aspects of their jobs. Other OSHA standards make it the employer's responsibility to limit certain job assignments to employees who are "certified," "competent," or "qualified," meaning that they possess the knowledge to perform the job safely. This knowledge may have been obtained at work or outside of the workplace. OSHA also utilizes the term "designated" personnel, meaning an employee selected or assigned by the employer as being qualified to perform specific duties. These requirements reflect OSHA's belief that training is an essential part of every employer's program for providing workers with a safe and healthy workplace. Many researchers conclude that those who are new on the job have a higher rate of accidents and injuries than more experienced workers. If ignorance of specific job hazards and of proper work practices is even partly to blame for higher injury rates, then training may help provide a solution.

The following issues will be covered in detail in this chapter to assist you in developing an effective training program:

- ❖ Value of Safety and Health Training
- ❖ Training Required by the Regulations
- ❖ Recommended Training
- ❖ The Training Process.
- ❖ Recordkeeping

VALUE OF SAFETY AND HEALTH TRAINING

An obvious benefit of an effective safety and health training program is that employees become better informed regarding the particular hazards present in their work area and in the job process itself. The greater the recognition of hazards, the more likely the employee will be able to avoid unsafe work practices and conditions. In return, accident and illness rates will decrease.

However, there are several other direct and indirect values obtained by a well managed and instituted training program. The first of these is that compliance with OSHA standards will be achieved, meaning that citations stemming from an OSHA inspection can be eliminated or significantly decreased. Compliance with the standards means that financial resources can be put to uses other than paying substantial fines. More importantly, decreased accidents and injuries will result in a better bottom line due to reduced insurance costs, less time off of work, and other related benefits.

An effective training program also has indirect effects on financial resources. Research shows that well-trained employees are better motivated and are more likely to feel that the employer cares about their well-being. This in turn leads to improved productivity which also has a positive effect on "the bottom line." The significance of these indirect effects should not be underestimated. A well-managed program can mean the difference between employees who are loyal, happy, productive, and cooperative and those who are unhappy, unproductive, and resistant to management involvement. Although meeting the OSHA training requirements fulfills management's obligation under the standards, taking an extra step by providing timely, quality training will result in a substantial array of benefits.

TRAINING REQUIRED BY THE REGULATIONS

The training requirements set forth in OSHA's general industry standards are numerous and often complex. With this in mind, an attempt is made in this section to simplify the process of identifying your training obligations. Rather than discuss every requirement of each standard, the standards and requirements are addressed in Table 4-1 located at the end of the Chapter. Before using this Table, there are several items to consider.

- ❖ The OSHA standards are constantly changing and it is important to review new and revised standards to ensure continual compliance.

- ❖ This Table does not include references to "competent," "qualified," "certified," "designated," or similar terms, because these terms do not necessarily require that the employee receive formal training at the workplace.

- ❖ If the frequency of training is listed as "unspecified," the standard in question does not indicate when the employee should receive training. However, to provide the highest effectiveness, training should always occur before the employee is assigned to the task requiring the training.

- ❖ An asterisk indicates that the standard either expands on what must be included in the training program or provides additional relevant information.

RECOMMENDED TRAINING

In addition to the training currently required by OSHA regulations, there are several other topics which management should consider integrating into the training program. These include ergonomics, heat stress and cold stress, non-ionizing radiation, environmental rules and regulations, first aid/cardiopulmonary resuscitation (CPR), and general occupational health and safety topics.

Ergonomics

Ergonomic hazards are common in today's world of video display terminals and fast paced, repetitive motion production lines. OSHA has developed voluntary guidelines for the meatpacking industry, which has a high rate of cumulative trauma disorders of the arms and wrists. A proposed rule for general industry is expected to be issued in the future. Several state and local governments have also issued guidelines. Combined with engineering modifications, training the worker to use proper body positioning can greatly decrease the incidence of injury.

Heat and Cold Stress

Although OSHA has not issued a standard on temperature extremes, and one is not expected, the agency may still issue a citation under the "general duty clause" of the Occupational Safety and Health Act. Workers properly trained in the hazards of heat and cold stress will tolerate these environments with less illness.

Non-Ionizing Radiation

Currently, there are no established training requirements for employees working with non-ionizing radiation. However, sources such as lasers can present serious hazards if employees are not familiar with appropriate procedures and protective devices. Training should be specific for the type of non-ionizing radiation utilized.

Environmental Rules and Regulations

The OSHA Hazardous Waste Operations and Emergency Response Standard (29 CFR 1910.120) does require that all employees receive "first responder" training at an awareness level which provides the employee with the proper procedures to follow if a spill or leak of a material is encountered. However, there are currently no OSHA standards which require the employer to provide general worker training on chemical waste treatment and disposal. The Resource Conservation and Recovery Act, under the EPA, may require training for certain workers handling hazardous waste. Because environmental regulations are often very complex and non-compliance can be very costly, it is imperative that employees receive training on the proper procedures for handling, collecting, and disposing of chemical waste resulting from materials with which they work. Failing to do so can result in severe liability for many years to come.

First Aid/CPR

First aid and CPR should be offered to all employees interested in receiving training in this area. Not only can it prove invaluable in the work setting, it can also be of great value to employees away from work.

General Occupational Safety and Health Training

It is also recommended that general information on how safety and health issues are resolved be included in the training agenda. Many employees can develop a greater understanding of a "safe and healthy" work place if some discussion time is devoted to explaining the interaction between the Occupational Safety and Health Act of 1970, OSHA, and the National Institute for Occupational Safety and Health (NIOSH). Employees should also be made familiar with the terminology used by safety and health professionals. Many employees have great difficulty understanding terms such as "part per million" or "time-weighted average." By providing this type of information, employees who are uncertain about the hazards on the job will be able to put things into better perspective.

THE TRAINING PROCESS

In order to maximize the effectiveness of any training program, you must incorporate each of the following steps:

❖ Identify training needs
❖ Identify goals and objectives
❖ Develop the training material
❖ Conduct the training
❖ Evaluate training effectiveness.

Failing to complete each of these steps in the sequence described can make any training program less effective.

Identifying Training Needs

The first step in instituting a training program is identifying the training needs. One way to identify the needs is to review the required and recommended training previously outlined. Obviously, if a potential hazard is present in the plant for which OSHA specifically requires training, developing and conducting training for that particular hazard should be a top priority. Additionally, even if training has been conducted in the past, refresher training may also be required under the regulations.

The following situations indicate that refresher training should be conducted:

❖ New equipment or processes are introduced
❖ Standard operating procedures have been revised
❖ Employee performance needs improvement
❖ Employees have difficulty remembering important information.

Many regulations do not specify a time-frame for conducting refresher training, although it is commonly done on an annual basis. The following indicators assist in identifying training needs where the interval for refresher training is not specified:

❖ Increased injury and illness incidence rates
❖ Increase in observation of unsafe work practices
❖ Increase in near-misses

❖ Revised safety and health policies.

The hazards, injury and illness statistics, training records, and employee comments for all plant operations should be reviewed when attempting to identify and prioritize training needs.

Program Goals and Objectives

Once the training needs have been identified and prioritized, you can then begin preparing the training goals and objectives. Effective, clearly stated objectives specifically describe what employees are expected to do, to do better, or to discontinue doing after completion of the training. Although objectives do not necessarily need to be written, this is the preferred method. Written objectives provide for consistency in training from employee to employee.

Objectives should also provide a means for demonstrating training effectiveness. Review the following examples:

Objective A— The employee will understand how to use a fire extinguisher.

Objective B— The employee will be able to describe how a fire extinguisher works and on what types of fires A, B, and C extinguishers should be used.

Objective B uses specific, action-oriented wording which will allow instructors to measure the competence of the trainees.

Developing the Materials

After clear objectives have been developed, development of the training material can follow. The training material should simulate the job or activity as closely as possible. Additionally, consider the following during development of the training material:

The *number of people* to be included in the training session may influence the type of training utilized. For example, training a large number of employees may be more conducive to using an outside source for conducting the training and developing more elaborate training materials, such as slides and films. Small groups of people or one-on-one training, which often includes job specific training, may best be done using plant safety and health professionals or supervisory personnel.

The *nature of the skills or knowledge to be gained* from the training can also influence the types of material used. Training in physical skills normally requires more "hands-on" training. Knowledge based training normally requires more lecture materials. However, using both the "hands-on" and lecture methods provides for the optimum result.

The variety of types of training materials are numerous. These include printed materials, overhead transparencies, slides, slide/tape formats, films, interactive computer programs, and other media. Using a combination of training methods assists in retaining employee interest. There are a number of excellent and inexpensive sources of training programs for employees. Many states offer programs for rent or purchase as part of their Workers' Compensation or state inspection services. The U.S. National Audiovisual Center offers a series of training programs developed by OSHA, NIOSH, and others. Trade associations, including the National Manufacturers Association, routinely sponsor various training seminars. Finally, professional associations, such as the National Safety Council, American Society of Safety Engineers, and the American Industrial Hygiene Association, all offer training programs and materials on a fee basis. However, avoid the prolonged use of videotapes and other similar types of visual aids, as these tend to shorten the attention span of the trainee.

Conducting the Training

The steps taken thus far will undoubtedly assist in providing interesting and informative training for your employees. However, even the most well-developed training material can be ineffective if not properly utilized and communicated. Below are important factors to consider when conducting the training or selecting the trainer:

❖ The organization and purpose of the training must be made clear to the trainee.

❖ New information must be related to the trainee's job or situation.

❖ Summarizing the objectives reinforces the learning process.

❖ The importance, relevance, and benefits of the training must be made clear.

❖ Participation of the trainee reinforces the learning process.

Evaluating Effectiveness

Perhaps the most critical part of the training process is that of evaluation. Only through evaluation can training needs be confirmed, objectives improved, and the training material be made more effective. Ideally, all aspects of the training process should be evaluated by the trainee either during the training session or as soon afterwards as possible. This can include paper performance or behavioral tests. The supervisors of the employees attending training sessions should also evaluate the training by observing the performance (changes in behavior) of those employees. And, just as injury and illness statistics may indicate the need for training, a decrease in those rates can also indicate the effectiveness of a training program.

Those parts of the program receiving poor evaluations should be reviewed and revised promptly. It is also important to determine what *was* effective so that the good elements of a program can be incorporated into future programs.

RECORDKEEPING

Federal, state, and local regulations which mandate training will also usually require that the employer be able to prove that the training was given and may require proof that a certain level of proficiency was achieved. Many employers have been faced with severe fines or liability because they were unable to prove that the employee took the training or understood it. All training that is provided should be documented in that employee's records. Many employers accomplish this by using special charge codes for time, as well as maintaining a sign-in log.

It is recommended that each training course given for which recordkeeping is important include a sign-in log (typed name and signature). It is also recommended that the course syllabus or schedule and any paper exercises be retained for your records. These can be important when the employee tells the inspector that he or she does not remember having taken the training in question.

Table 4-1
Training Required by the Regulations

Standard	Requirement	Frequency of Training
Subpart E - Means of Egress		
Employee Emergency Plans and Fire Prevention Plans (1910.38)	The employer shall designate and train a sufficient number of persons to assist in the safe and orderly emergency evacuation of employees	Before Plan Implementation
	The employer shall review the plan with each employee covered by the plan	Upon Plan Development Change in Responsibilities Change in the Plan
	The employer shall review with each employee those parts of the plan which the employee must know to protect the employee in the event of an emergency	Upon Initial Assignment
	The employer shall apprise employees of the fire hazards of the materials and processes to which they are exposed	Not Specified
	The employer shall review with each employee those parts of the fire prevention plan which the employee must know to protect the employee in the event of an emergency	Upon Initial Assignment
Subpart F - Powered Platforms, Manlifts, and Vehicle-Mounted Platforms		
Powered Platforms for Building Maintenance (1910.66)	All employees who operate working platforms shall be trained*	Unspecified

Table 4-1 (cont'd)

Subpart G - Occupational Health and Environmental Control

Ventilation (1910.94)	All employees working in and around open-surface tank operations must be instructed*	Unspecified
	If in emergencies, it is necessary to enter a tank which may contain a hazardous atmosphere, at least one trained standby employee shall be present	Unspecified
Occupational Noise Exposure (1910.95)	The employer shall institute a training program for all employees who are exposed to noise at or above an 8-hour Time-Weighted Average of 85 dB, and shall ensure employee participation in such program*	Annually
Ionizing Radiation (1910.95)	All employees whose work may necessitate their presence in an area covered by the immediate evacuation warning signal shall be made familiar with the actual sound of the signal by actual demonstration	Before System Operation
	All individuals working in or frequenting any portion of a radiation area shall be informed and instructed*	Unspecified

Subpart H - Hazardous Materials

Flammable and Combustible Liquids (1910.106)	Station operators and other employees depended upon to carry out flood emergency instructions shall be thoroughly informed*	Unspecified
Explosives and Blasting Agents (1910.109)	Vehicles transporting explosives shall only be driven by, and be in the charge of, a driver who is familiar with the traffic regulations, State laws, and the provisions of the standard	
	The driver of every motor vehicle transporting any quantity of Class A or Class B explosives shall have been instructed in the measures and procedures to be followed in order to protect the public from danger and shall be trained to move the vehicle when required	Before Assignment

Table 4-1 (cont'd)

	Operators of bulk delivery vehicles shall be trained in the safe operation of the vehicle together with its mixing, conveying, and related equipment, and shall be familiar with the commodities be delivered and the general procedure for handling emergency situations	Unspecified
	The driver of vehicles transporting blasting agents shall be familiar with the States' vehicle and traffic laws	Unspecified
	The driver of vehicles used over public highways for the bulk transportation of water gels or of ingredients classified as dangerous commodities shall meet the training requirements as described above	Before Assignment
Storage and Handling of Liquefied Petroleum Gases (1910.110)	Personnel performing installation, removal, operation, and maintenance work shall be properly trained in such function	Unspecified
	When standard watch service is provided, it shall be extended to the Liquefied Petroleum Gas installation, and personnel properly trained	Unspecified
Storage and Handling of Anhydrous Ammonia (1910.111)	Personnel required to handle ammonia should be trained in safe operating practices and in the proper action to take in the event of emergencies	Unspecified
	The employer shall insure that unloading operations are performed by reliable persons properly instructed	Unspecified
Hazardous Waste Operations and Emergency Response (1910.120)	All employees working on site exposed to hazardous substances, health hazards, or safety hazards, and their supervisors and management responsible for the site shall receive training before engaging in hazardous waste operations*	Before Assignment
	General site workers engaged in hazardous substance removal or other activities which expose workers to hazardous substances and health hazards shall receive a minimum of 40 hours of instruction, and a minimum of three days actual field experience	Before Assignment

Table 4-1 (cont'd)

Workers on site only occasionally for a specific limited task and who are unlikely to be exposed over permissible exposure limits and published exposure limits shall receive a minimum of 24 hours of instruction, and a minimum of one day actual field experience	Before Assignment
Workers regularly on site who work in areas which have been monitored and fully characterized, and the characterization indicates that there are no health hazards or the possibility of an emergency developing, shall receive a minimum of 24 hours of instruction and the minimum of one day actual field experience	Before Assignment
Workers with 24 hours of training who are covered as above, and who become general site workers or who are required to wear respirators, shall have 16 hours and 2 days training to total the training for a general site worker	As Required
On-site management and supervisord directly responsible for, or who supervise employees engaged in, hazardous waste operations shall receive 40 hous initial training, and three days of supervised field experience	Before Assignment
Employees who are engaged in responding to hazardous waste clean-up sites that may expose them to hazardous substances shall be trained in how to respond to such expected emergencies	Before Assignment
Employees shall receive eight hours of refresher training	Annually
Employees who, in the course of their regular job duties, work with and are trained in the hazards of specific hazardous substances, and who will be called upon to provide technical advice or assistance at a hazardous substance release incident to the individual in charge, shall receive training or demonstrate competency in the area of their specialization	Annually

Table 4-1 (cont'd)

Subpart I - Personal Protective Equipment		
	Employees who participate, or are expected to participate, in emergency response, shall be given training*	Before Assignment
	Those employees who are trained to participate in emergency response shall receive refresher training	Annually
Respiratory Protection (1910.134)	The user shall be instructed and trained in the proper use of respirators and their limitations	Unspecified
	Personnel shall be familiar with the procedures covering safe use of respirators in dangerous atmospheres that might be encountered in normal operations or in emergencies	Unspecified
	Supervisors and workers shall be instructed in the selection, use, and maintenance of respirators*	Unspecified
	Every respirator wearer shall receive fitting instructions including demonstrations and practice*	Unspecified
Subpart J - General Environmental Controls		
Specifications for Accident Prevention Signs and Tags (1910.145)	All employees shall be instructed that danger signs indicate immediate danger and that special precautions are necessary	Unspecified
	All employees shall be instructed that caution signs indicate a possible hazard against which proper precautions should be taken	Unspecified

Table 4-1)cont'd)

Confined Spaces (1910.146)	*Authorized entrants, attendants, entry supervisors, and rescue and emergency service personnel must be trained to acquire the understanding, knowledge and skills necessary for the safe performance of duties assigned*	*Before being assigned duty* *When changes are made* *When needed to improve proficiency* *Annually for rescue service*
The Control of Hazardous Energy (Lockout/Tagout) (1910.147)	*The employer shall provide training to ensure that the purpose and function of the energy control program are understood by employees and that the knowledge and skills required are acquired by employees**	*Upon Initial Assignment* *Change in Assignments* *New Hazards* *Change in Control Procedures*
Subpart L - Fire Protection		
Fire Brigades (1910.156)	*The employer shall provide training and education for all fire brigade members commensurate with those duties and functions that fire brigade members are expected to perform**	*Before Performance* *For Satisfactory Performance* *At Least Annually*
	Fire brigade members who are expected to perform interior structural fire fighting shall be provided with an education session or training	*Quarterly*
Portable Fire Extinguishers (1910.157)	*Where the employer has provided portable fire extinguishers for employee use in the workplace, the employer shall provide an educational program**	*Upon Initial Assignment* *Annually*
	The employer shall provide employees who have been designated to use fire fighting equipment as part of an emergency action plan with training*	*Upon Initial Assignment* *Annually*
Fire Extinguishing Systems (1910.160)	*The employer shall train employees designated to inspect, maintain, operate, or repair fixed extinguishing systems*	*Upon Initial Assignment* *Annually*

Table 4-1 (cont'd)

Subpart N - Materials Handling and Storage		
Servicing of Single Piece and Multi-Piece Rim Wheels (1910.177)	The employer shall provide a training program to train and instruct all employees who service multi-piece rim wheels*	Before Assignment As Necessary to Assure Proficiency
Powered Industrial Trucks (1910.178)	Only trained and authorized operators shall be permitted to operate a powered industrial truck. Methods shall be devised to train operators in the safe operation of powered industrial trucks.	Unspecified
Subpart O - Machinery and Machine Guarding		
Mechanical Power Presses (1910.217)	The employer shall train and instruct the operator in the safe method of work	Before Assignment
Forging Operations (1910.218)	It shall be the responsibility of the employer to train personnel for the proper	Unspecified
Subpart Q - Welding, Cutting and Brazing		
Welding, Cutting and Brazing (1910.252)		

CHAPTER
5
RECORDKEEPING

This chapter is intended as a guide for collecting, recording, and analyzing safety and health information. It approaches recordkeeping as a tool for the proactive manager to assist him or her in making decisions for the prevention of accidents in the manufacturing environment, as well as meeting reporting and recordkeeping requirements of the OSH Act. The chapter is divided into three sections: Occupational Injury and Illness Records, Miscellaneous Safety and Health Records, and Recordkeeping Systems - A Management Tool. Appendix 5-A at the end of the chapter contains a detailed description of recordkeeping requirements under OSHA organized by major sections of the regulations.

OCCUPATIONAL INJURY AND ILLNESS RECORDS

Recordkeeping requirements of the OSH Act apply to nearly all employers. The purpose of the recordkeeping requirements of the Act is to enable OSHA to develop statistical information concerning injuries and illnesses. Injury and illness reporting provides OSHA with information on occupational injuries or illnesses: when they occur, their causes(s), types, severity, and frequency.

OSHA compiles this data to develop inspection schedules, to determine needs for additional regulations, and to participate in the investigations of fatalities and major accidents. Since management has the obligation to collect, record, and maintain injury and illness information for OSHA and similar state agencies, it is only logical that the proactive manager employ the same information for controlling accidents and their associated costs in his or her facility. Later, this chapter discusses integrating required injury/illness data into a management information system.

There are two reasons for managers to maintain occupational injury and illness records on a timely and accurate basis. First, it is an OSHA requirement. Businesses that have falsified or failed to maintain these records have received willful citations with fines some of which have been in excess of $1 million dollars. Secondly, it is a good business tool. The data from these records provides valuable information for a manager to develop action plans to prevent the recurrence of injuries and illnesses and ultimately reduce workers' compensation costs.

The forms required for OSHA recordkeeping are (1) Log and Summary of Occupational Injuries and Illnesses, OSHA No. 200 and (2) the Supplementary Record of Occupational Injuries and Illnesses, OSHA No. 101. These forms will be discussed in detail later in this chapter.

Who Must Keep OSHA Records?

There are three categories of employers: employers exempt from OSHA required recordkeeping requirements, employers who infrequently must keep OSHA records, and employers that must keep OSHA records. The reader is cautioned that individual state or local agencies may also have requirements for reporting accidents and illnesses that are not addressed in this guide.

Employers exempt from OSHA recordkeeping requirements include the following:

❖ Self-employed individuals

❖ Partnerships with no employees

❖ Employers of employees engaged in domestic services.

❖ Employers engaged in religious activities. However, records of accidents and illnesses of employees engaged in secular activities must be kept.

Employers who infrequently must keep OSHA records are employers in manufacturing industries with 10 or fewer full- or part-time employees at one time in the previous calendar year and employers in the retail trade, finance, insurance, real estate, and services industries. A sample of employers in this category is randomly selected each year and required to keep occupational injury and illness records. It is mandatory for employers in this category to participate in the sampling program. OSHA contacts the selected employers prior to the recordkeeping year and provides the necessary forms for recording accident and illness data and instructs the employer on how to record the data.

Employers who must keep OSHA records include all employers having 11 or more full- or part-time employees at any one time in the previous calendar year and engaged in the following industries:

❖ Agriculture, forestry, and fishing

❖ Oil and gas extraction

❖ Construction

❖ Manufacturing

❖ Transportation and public utilities

❖ Wholesale trade

❖ Building materials and garden supplies

❖ General merchandise and food stores

❖ Hotels and other lodging places

❖ Repair services

❖ Amusement and recreation services

❖ Health services

Location, Retention, and Maintenance of Records

Injury and illness records must be kept by employers for each of their establishments. OSHA defines an establishment as "a single physical location where business is conducted or where services or industrial operations are performed." In addition, OSHA specifies that distinctly separate activities performed at the same physical location shall be considered separate establishments for recordkeeping purposes. The following list of conditions would be considered separate establishments by OSHA and would require separate recordkeeping systems:

❖ Production of dissimilar products

❖ Different facilities

❖ Different kinds of operational procedures

❖ Separate management organizations (payroll, personnel or support staff)

Under the OSHA regulations, the location where occupational injury and illness records are kept depends on whether or not the employees are associated with a fixed or nonfixed establishment. Fixed establishments remain at a given location for long term or permanent time. OSHA considers "long term" as any period of time exceeding one year. Nonfixed establishments usually operate at a single site for a relatively short period of time.

Records for employees associated with fixed establishments should be located as follows:

❖ Records for employees working at fixed locations should be kept at the work location.

❖ Records for employees that report to a fixed location but perform their work tasks elsewhere should be kept at the location where they *report* each day.

❖ Records for employees whose payroll and personnel records are maintained at a fixed location, but who do not report regularly at that location should be kept at the site of the business's payroll and/or personnel offices.

Records for employees associated with nonfixed establishments should be located as follows:

❖ Records should be kept at the field office or mobile base of operations.

❖ Records may be kept at an established location, providing:

1) The address and phone number of the location at which the records are kept is available at the work site.

2) Someone at that location during normal business hours can provide information from the records.

OSHA requires employers to complete and maintain two specific injury and illness forms: the OSHA Form 200 and OSHA Form 101 or its equivalent. Occasionally an employer will be sent an OSHA 200S Form that must be completed. The 200S form is a summary of the 200 form. Randomly selected businesses are sent the summary form.

LOG AND SUMMARY OF OCCUPATIONAL INJURIES AND ILLNESSES: OSHA FORM 200

The OSHA 200 log is used for recording and classifying occupational injuries and illnesses and for documenting the extent of each case. Employers may use the log provided by OSHA (Figure 5-1) or use an equivalent form including computer generated forms, providing the form contains the same information as the OSHA form. A brief overview of the information required on the 200 log follows:

(1) Case or file number

This entry must coincide with the case or file number on OSHA Form 101.

(2) Name of injured employee

(3) Date of injury or illness

(4) Occupation (job title) of injured employee

(5) Department or description of normal workplace

(6) Description of injury or illness and extent and outcome of injury:

  (a) Fatality

  (b) Injuries involving lost and/or restricted workdays

  (c) Injuries involving days away from work

  (d) Number of days away from work

  (e) Number of days of restricted work

  (f) Injuries without lost workdays

  (g) Types of illnesses

      (i) Skin diseases or disorders

      (ii) Dust diseases of the lungs

      (iii) Respiratory conditions due to toxic materials

      (iv) Poisoning (Systemic effects of toxic materials)

      (v) Disorders due to physical agents

      (vi) Disorders associated with repeated trauma

      (vii) All other occupational illnesses

Extent and outcome of illnesses:

  (h) Fatality

  (i) Illnesses involving lost and/or restricted workdays

  (j) Illnesses involving days away from work

(k) Number of days away from work

(l) Number of days of restricted work

(m) Illnesses without lost workdays.

Care must be taken to ensure that the 200 log is accurate, up-to-date, and that over- or under-reporting of accidents and illnesses is avoided. Over-reporting can cause a business to be targeted for an OSHA inspection while under-reporting can lead to OSHA citations.

The persons maintaining the records must analyze each case and decide if the case is REPORTABLE. The decision process consists of four steps:

1. Determine whether a case has occurred. Has there been a fatality, injury, or illness?

2. Establish that the case was work-related.

3. Determine if the case is an injury or an illness.

4. If the case is a work-related illness, enter the required information in the illness category section of the log.

5. If the case is a work-related injury, enter the required information in the injury category section of the log.

Figure 5-2 presents this scheme in graphic form.

DETERMINING IF A REPORTABLE CASE HAS OCCURRED

Determination that a work related injury or illness case has occurred is not as straight forward as it first appears. There are many instances in which an injury or illness case is not reportable. Examples of cases that are not reportable include but are not limited to:

❖ Accidents in employer parking lots, provided the injured employee is not performing a work related assignment.

❖ Injuries sustained in company sponsored activities such as athletic games or picnics if such activities are not made mandatory by the employer. Such injuries may be compensable but they are not reportable.

❖ Instances when employees are sent to the hospital, providing they do not receive medical attention other than first aid treatment or diagnostic treatment.

❖ When there are recurring symptoms from a previous injury or illness. The recurrence of symptoms from diseases such as silicosis which have prolonged effects would not be considered a new case. The recurrence of symptoms may require the record keeper to adjust the original entries on the 200 log. Aggravation of previous injuries generally results from a new incident such as a slip, fall, or strain. In such cases a new entry would be required.

Figure 5-3 is a general guide for determining if an injury or illness is work-related.

Figure 5-1
Log and Summary of Occupational Injuries and Ilnesses
OSHA No. 200

Figure 5-1 (cont'd)

Instructions for OSHA No. 200

I. Log and Summary of Occupational Injuries and Illnesses

Each employer who is subject to the recordkeeping requirements of the Occupational Safety and Health Act of 1970 must maintain for each establishment a log of all recordable occupational injuries and illnesses. This form (OSHA No. 200) may be used for that purpose. A substitute for the OSHA No. 200 is acceptable if it is as detailed, easily readable, and understandable as the OSHA No. 200.

Enter each recordable case on the log within six (6) workdays after learning of its occurrence. Although other records must be maintained at the establishment to which they refer, it is possible to prepare and maintain the log at another location, using data processing equipment if desired. If the log is prepared elsewhere, a copy updated to within 45 calendar days must be present at all times in the establishment.

Log must be maintained and retained for five (5) years following the end of the calendar year to which they relate. Logs must be available (normally at the establishment) for inspection and copying by representatives of the Department of Labor, or the Department of Health and Human Services, or States accorded jurisdiction under the Act. Access to the log is also provided to employees, former employees, and their representatives.

II. Changes in Extent of or Outcome of Injury or Illness

If, during the 5-year period the log must be retained, there is a change in an extent and outcome of an injury or illness which affects entries in columns 1, 2, 6, 8, 9, or 13, the first entry should be lined out and a new entry made. For example, if an injured employee at first required only medical treatment but later lost workdays away from work, the check in column 6 should be lined out, and checks entered in columns 2 and 3 and the number of lost workdays entered in column 4.

In another example, if an employee with an occupational illness lost workdays, returned to work, and then died of the illness, any entries in columns 9 through 12 should be lined out and the date of death entered in column 8.

The entire entry for an injury or illness should be lined out if later found to be nonrecordable. For example, an injury which is later determined not to be work related, or which was initially thought to involve medical treatment but later was determined to have involved only first aid.

III. Posting Requirements

A copy of the totals and information following the fold line of the last page for the year must be posted at each establishment in the place or places where notices to employees are customarily posted. This copy must be posted no later than *February 1 and must remain in place until March 1*.

Even though there were no injuries or illnesses during the year, zeros must be entered on the totals line, and the form posted.

The person responsible for the *annual summary totals* shall certify that the totals are true and complete by signing at the bottom of the form.

IV. Instructions for Completing Log and Summary of Occupational Injuries and Illnesses

Column A – CASE OR FILE NUMBER. Self-explanatory.

Column B – DATE OF INJURY OR ONSET OF ILLNESS.
For occupational injuries, enter the date of the work accident which resulted in injury. For occupational illnesses, enter the date of initial diagnosis of illness, or, if absence from work occurred before diagnosis, enter the first day of the absence attributable to the illness which was later diagnosed or recognized.

Columns C through F – Self-explanatory.

Columns 1 and 8 – INJURY OR ILLNESS-RELATED DEATHS. Self-explanatory.

Columns 2 and 9 – INJURIES OR ILLNESSES WITH LOST WORKDAYS. Self-explanatory.

Any injury which involves days away from work, or days of restricted work activity, or both must be recorded since it always involves one or more of the criteria for recordability.

Columns 3 and 10 – INJURIES OR ILLNESSES INVOLVING DAYS AWAY FROM WORK. Self-explanatory.

Columns 4 and 11 – LOST WORKDAYS—DAYS AWAY FROM WORK.
Enter the number of workdays (consecutive or not) on which the employee would have worked but could not because of occupational injury or illness. The number of lost workdays should not include the day of injury or onset of illness or any days on which the employee would not have worked even though able to work.
NOTE: For employees not having a regularly scheduled shift, such as certain truck drivers, construction workers, farm labor, casual labor, part-time employees, etc., it may be necessary to estimate the number of lost workdays. Estimates of lost workdays shall be based on prior work history of the employee AND days worked by employees, not ill or injured, working in the department and/or occupation of the ill or injured employee.

Columns 5 and 12 – LOST WORKDAYS—DAYS OF RESTRICTED WORK ACTIVITY.
Enter the number of workdays (consecutive or not) on which because of injury or illness:
(1) the employee was assigned to another job on a temporary basis, or
(2) the employee worked at a permanent job less than full time, or
(3) the employee worked at a permanently assigned job but could not perform all duties normally connected with it.

The number of lost workdays should not include the day of injury or onset of illness or any days on which the employee would not have worked even though able to work.

Columns 6 and 13 – INJURIES OR ILLNESSES WITHOUT LOST WORKDAYS. Self-explanatory.

Columns 7a through 7g – TYPE OF ILLNESS.
Enter a check in only *one* column for each illness.

TERMINATION OR PERMANENT TRANSFER–Place an asterisk to the right of the entry in columns 7a through 7g (type of illness) which represented a termination of employment or permanent transfer.

V. Totals

Add number of entries in columns 1 and 8.

Add number of checks in columns 2, 3, 6, 7, 9, 10, and 13.

Add number of days in columns 4, 5, 11, and 12.

Yearly totals for each column (1-13) are required for posting. Running or page totals may be generated at the discretion of the employer.

If an employee's loss of workdays is continuing at the time the totals are summarized, estimate the number of future workdays the employee will lose and add that estimate to the workdays already lost and include this figure in the annual totals. No further entries are to be made with respect to such cases in the next year's log.

VI. Definitions

OCCUPATIONAL INJURY is any injury such as a cut, fracture, sprain, amputation, etc., which results from a work accident or from an exposure involving a single incident in the work environment.

OCCUPATIONAL ILLNESS of an employee is any abnormal condition or disorder, other than one resulting from an occupational injury, caused by exposure to environmental factors associated with employment. It includes acute and chronic illnesses or diseases which may be caused by inhalation, absorption, ingestion, or direct contact.

NOTE: Conditions resulting from animal bites, such as insect or snake bites or from one-time exposure to chemicals, are considered to be injuries.

The following listing gives the categories of occupational illnesses and disorders that will be utilized for the purpose of classifying recordable illnesses. For purposes of information, examples of each category are given. These are typical examples, however, and are not to be considered the complete listing of the types of illnesses and disorders that are to be counted under each category.

7a. **Occupational Skin Diseases or Disorders**
Examples: Contact dermatitis, eczema, or rash caused by primary irritants and sensitizers or poisonous plants; oil acne; chrome ulcers, chemical burns or inflammations; etc.

7b. **Dust Diseases of the Lungs (Pneumoconioses)**
Examples: Silicosis, asbestosis and other asbestos-related diseases, coal worker's pneumoconiosis, byssinosis, siderosis, and other pneumoconioses.

7c. **Respiratory Conditions Due to Toxic Agents**
Examples: Pneumonitis, pharyngitis, rhinitis or acute congestion due to chemicals, dusts, gases, or fumes; farmer's lung; etc.

7d. **Poisoning (Systemic Effect of Toxic Material)**
Examples: Poisoning by lead, mercury, cadmium, arsenic, or other metals; poisoning by carbon monoxide, hydrogen sulfide, or other gases; poisoning by benzol, carbon tetrachloride, or other organic solvents; poisoning by insecticide sprays such as parathion, lead arsenate; poisoning by other chemicals such as formaldehyde, plastics, and resins; etc.

7e. **Disorders Due to Physical Agents (Other than Toxic Materials)**
Examples: Heatstroke, sunstroke, heat exhaustion, and other effects of environmental heat; freezing, frostbite, and effects of exposure to low temperatures; caisson disease; effects of ionizing radiation (isotopes, X-rays, radium); effects of nonionizing radiation (welding flash, ultraviolet rays, microwaves, sunburn); etc.

7f. **Disorders Associated With Repeated Trauma**
Examples: Noise-induced hearing loss; synovitis, tenosynovitis, and bursitis; Raynaud's phenomena; and other conditions due to repeated motion, vibration, or pressure.

7g. **All Other Occupational Illnesses**
Examples: Anthrax, brucellosis, infectious hepatitis, malignant and benign tumors, food poisoning, histoplasmosis, coccidioidomycosis, etc.

MEDICAL TREATMENT includes treatment (other than first aid) administered by a physician or by registered professional personnel under the standing orders of a physician. Medical treatment does NOT include first-aid treatment (one-time treatment and subsequent observation of minor scratches, cuts, burns, splinters, and so forth, which do not ordinarily require medical care) even though provided by a physician or registered professional personnel.

ESTABLISHMENT: A single physical location where business is conducted or where services or industrial operations are performed (for example, a factory, mill, store, hotel, restaurant, movie theater, farm, ranch, bank, sales office, warehouse, or central administrative office). Where distinctly separate activities are performed at a single physical location, such as construction activities operated from the same physical location as a lumber yard, each activity shall be treated as a separate establishment.

For firms engaged in activities which may be physically dispersed, such as agriculture; construction; transportation; communications; and electric, gas, and sanitary services, records may be maintained at a place to which employees report each day.

Records for personnel who do not primarily report or work at a single establishment, such as traveling salesmen, technicians, engineers, etc., shall be maintained at the location from which they are paid or the base from which personnel operate to carry out their activities.

WORK ENVIRONMENT is comprised of the physical location, equipment, materials processed or used, and the kinds of operations performed in the course of an employee's work, whether on or off the employer's premises.

Figure 5-2
Guide to Recordability of Cases Under the OSH Act

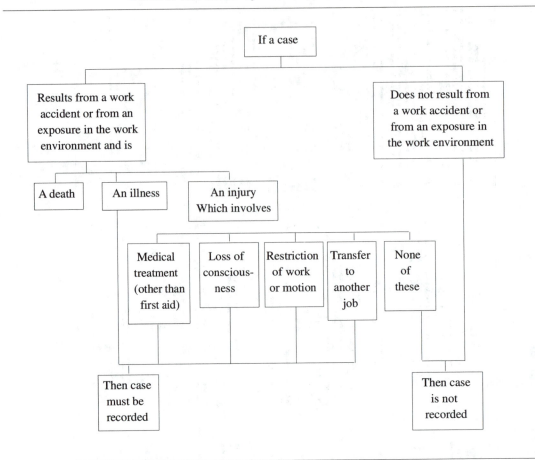

MEDICAL AND FIRST AID TREATMENT

The kinds of treatment that an employee receives for an injury determines whether or not a case is reportable. The following guidelines can be used to determine if an accident or illness is reportable:

1. Removal of foreign objects from the eye.

    ❖ If object is not imbedded in the eye   –NOT REPORTABLE
    ❖ If object is imbedded in the eye   –REPORTABLE

2. Burns regardless of size.

    ❖ First degree burns   –NOT REPORTABLE
    ❖ Second and third degree burns   –REPORTABLE

3. Administering of prescription medication.

Figure 5-3
Guidelines for Establishing Work Relationship of Injuries and Illnesses

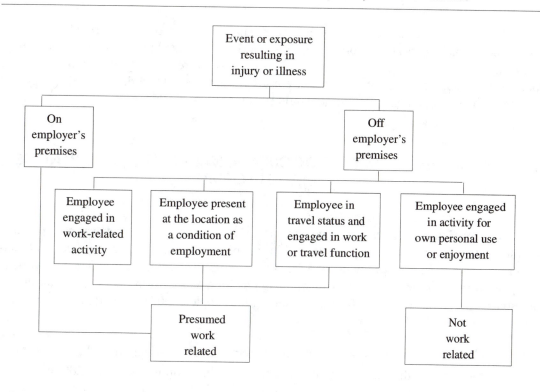

❖ Single dose on first visit for
   minor discomfort   –NOT REPORTABLE

❖ Preventative shots and boosters,
   such as tetanus   –NOT REPORTABLE

❖ All other prescription medications –REPORTABLE

4. Application of sutures or steri-strip bandages.   –REPORTABLE

5. Treatment involving the use of antiseptics heat, soaking, whirlpool bath.

   ❖ First visit only   –NOT REPORTABLE
   ❖ More than one visit   –REPORTABLE

6. The cutting away of dead skin (debridement).   –REPORTABLE

7. Loss of consciousness.   –REPORTABLE

8. Restriction of work or motion.   –REPORTABLE

9. Transfer to another job, such as in cases
   involving allergic reactions to chemicals.   –REPORTABLE

DETERMINING WHETHER A CASE IS AN INJURY OR AN ILLNESS

Determining if a case is an injury or illness usually does not present a problem. Illnesses are defined by OSHA as any abnormal condition or disorder other than an occupational injury. Acid burns would be considered injuries while disorders of the respiratory system from repeated long term exposures to acid vapors would be considered an illness. Disorders caused by repeated trauma such as carpal tunnel syndrome would be reported as an illness.

SUPPLEMENTARY RECORD OF OCCUPATIONAL INJURIES AND ILLNESSES: (OSHA 101 FORM)

For every injury or illness recorded on your OSHA 200 log you must have a corresponding supplemental record of the injury or illness. The supplemental record describes pertinent facts about the injury. Managers may use the OSHA 101 form, shown in Figure 5-4, the workers compensation form, or a form of original design as long as the form contains the minimum required information. Whatever form is chosen for the supplemental record, it must contain the following minimum information:

1. Case or File Number—The case or file number on the supplemental form must be the same as the case or file number recorded in column (A) on the 200 log. This requirement ties the information on the 200 log to the corresponding information on the supplementary report for each reportable incident. Failure to have corresponding case or file numbers may cause you to be cited and fined by OSHA.

2. Employer Information—The name and the mailing address of the employer and the location address of the employer if different than the mailing address.

3. Employee Information—Personal data concerning the injured employee:

   ❖ Name, address, age, sex and social security number

   ❖ Occupation—Record the employee's regular job title, not the specific task being done at the time of the injury. Example: Truck driver as opposed to loading truck or Repairman as opposed to parts cleaning.

   ❖ Department—Record the employee's regular assigned department, not the department where the injury took place.

   ❖ Location of the accident or exposure—Record the address of the location where the incident took place and whether or not the incident took place on the employer's premises.

4. Describe how the injury or illness occurred—The following are examples of accident descriptions:

Example One   The employee was taking 50 pound bags of fertilizer from a pallet and loading them into the back of a delivery van. After placing the second bag in the back of the truck the employee felt a sharp pain in his lower back. The employee was not using any material handling equipment.

Example Two   The employee was cleaning grease from gear parts in the maintenance shop degreasing machine. The degreasing machine has an electrical hoist for inserting machine parts into the cleaning solution and for removal following the cleaning cycle. The cleaning chamber has a ventilation hood for removal of solvent vapors. All equipment was operational at the time of the incident. The employee had lowered the loaded cleaning basket and started the cleaning cycle. At this time he discovered he had failed to place an eight-inch diameter gear into the basket. He leaned over the edge of the cleaning chamber and dropped the gear into the basket. Hot solvent 1,1,1-trichloroethane splashed up and struck the employee in the face. He was wearing standard safety glasses, but no face shield. The employee broke out in a rash after leaving work at the end of the shift, and went to the local hospital for an examination.

5. List the objects or substances involved—Only list the objects or substances that have an influence on the incident.

Example One   50 pound bags of fertilizer.

Example Two   1,1,1 - trichloroethane.

6. Describe the injury and illness in detail

Example One   The employee was diagnosed as having a muscle strain and missed 10 days of work starting on $\frac{1}{14}$ and returned to work on $\frac{1}{28}$.

Example Two   The employee was diagnosed as having contact dermatitis on his forehead; he was treated and released for work without loss of work or restrictions.

NOTE: OSHA does not specify that the dates employees miss work or are on work restriction be placed on the supplemental record. However, including this information may prove helpful during an OSHA recordkeeping inspection.

7. Other information

❖ Date of injury or diagnosis of illness
❖ Name and address of hospital if the employee is admitted
❖ If injury or illness resulted in a fatality
❖ Name and address of physician
❖ Date of report, signature, and title of person filling out the report

Reporting of Injuries and Illnesses

Employers are required to report all fatalities and the hospitalization of five or more employees to the nearest office of the Area Director of the Occupational Safety and Health Administration or to the local state agency for states with approved state plans. For notification

Figure 5-4
Supplementary Record of Occupational Injuries and Illnesses, OSHA 101 Form

SUPPLEMENTARY RECORD OF OCCUPATIONAL INJURIES AND ILLNESSES

To supplement the Log and Summary of Occupational Injuries and Illnesses (OSHA No. 200), each establishment must maintain a record of each recordable occupational injury or illness. Worker's compensation, insurance, or other reports are acceptable as records if they contain all facts listed below or are supplemented to do so. If no suitable report is made for other purposes, this form (OSHA No. 101) may be used or the necessary facts can be listed on a separate plain sheet of paper. These records must also be available in the establishment without delay and at reasonable times for examination by representatives of the Department of Labor and the Department of Health and Human Services, and States accorded jurisdiction under the Act. The records must be maintained for a period of not less than five years following the end of the calendar year to which they relate.

Such records must contain at least the following facts:

1) *About the employer*—name, mail address, and location if different from mail address.

2) *About the injured or ill employee*—name, social security number, home address, age, sex, occupation, and department.

3) *About the accident or exposure to occupational illness*—place of accident or exposure, whether it was on employer's premises, what the employee was doing when injured, and how the accident occurred.

4) *About the occupational injury or illness*—description of the injury or illness, including part of body affected; name of the object or substance which directly injured the employee; and date of injury or diagnosis of illness.

5) *Other*—name and address of physician; if hospitalized, name and address of hospital; date of report; and name and position of person preparing the report.

SEE *DEFINITIONS* ON THE BACK OF OSHA FORM 200.

Bureau of Labor Statistics
Supplementary Record of
Occupational Injuries and Illnesses

U.S. Department of Labor

This form is required by Public Law 91-596 and must be kept in the establishment for 5 years.
Failure to maintain can result in the issuance of citations and assessment of penalties.

Case or File No. _____

Form Approved
O.M.B No. 1220-0029

Employer

1. Name

2. Mail address *(No. and street, city or town, State, and zip code)*

3. Location, if different from mail address

Injured or Ill Employee

4. Name *(First, middle, and last)* Social Security No.

5. Home address *(No. and street, city or town, State, and zip code)*

6. Age

7. Sex: *(Check one)* Male ☐ Female ☐

8. Occupation *(Enter regular job title, not the specific activity he was performing at time of injury.)*

9. Department *(Enter name of department or division in which the injured person is regularly employed, even though he may have been temporarily working in another department at the time of injury.)*

The Accident or Exposure to Occupational Illness

If accident or exposure occurred on employer's premises, give address of plant or establishment in which it occurred. Do not indicate department or division within the plant or establishment. If accident occurred outside employer's premises at an identifiable address, give that address. If it occurred on a public highway or at any other place which cannot be identified by number and street, please provide place references locating the place of injury as accurately as possible.

10. Place of accident or exposure *(No. and street, city or town, State, and zip code)*

11. Was place of accident or exposure on employer's premises? Yes ☐ No ☐

12. What was the employee doing when injured? *(Be specific. If he was using tools or equipment or handling material, name them and tell what he was doing with them.)*

13. How did the accident occur? *(Describe fully the events which resulted in the injury or occupational illness. Tell what happened and how it happened. Name any objects or substances involved and tell how they were involved. Give full details on all factors which led or contributed to the accident. Use separate sheet for additional space.)*

Occupational Injury or Occupational Illness

14. Describe the injury or illness in detail and indicate the part of body affected. *(E.g., amputation of right index finger at second joint; fracture of ribs; lead poisoning; dermatitis of left hand, etc.)*

15. Name the object or substance which directly injured the employee. *(For example, the machine or thing he struck against or which struck him; the vapor or poison he inhaled or swallowed; the chemical or radiation which irritated his skin; or, in cases of strains, hernias, etc., the thing he was lifting, pulling, etc.)*

16. Date of injury or initial diagnosis of occupational illness

17. Did employee die? *(Check one)* Yes ☐ No ☐

Other

18. Name and address of physician

19. If hospitalized, name and address of hospital

Date of report Prepared by Official position

OSHA No. 101 (Feb. 1981)

purposes hospitalization means *admitted* to the hospital *not* hospital treatment alone. Notification to the OSHA Area Director must be within forty-eight hours following the accident that caused the fatality and/or hospitalizations. It is advisable to check local state laws for reporting requirements, as they can vary in some states.

It is recommended that the notification be made by telephone and or fax followed by a written report reiterating the verbal notification. The following information must be included in the report:

❖ Circumstances surrounding the accident. Included in this portion of the report would be the type of work being performed, any circumstances not normally associated with the work and the probable cause(s) of the accident, if known.

❖ The number of fatalities and/or the number of persons hospitalized.

❖ The extent of any injuries.

The Area Director may require additional information on the accident or incident.

Retention and Access to OSHA Records

OSHA accident and illness records (OSHA 200 log and 101 forms) must be kept for a period of not less than 5 calendar years following the end of the year to which they relate. When a business changes owners, the responsibility for retaining these records transfers to the new owner. The new owner is not responsible for updating the records of the previous owner. OSHA accident and illness records (OSHA 200 log and 101 forms) should be available for inspection and copying by all approved federal and state government officials. Employees, former employees, and their representatives are mandated access to only the 200 log unless otherwise provided by state law.

MISCELLANEOUS SAFETY AND HEALTH RECORDS

In addition to injury and illness recordkeeping requirements there are many other OSHA regulations that require employers to compile, record, and maintain data. A brief summary of the scope of these miscellaneous recordkeeping requirements is included in Appendix 5-A at the end of this chapter. Remember that failure to meet the requirements in this Appendix can lead to OSHA citations and penalties. Always review the latest edition of the standard for all applicable requirements.

TRAINING AND INSPECTION RECORDS

As a manager you will be responsible for providing specified safety and health training for your employees and for conducting safety inspections of equipment that does not have specific recordkeeping requirements. If you are challenged by an OSHA compliance officer, insurance inspector or for other reasons, it is important that you have accurate and timely records.

The prudent manager can save valuable time and avoid adverse consequences by instituting a recordkeeping system for all safety and health related training and inspections. Such records do not have to be complex but should contain enough information to establish what was done. Records for training should contain the following elements:

- ❖ Employee's name
- ❖ Subject of the training
- ❖ Outline of the training program
- ❖ Date training was given
- ❖ Length of the training program (30 minutes, 1 hour, etc.)
- ❖ Instructor's name and title

Records for inspections should be tailored to meet the requirements of the regulation requiring the inspection. All inspections should include certain basic information:

- ❖ Name and title of inspector
- ❖ Inspection date
- ❖ Inspection schedule
- ❖ Components or elements inspected
- ❖ Deficiencies observed

One of the most difficult tasks a manager may face is establishing a records retention schedule. OSHA and other government agencies provide minimum records retention schedules for mandatory recordkeeping. Managers must determine a records retention schedule that will meet their needs to verify compliance and to provide company protection and yet not create a records glut. A good guide for records retention is to maintain records just long enough to demonstrate that you are providing training and conducting inspections at the required interval. As a rule of thumb, keeping records of two previous inspection or training cycles should be enough to demonstrate compliance. You should consult with your corporate attorney to see if your situation requires a longer retention period.

RECORDKEEPING SYSTEMS: A MANAGEMENT TOOL

Managers know that they must keep records to satisfy governmental requirements. Government agencies use these records to evaluate safety and health programs by comparing the number of accidents and illnesses in a business against industrial averages, plotting the type, cause, and severity of accidents and illnesses, and to evaluate overall safety performance. It is therefore prudent that you do the same. The development of a computerized safety and health recordkeeping database is the most practical way to generate meaningful data that you may use for assisting in the decision making process.

First, it is necessary to determine exactly what information you need for the safety information system (SIS). Do you want to keep employee training records, accident and illness data, safety inspection records, workers' compensation costs with personal employee information, or any part of this information? Your decision will be influenced by the size of your computer system, the capabilities of existing software or the software you must obtain to drive the SIS. Small businesses utilizing a personal computer with software capable of data sorting can develop an effective SIS.

A developed SIS can plot injury and illness trends, pinpoint the areas that need management attention for controlling accidents, identify departments and individuals that have high accident rates, and derive other important information. Study the simplistic example of an SIS below and see how it can be used to develop information for decision-making.

SIS Example   The manager of the Boulder Co. has requested information about the overall safety program. The manager has four questions. Is safety performance going up or down (Table 5-1)? Where are accidents happening (Table 5-2)? Who is having the accidents (Table 5-3)? What are the causes of the accidents (Table 5-4)? Tables 5-1 through 5-4 provide the answers. All of the information in these tables was taken from the OSHA 200 log and 101 forms.

Table 5-1
Monthly Plant Injury and Illness Cases

Month	JAN	FEB	MAR	APR	MAY	JUN
No.	05	04	11	09	12	14

Through the first six months of the year there appears to be an unfavorable trend.

Table 5-2
Monthly Injury and Illness Cases by Department

No./ Dept.	JAN	FEB	MAR	APR	MAY	JUN	YTD
101	01	00	01	01	00	01	04
104	04	04	06	04	08	10	36
200	00	00	02	03	01	02	08
201	00	00	02	01	03	01	07

Department 104 should be targeted for immediate action. Although the other departments do not have as many accidents recorded, they are showing an unfavorable trend and action should be planned.

Table 5-3
Department 104 Personnel Injury and Illness Cases

Cases/ Employee	JAN	FEB	MAR	APR	MAY	JUN	YTD
Brown	00	01	00	01	01	00	03
Collins	01	01	02	00	02	03	09
Harper	01	00	01	00	01	01	04
Horn	00	00	00	00	00	01	01
Kern	00	00	01	00	01	01	03
Lot	00	01	01	01	02	02	07
Muse	01	00	00	01	01	01	04
Post	01	00	00	00	00	01	02
Smith	00	00	01	00	00	00	01
Zonk	00	01	00	01	00	00	02

Collins and Lot should be sceduled for individual counseling to determine why they are having so many accidents. Secondly, a safety awareness session for the entire department should be considered.

Table 5-4
Department 104 Injury Causes

Cause	Slip/Fall	Lifting	Laceration	Striking Object
Number	16	04	8	8

An inspection of the work area is dictated with an emphasis on walking and working surfaces. The injury causes suggest that such conditions as housekeeping and aisle clearances are contributing to the high accident rates.

Appendix 5-A

RECORDKEEPING REQUIREMENTS

The requirements that follow are identified by SUBPART and PARAGRAPH of 29 CFR 1910, OSHA's Occupational Safety and Health Standards.

SUBPART C
General Safety and Health Provisions

1910.20   Access to employee exposure and medical records.   Employers shall assure that all employee medical and exposure records and their analyses are preserved, and access to such records is made available to employees or their representatives. Employers are required to inform all new employees of their right to access of medical and exposure records and the procedure for exercising this right. All employees are to be informed of this right on an annual basis.

SUBPART F
Powered Platforms, Manlifts and Vehicle Mounted Work Platforms

1910.66(g)(2)(iii)   Building owners must keep records of powered platform inspections, prior to being placed in service (new installations), following alterations, and at intervals not to exceed 12 months.

1910.66(g)(3)(ii)   Building owners must keep certified records of cyclical inspections and tests for each platform.

1910.66(i)(1)(v)   Employers must keep records certifying that employees have been trained in operating and inspecting work platforms. The training records must be kept for the duration of the time the employee is employed and contain the name of the employee trained, the date of the training and the signature of the trainer.

1910.68(e)(3)   Employers must keep certified records of manlift inspections. General safety inspections shall not exceed 30 days. Limit switches shall be checked weekly.

SUBPART G
Occupational Health and Environmental Control

1910.95(g)(5)   Employers must establish valid baseline audiograms for employees exposed to noise levels at or above the noise action level for comparison against subsequent audiograms.

1910.95(m)(1)   Employers must retain all employee noise exposure measurements.

1910.95(m)(2)(i)   Employers must retain all employee audiometric test records pursuant to this paragraph.

(A) Employee name and job classification

(B) Dates of audiograms

(C) The examiner's name

(D) Date of last acoustic or exhaustive calibration of audiometer

(E) Employee's most recent noise exposure assessment

(F) Measurements of background sound pressure levels in audiometric testing rooms.

1910.95(m)(3)   Employers must retain the records in this paragraph as follows:

(i) Noise exposure measurements—2 years.

(ii) Audiometric test records—for the duration of the employee's employment.

1910.95(m)(5)   If the employer ceases to do business, the employer shall transfer to the successor all records in this section and the successor shall retain them for the remainder of the prescribed times.

1910.96(b)(2)(iii)   Employers must maintain adequate past and current whole body radiation dose exposure records for all employees to show that any addition of such a dose does not exceed permissible levels.

1910.96(n)(1)   When personnel monitoring for radiation exposures is required, employers must retain records of the radiation exposure.

SUBPART H
Hazardous Materials

1910.120(f)(8)(i) Employers engaged in emergency response must maintain records of their employee's medical surveillance examinations that include:

(A) Employee name and social security number

(B) Physician's written opinions, recommended limitations, and results of examinations and tests

(C) Employee complaints related to hazardous material exposures

(D) A copy of the required information provided to the physician with the exception of the standard and its appendices.

Medical records of employees must be retained for a period of 30 years following termination of employment. Records for employees that have worked less than one year need not be retained providing any such records are given the employee at the termination of employment.

1910.120(p)(8)(iii)(C)   Employers must maintain records certifying that all employees participating in their emergency response plan have been trained or have demonstrated competency annually.

SUBPART I
Personal Protective Equipment

1910.134(f)(1)(ii)   Employers must keep records of all required inspections of respirators maintained for emergency use.

SUBPART J
CONFINED SPACES

1910.146 (e)(6)   Employers must retain cancelled permits for at least 1 year to facilitate review of the permit required confined space program and permits required by paragraph (d)(14).

1910.146 (g)(4)   Certification of training is required including name, identity of trainers and training dates. Certification must be available to employees and their authorized representatives.

1910.146 (c)(7)(iii)   Certification that all hazards in a permit space have been eliminated [allowing reclassification from permit to a non-permit space] must contain the date, location, and signature of the person making the determination.

SUBPART L
Fire Protection

1910.157(e)(3)   Employers must keep records of required maintenance checks for all portable fire extinguishers. Maintenance checks are required annually and records must be kept one year following the shelf life of the extinguisher.

1910.157(f)(16)   Employers must keep records of required hydrostatic tests for fire extinguishers. These records must be maintained for the period between hydrostatic tests, as listed in table L-1 of the standard, or until the extinguisher is removed from service.

1910.159(c)(11)   Records must be maintained for hydraulically designed automatic systems when these systems are not identified by signs.

1910.160(b)(9)   Employers having fixed extinguishing systems must record inspection and maintenance dates for such systems, on the container or in a central location. In addition a record of the last semi-annual check must be retained until the container is rechecked or removed from service.

SUBPART N
Materials Handling and Storage

1910.179(j)(2)(iii) A daily record of inspections of hooks for overhead and gantry cranes must be maintained that includes the date of inspection, signature of the inspector, and the serial number or other identifier for the hook.

1910.179(j)(2)(iv) A daily record of inspections of chains and end connectors for overhead and gantry cranes must be maintained that includes the date of inspection, signature of the inspector, and an identifier for the chain and end connector inspected.

NOTE: Daily inspections need not be made for hooks and chains when the overhead or gantry crane is not in daily service. These inspections are to be performed prior to the use of the crane. Similar records must be maintained of the inspections.

1910.179(m)(1) All running ropes for overhead and gantry cranes must be inspected at least monthly. Records of these inspections must be maintained that include the date of inspection, the signature of the inspector, and an identifier for the rope.

1910.179(m)(2) All installed ropes on overhead and gantry cranes that have been idle for longer than one month must be inspected prior to being placed in service. Records for these inspections must be maintained that include the date of inspection, signature of the inspector, and an identifier for the rope.

1910.180(d)(6) Records of monthly inspections of all critical parts, such as brakes, crane hooks, and ropes, on crawler locomotive and truck cranes must be maintained. The records are to include date of inspection, signature of inspector, and serial number or other identifier of the crane.

1910.180(g)(1) All running ropes for crawler locomotive and truck cranes must be inspected at least monthly. Records of these inspections must be maintained that include the date of inspection, the signature of the inspector, and an identifier for the rope.

1910.180(g)(2)(ii) All installed ropes on crawler locomotive and truck cranes that have been idle for longer than one month must be inspected prior to being placed in service. Records for these inspections must be maintained that include the date of inspection, signature of the inspector, and an identifier for the rope.

1910.181(g)(1) All running ropes for crawler derricks must be inspected at least monthly. Records of these inspections must be maintained that include the date of inspection, the signature of the inspector, and an identifier for the rope.

1910.181(g)(3) All installed ropes on derricks that have been idle for longer than one month must be inspected prior to being placed in service. Records for these inspections must be maintained that include the date of inspection, signature of the inspector, and an identifier for the rope.

1910.184(e)(3)(i) & (ii) Records of alloy steel chain sling inspections must be maintained. Inspection can not be at intervals exceeding 12 months.

1910.184(e)(4)   Employers must retain a certificate of proof testing for all new, repaired, or reconditioned alloy steel chain slings.

1919.184(f)(4)(ii)   Employers must retain a certificate of proof testing for all welded end attachments for wire rope slings.

1910.184(g)(8)(ii)   Records of repairs to wire mesh slings must be maintained. Permanently marking or tagging the sling with the date, nature of the repair, and the identity of the person or firm making the repair may be done in place of a written record.

1910.184(i)(8)(ii)   Employers must retain a certified copy of proof testing for all repaired synthetic web slings.

SUBPART O
Machinery and Machine Guarding

1910.217(c)(iv)(d)   Records must be kept of inspections of pull out devices for power presses. Inspections are to be performed prior to the start of each shift, following new die set ups, and when operators are changed. Inspection records are to include the date of inspection, signature of the inspector, and the serial number or other identifier of the press.

1910.217(e)(1)(i)   Employers must establish and follow a program of periodic and regular inspections of power presses that certify that all parts, auxiliary equipment, and safe guards are in a safe operating condition and in proper adjustment. Records of these inspections must be maintained. The inspection records must include the date of inspection, signature of the inspector, and the serial number or other identifier of the press.

1910.217(e)(1)(ii)   Weekly inspections and tests of power presses must be conducted including clutch/brake mechanisms, anti-repeat features, and single stroke mechanisms and records of these inspections maintained that include date of inspection, signature of the inspector, and the serial number or other identifier of the press.

1910.217(h)(11)(B)(iv)   Records of installation certification and validation and the most recent recertification and revalidation of power press presence sensing device initiation (PSDI) systems must be maintained. These records must include the manufacturer and model number of each component and subsystem, the calculation of stopping distance, and the stopping time measurement.

1910.217(h)(13)(i)   Training records for operators of power presses equipped with PSDI systems must be maintained. Operators must be trained prior to working with presses having PSDIs and as necessary to maintain competence, but not less than annually thereafter.

1910.218(a)(2)(i)   Employers having forging machines must establish periodic and regular maintenance safety checks and keep certified records of these inspections that include the date of inspection, signature of the inspector, and the serial number or other identifier of the forging machine.

1910.218(a)(2)(ii)   Employers must schedule and record inspections of guards and point of operation guarding devices for forging machines at frequent and regular intervals. The records must include the date of inspection, signature of the inspector, and the serial number or other identifier of the equipment inspected.

SUBPART R
Special Industries

1910.268(c)   New employees in the telecommunications industry must be trained or certified by the employer as being trained during prior employment in the various precautions and safe practices in this section. Training records must be maintained for the duration of the employee's employment. Training records must include the date of training or certification as being trained, the name of the person trained, and the signature of the trainer.

1910.272(l)(3)   Records of regularly scheduled inspections of applicable mechanical and safety control equipment in grain handling facilities must be kept.

SUBPART T
Commercial Diving Operations

1910.425(d)(1)   Commercial diving operations require post-dive records be kept for each diving operation.

1910.430(a)(1)   Each equipment modification, repair, test, calibration or maintenance service shall be recorded by a means of a tagging or logging system.

1910.440(a)(2)   Employers must keep records of any diving related injury or illness which requires the hospitalization of any dive team member for longer than 24 hours specifying the circumstances of the incident and the extent of injuries or illnesses.

1910.440(b)(2)   Employers must retain the following records for ready inspection:

    1910.420   Safe practice manuals
    1910.422   Depth-time profiles
    1910.422   Recordings of dives
    1910.423   Decompression procedure assessment evaluations.

Employers must supply the records in **430** and **440** to employees or their representatives when requested.

1910.440(b)(3)   Record retention periods are as follows:

(i) Dive team medical records—5 years

(ii) Safe practice manual—Current manual only

(iii) Depth-time profiles until completion of the recording of dive or until completion of decompression as applicable

(iv) Recording of dive—1 year, except 5 years in cases of incidents having decompression sickness

(v) Decompression procedure assessments—5 years

(vi) Equipment testing & test records—current entry or tag

(vii) Records of hospitalization—5 years.

1910.440(b)(4) After the expiration of the retention period of any record required to be kept for 5 years, the employer shall forward such records to NIOSH, Department of Health and Human Services.

SUBPART Z
Toxic and Hazardous Substances

Employers are required to maintain records of all monitoring data associated with hazardous chemicals and medical examinations for employees exposed or that may be exposed to hazardous chemicals to meet the requirements of 1910.20.

In addition to the general recordkeeping requirement there are recordkeeping requirements for specific hazardous chemical compounds that require the employer to maintain monitoring data, employee training records, and medical examinations. A list of these chemical compounds follows.

- ❖ **1910.1001** ASBESTOS, TREMOLITE*, ANTHOPHYLLITE*, and ACTINOLITE*
- ❖ **1910.1003** 4-NITROBIPHENYL
- ❖ **1910.1004** ALPHA-NAPHTHYLAMINE
- ❖ **1910.1006** METHYL CHLOROMETHYL ETHER
- ❖ **1910.1007** 3,3'-DICHLOROBENZIDINE (and its salts)
- ❖ **1910.1008** BIS-CHLOROMETHYL ETHER
- ❖ **1910.1009** BETA-NAPHTHYLAMINE
- ❖ **1910.1010** BENZIDINE
- ❖ **1910.1011** 4-AMINODIPHENYL
- ❖ **1910.1012** ETHYLENEIMINE
- ❖ **1910.1013** BETA-PROPIOLACTONE
- ❖ **1910.1014** 2-ACTYLAMINOFLUORENE
- ❖ **1910.1015** 4-DIMETHYLAMINOAZOBENZENE
- ❖ **1910.1016** N-NITROSODIMETHYLAMINE
- ❖ **1910.1017** VINYL CHLORIDE
- ❖ **1910.1018** INORGANIC ARSENIC
- ❖ **1910.1025** LEAD

*   Recordkeeping requirements on hold for these compounds at this time.

- ❖ **1910.1028**   BENZENE
- ❖ **1910.1029**   COKE OVEN EMISSIONS
- ❖ **1910.1030**   BLOODBORNE PATHOGENS
- ❖ **1910.1043**   COTTON DUST
- ❖ **1910.1044**   1,2-DIBROMO-3-CHLOROPROPANE
- ❖ **1910.1045**   ACRYLONITRILE
- ❖ **1910.1047**   ETHYLENE OXIDE
- ❖ **1910.1048**   FORMALDEHYDE
- ❖ **1910.1101**   ASBESTOS

1910.1200   Toxic and Hazardous Substances Hazard Communication Standard

(e)(1)   Employers must develop, implement, and maintain a written hazard communication program.

(e)(1)(i)   Provide and maintain a list of hazardous chemicals known to be present in the workplace.

(g)(1)   Obtain and keep a material safety data sheet (MSDS) for each hazardous chemical present in the workplace.

1910.1450   Occupational exposure to hazardous chemicals in laboratories

(e)   Employers must develop and maintain a written chemical hygiene plan.

(j)(1)   Employers must retain records of all monitoring data and any medical consultations and examinations including tests or written opinions required by the standard.

PART TWO

PREVENTING INJURY AND ILLNESS IN MANUFACTURING PLANTS

CHAPTER
6
PLANT DESIGN AND LAYOUT FOR SAFETY

Accident prevention can be enhanced in large measure before a plant is built, i.e., at the planning stage. Proper design and layout of plants and facilities can significantly reduce the potential for hazards. Ideally, safety should be an integral part of the planning process with a hazard and operability (Haz-Op) study completed before construction is started.

PREPARING A HAZARD AND OPERABILITY (HAZ-OP) STUDY

A Haz-Op study is the systematic evaluation of facilities, processes, and equipment to assess their hazard potential under expected operating conditions and during malfunctions. Assessing potential hazards during the design stage permits changes to be incorporated at the least costly point. Retrofitting, renovation, or process changes after construction are always more difficult to achieve and much more costly. Major safety, health, or environmental requirements which will influence, if not mandate, the type, shape, and size of structures or buildings are based on the process and materials involved, mechanical equipment used, and working conditions (e.g., wet, dry, dusty, cleanliness required).

Some examples of the many factors impacting on safety that need to be considered in the initial analysis of the plant building design include:

- ❖ Site location relative to other companies and emergency services
- ❖ Climate as it relates to emergency planning for severe weather
- ❖ Traffic patterns and flows
- ❖ Raw materials storage requirements

❖ Flammability and reactivity of raw materials and products

❖ Materials flow for the process

❖ Mechanical equipment required

❖ Electrical requirements and hazards

❖ Noise generation from operating equipment

❖ Solid and hazardous waste storage requirements

❖ Pollution control devices and efficiencies

The safety aspects within the facility itself must also be considered. Modifications after final building prints are drawn are expensive, but not as expensive as retrofitting once the plant is built. The following list includes examples of many of the common interior aspects that need to be considered during planning:

❖ Exits and other wall openings

❖ Floors, walkways, stairs, ramps, platforms

❖ Storage facilities, including those for explosives and flammable materials, harmful substances, finished products, and yard storage

❖ Electric wiring runs

❖ Illumination

❖ Mechanical handling equipment: cranes, conveyors, industrial trucks

❖ Elevators

❖ Boilers and other pressure equipment

❖ Ventilation, heating, and air conditioning

❖ Fire prevention and extinguishment

❖ Personal service facilities: parking, food service, rest rooms, employment, training

CODES AND STANDARDS

Codes and standards deserve special consideration in the planning of operations. Many of these are ordinances and local or state laws governing means of egress (doors and emergency escapes), emergency lighting, fire alarms, fires sprinklers, and exhaust and ventilation systems. Additionally, city, county, state, and federal agencies may have specific standards for sanitation, use of radiation devices (ionizing and non-ionizing), building construction, and pollution prevention requirements. In some cases, operating permits may be required on a local, regional, state, and federal level for equipment (e.g., spray paint booth) or operations (e.g., spray painting).

There are a variety of "consensus" or voluntary safety codes and standards which in some cases have been incorporated into law. One well-known example of a standard setting group is the American National Standards Institute (ANSI). ANSI has standards for such items as building load design, sanitation, stairs, construction, floor and wall openings, ladders, window cleaning, marking hazards, identification of piping, and accident prevention signs. Other example organizations include the Illuminating Engineering Society, National Fire Protection Association, and the American Society of Mechanical Engineers. The Illuminating Engineering

Society (IES) has published practices on industrial lighting while electrical wiring and installations are covered by the National Electrical Code (NEC) which is published by the National Fire Protection Association (NFPA). Code requirements for fire extinguishers and equipment, flammable liquids and gases, combustible solids, dusts, chemicals, and explosives are also published by the NFPA. The American Society of Mechanical Engineers (ASME) has specific codes on boilers, pressure vessels and other mechanical equipment (see Chapter 23, Pressure Vessels). As comprehensive as some of these codes are, they should only be considered as starting points for safe plant design. Names and addresses of some of the key code and standards setting organizations are listed at the end of this chapter.

Actual design and layout of the manufacturing process is well beyond the scope of this book, however, some key design aspects to consider are briefly discussed in the remainder of this chapter. This includes the location of buildings and structures, flow diagrams, equipment layout, electrical equipment, heating, ventilation and air conditioning, storage, illumination, signs, color coding, aisles and walkways, sanitation, personal facilities such as drinking fountains and washrooms, and purchasing requirements. Your design engineer or contractor should be asked about specific requirements for your location.

PHYSICAL PLANT LAYOUT

The prime consideration for location, size, shape, and layout of the plant buildings should be the safest and most efficient utilization of materials, processes, and methods. In past years, multistoried plants often moved a partially finished product up and down floors using various elevators and ramps with little regard for process flow. The more efficient and usually safer way is a larger single floor factory where raw materials enter at one end of the plant and finished product is shipped at the other.

Location of Buildings and Structures

Fire and explosion hazards are important considerations for raw materials and finished product storage and process areas. For example, highly flammable liquids and gases such as liquid propane gas (LPG) may have to be stored separately. These considerations are addressed in Chapter 15, Materials Handling and Storage, and Chapter 19, Fire Safety.

Bulk storage of liquids will usually require diking with separation from other incompatible bulk storage (e.g., acids and bases). Pressurized storage will require specially designed containers and appropriate relief venting systems. Many of these locations and design requirements are contained in national design codes such as those previously discussed and referenced at the end of this chapter (see Chapter 23 for a discussion on pressure vessels).

Flow Diagrams

Use of process flow sheets or diagrams showing the movement of materials through the plant will help to emphasize locations of hazardous materials and operations and of potential inefficiencies or unsafe practices during the planning process. Locations presenting potential fire hazards can also be identified and provided with automatic sprinklers or other fire suppression systems based on the process flow layout. Types of fire construction (fire resistant structures) and location of emergency egress should also be noted. Guidance on fire suppression requirements is available from NFPA.

Equipment Layout

One method to help determine optimum equipment layout is the flat (paper) two-dimensional scale layout using equipment templates. This technique allows equipment to be "moved around" to evaluate fit and most efficient location. These paper simulations can be automated to allow for more rapid manipulation using computer-aided design (CAD) techniques.

Building a three-dimensional model is ultimately the best technique. This has traditionally been done with wood, plastic, or paper scale models; however, with the isometric, three dimensional possibilities of the current versions of CAD this can be accomplished on the computer. This technique can indicate potential congestion areas and the best traffic flow patterns for fork lift trucks and personnel. Aisle and walkway needs can also be derived from computer models and simulations.

Electrical Equipment

Electrical design and layouts are governed by equipment requirements and other factors. The National Fire Protection Association publishes the National Electrical Code which lists consensus standards for electrical wiring and design. These are discussed in Chapter 21, Electrical Safety.

Heating, Ventilation, and Air Conditioning (HVAC)

HVAC systems require special design attention to avoid creating or adding to potential health, safety, and environmental problems. For example, in factories where direct fired gas space heaters are used, special consideration must be given to the possibility of burning contaminated plant air and recirculating it back into the workers' environment. The American Society of Heating, Refrigeration, and Air Conditioning Engineers (ASHRAE) has established criteria and guidelines for HVAC of factory and office environments. Overall, factory HVAC systems should be balanced from room to room and area to area so that "wind tunnel" effects are not created. Excess drafts occur if systems are unbalanced or severely starved for makeup air. This can also cause odors and contaminants to be pulled into other areas, including offices, causing indoor air quality complaints.

Maintenance and accessibility must also be considered for all ventilation systems but particularly for local exhaust ventilation systems that control dusts, fumes, or solids. This includes planning for access doors for cleanout and accessibility for repair.

Storage

A caveat is that factories or plants usually have insufficient storage space. Besides adequate raw material and finished product storage, consideration must be given to the need for items like janitorial rooms or closets which need to be strategically located in work areas. Additionally, storage areas for maintenance supplies are also important and need to be considered in the planning stages. This topic is discussed in Chapter 15, Materials Handling and Storage.

ILLUMINATION

Lighting or illumination can be provided by natural daylight. This can be done through windows and skylights in plants where the distance from the outer walls to the center of the plant is not great. However, even for smaller scale buildings, supplemental electric lighting will need to be used for night operations. Even during the day, natural light is usually too variable to use as the sole source of illumination. Additional high intensity supplementary or localized lighting may also be necessary for specialized tasks such as product inspection.

Care must be taken with lighting design so that the lighting does not interfere with work or cause visual fatigue. For example, direct or reflected glare with high contrast can contribute to accidents and should be minimized.

The required quantity or amount of illumination (normally measured in units called footcandles or lux) will vary with the task or job function. "Poor illumination" is commonly listed as a cause of accidents. Minimum levels of illumination for various industrial areas and tasks are given in ANSI Standard RP7 "Practice for Industrial Lighting." Table 6-1 has been extracted from that standard.

Table 6-1
Minimum Levels for Industrial Lighting

Area	Foot-Candles
Assembly-rough, easy seeing	30
Assembly-medium	100
Building construction-general	10
Corridors	20
Drafting rooms-detailed*	200
Electrical equipment, testing	100
Elevators	20
Garages-repair areas	100
Garages-traffic areas	20
Inspection, ordinary	50
Inspection, highly difficult	200
Loading platforms	20
Machine shops-medium work	100
Materials-loading, trucking	20
Offices-general areas*	100
Paint dipping, spraying	50
Service spaces-wash rooms, etc.	30
Sheet metal-presses, shears	50
Storage rooms-inactive	5
Storage rooms-active, medium	20
Welding-general	50
Woodworking-rough sawing	30

*From *Practice for Office Lighting*, ANSI/IES RP1-1982. Others from ANSI/IES RP7-1983.

Color Codes, Signs and Warnings

Color is used effectively as an inherent means of warning throughout our daily lives. For example, the color red is often associated with stop, danger, or fire whereas most people perceive that green means "go" or "safe." The ANSI Standard Z53.1, Safety Color Code for Marking Physical Hazards and Identification of Certain Equipment, recommends color use to identify the following categories of hazard:

RED	Fire hazards, fire protection, danger, emergency stops on equipment.
YELLOW	Hazards from slipping, falling, etc.; hazards from material handling equipment such as cranes and lift trucks; radiation hazards (along with magenta on black).
GREEN	First aid and safety equipment locations.
BLACK AND WHITE	Housekeeping and traffic markers.
ORANGE	Dangerous parts of machinery such as energized equipment, exposed cutting devices, insides of guards.
MAGENTA (REDDISH-PURPLE)	Radiation protection.
BLUE	Information that is not safety related.

There is also a similar color code for piping systems identification. ANSI A13.1, Scheme for the Identification of Piping Systems provides:

Safety Red: Fire Protection
Safety Yellow: Dangerous
Safety Green: Safe Areas
Safety Blue: Protective Materials (Inert Gases).

Table 6-2 is a summary of the OSHA and ANSI color designations.

Signs

Accident prevention signs use several consensus conventions such as ANSI Z35.1, Specifications for Industrial Accident Prevention Signs, to convey messages or information. The ANSI specifications are:

DANGER	Red oval in top panel, black or red lettering on white backround in lower panel.
CAUTION	Yellow backround color with black lettering.
GENERAL SAFETY	Green backround on upper panel, black or green lettering on white backround on lower panel.
FIRE AND EMERGENCY	White letters on red backround.
INFORMATION	Blue letters on white backround.

Exit signs have separate requirements which are listed in the Life Safety Code, NFPA 101.

Table 6-2
Summary of OSHA and ANSI Safety Color Code*

Color	Designation
Red	Fire: Protection equipment and apparatus, including fire-alarm boxes, fire-blanket boxes, fire extinguishers, fire-exit signs, fire-hose locations, fire-hydrants, and fire pumps. Danger: Safety cans or other portable containers of flammable liquids, lights at barricades and at temporary obstructions, and danger signs. Stop: Stop buttons and emergency stop bars on hazardous machines.
Orange	Dangerous Equipment: Parts of machines and equipment that may cut, crush, shock, or otherwise injure.
Yellow	Caution: Physical hazards such as stumbling, falling, tripping, striking against, and being caught in between.
Green	Safety: First-aid equipment.
Blue	Warning: Caution limited to warning against starting, using, or moving equipment under repair.
Black or Yellow	Radiation: X-ray, alpha, beta, gamma, neutron, proton radiation.
Black and White	Boundaries of traffic aisles, stairways (risers, direction, and border limit lines), and directional signs.

*See full text under Section 1910.144 of Occupational Safety and Health Standards. For piping colors, see ANSI Standard, A13.1-1981, Scheme for the Identification of Piping Systems.

Aisles and Walkways

Aisles should be designed to allow for the safe and smooth flow of people and equipment inside a building and should be specifically marked. Walkways, by definition, are passages connecting two buildings and are typically outdoors, but may be totally enclosed or covered. Both aisles and walkways are discussed in Chapter 14, Walking and Working Surfaces. In brief, adequate entrance and exit space (usually 28 inches or greater in width) must be maintained for all working or occupancy areas of the plant.

HIGH HAZARD AREAS

Some areas or operations have an inherent high hazard potential. These include:

1. Manufacturing, use, or storage of flammable liquids

2. Explosives manufacturing, use, or storage

3. High voltage/high current electric equipment and areas

4. Confined spaces

5. Radiation area (such as in QA/QC inspection)

6. Robotic operational areas

7. Areas with process equipment of high energy movement through a mechanical, pneumatic, steam, electrical, or other power sources

These areas may require special precautions and planning, such as the need for physical barriers, electrical interlocks, alarms, and other special measures.

SANITATION AND PERSONAL SERVICE FACILITIES

In order for employees to work efficiently and to have their health protected, several sanitation or hygiene requirements must also be considered in the planning stages. These include:

1. Potable water for drinking and washing

2. Sewage, solid waste, and garbage disposal

3. Sanitary food service

4. Personal service facilities such as:

   ❖ Drinking fountains

   ❖ Washrooms and locker rooms

   ❖ Toilets

   ❖ Showers

5. Janitorial services

Drinking water quality is mandated by state regulations, as well as the U.S. EPA. Even though potable water is delivered to the plant property, there is no automatic assurance that the water at the employee drinking fountain will be completely potable. Because of the potential hazards of contamination, the integrity of the drinking water system must be maintained. No cross or direct connections are allowed without an approved backflow prevention device. Additionally, water fountains may need to be protected in dusty or dirty areas.

Specific requirements and methods for water supply, plumbing, sewage and solid waste disposal, and storm water drainage are normally governed by the state or local city or county health department. They should be contacted for the specific requirements for your local area.

Control of vermin such as insects and rodents may also need to be considered in the planning stages. Special construction practices and licensed exterminators may need to be used depending on what is produced (e.g., food products) and the geographical location (e.g., deep South) or climate.

Personal Facilities

Drinking fountains should be available at about one per 50 people to meet the requirements of ANSI Standard A112, Specifications for Drinking Fountains.

Washrooms and locker rooms are specified in ANSI Z4.1, Minimum Requirements for Sanitation in Places of Employment.

Approximately one lavatory per 15 employees is usually considered adequate. The number of toilets may vary from one per 15 employees to one per 35 employees for larger facilities. There can also be distance requirements. Some states require restrooms to be accessible within 200 feet (or less) of the work station.

Purchasing Requirements

Local, state, and federal codes must also be considered when purchasing specialized equipment for hazardous locations such as explosion-proof electric equipment or sealed pumps. Since these requirements can vary based on location, they should be discussed with your contractor.

STANDARDS ORGANIZATIONS

A list of standards organizations with addresses follows. Example publications from the listed organization are also given. These organizations should be contacted for additional information on their respective standards, codes, or recommendations.

American Society of Heating, Refrigerating, and Air Conditioning Engineers, 1791 Tullie Circle, N.E., Atlanta, GA 30329.

> *Guide and Data Book.*
> *Standard for Acceptable Indoor Air Quality, ASHRAE 62.*

American National Standards Institute, 1430 Broadway, New York, N.Y. 10018.

> *Manual on Uniform Traffic Control Devices for Streets and Highways,* ANSI D6.1.
> *Minimum Design Loads for Buildings and Other Structures,* A58.1.
> *Life Safety Code,* A9.1 (NFPA 101).
> *Minimum Requirements for Sanitation in Places of Employment,* Z4.1.
> *National Electrical Code,* ANSI/NFPA 70.
> *Practices for Industrial Lighting,* ANSI/IES RP7.
> *Requirements for Fixed Industrial Stairs,* A64.1.
> *Safety Requirements for Construction,* A10 Series.
> *Safety Requirements for Floor and Wall Openings, Railings, and Toeboards,* A12.1.
> *Safety Color Code for Marking Physical Hazards,* Z53.1.
> *Scheme for the Identification of Piping Systems,* A13.1.

Specifications for Accident Prevention Signs, Z35.1.

"Gas-Burning Appliances," Z21 Series.

Minimum Requirements for Non-Sewered Disposal Systems, Z4.3.

"Code for Pressure Piping," B31.

Specification for Drinking Fountains, A112.

Buildings and Facilities—Providing Accessibility and Usability for Physically Handicapped People, A117.1.

Illuminating Engineering Society, 345 E. 47th St., New York, N.Y. 10017.

Glare and Lighting Design.

Lighting Handbook.

Recommended Practice for Daylighting.

Practice for Industrial Lighting, ANSI/IES RP7.

National Sanitation Foundation, P.O. Box 1468, 3475 Plymouth Rd., Ann Arbor, MI 48106.

Various publications on food and drink related equipment.

American Society of Mechanical Engineers, 345 East 47th Street, New York, NY 10017.

ASME Boiler and Pressure Vessel Code.

National Fire Protection Association, 1 Batterymarch Park, P.O. Box 9101, Quincy, MA 02269.

Standard for Installation of Sprinkler Systems, NFPA-13.

Standard for Installation of Standpipe and Hose Systems, NFPA-14.

Standard for Flammable and Combustible Liquids, NFPA-30.

Standard for Fire Doors and Windows, NFPA-80.

Life Safety Code, NFPA-101.

CHAPTER
7
REDUCING ACCIDENTS AND ILLNESS

This chapter will define the term "accident," describe why preventing accidents is so important to business, and discuss how to reduce or prevent them. The concept of "near misses" will be discussed, including why these events are as important as an actual accident.

Before discussing how or why accidents (injuries and illnesses) occur, it is necessary to define what is meant by the term "accident." An important concept is that accidents are not really random or totally unpredictable. Inherent in the term is a fatalistic outlook—injuries are unpredictable and unpreventable. Perhaps a better term would be unwanted incidents or events; however, since the idea of "accidents" is so firmly implanted in the literature, this term will be used.

What is an accident? One authority (Frank E. Bird, Jr. in *Management Guide to Loss Control*, Institute Press 1978) has defined it as:

> An undesired event that results in physical harm to a person or damage to property. It is usually the result of contact with a source of energy (i.e., kinetic, electrical, chemical, thermal, etc.) above the threshold limit of the body or structure.

Why should management be concerned with accidents? Apart from obvious ethical reasons, accidents are an important bottom line consideration. Accidents cost money and it follows that the more accidents occur, the more money that is lost.

The National Safety Council has estimated that work accidents cost industry and the nation almost $64 billion in 1990. This included insurance administration costs of 10.3 billion, wage losses of 10.2 billion, and medical costs of 8.7 billion. The remaining 35 billion were for uninsured costs including the cost of lost time, fire losses, and vehicle accident losses. Its study also estimated an annual cost of $540 per American worker was needed to offset the cost of workplace injuries. Clearly, accidents are expensive for both business and the nation.

Another approach to evaluating the impact of accidents on the bottom line is to examine the amount of sales necessary to recover the costs of accidents. The average lost workday accident has been estimated to result in approximately $28,000 in direct and indirect costs. Assuming a workforce of 100 employees and an incidence rate of two accidents per year for this workforce (approximate U.S. average), you would expect a cost of $56,000 per year. If your company had a five percent net return on sales, you would need over one million dollars in sales just to pay for these accidents. If you had 1000 employees and the same incidence rate, you would need over ten million dollars in sales. Assume you were able to reduce your incidence rate to a value of one half the national average (rate of one per 100). With the same workforce, you would save approximately $280,000 in direct and indirect costs or the need for an additional 5.6 million in sales. For companies with incidence rates well below national averages, the contribution to profits (keeping profits) can be very significant. This is true for both large and small companies.

The remainder of this chapter discusses accident types, causes of accidents, including unsafe acts and conditions, workers' compensation and other costs of accidents, preventing accidents, reducing workers' compensation and accident costs, and recordkeeping and regulatory compliance.

ACCIDENT TYPES AND ACCIDENT RATIOS

Studies have shown that for every serious or disabling accident there are on the average at least ten minor injuries, thirty property damage accidents, and six hundred near-misses (an incident or near-miss accident with no visible injury or damage). This has been referred to as the 1:10:30:600 ratio. This means that there are approximately 600 near miss accidents in which no equipment is damaged and no one is hurt for every "true accident." This concept is illustrated in Figure 7–1. This is why the recognition and control of the near-miss accident is so important. Correction of the conditions causing near misses will prevent moving up the triangle. Investigation of near-miss accidents are described in more detail in the next chapter.

Figure 7–1
Accident Ratios

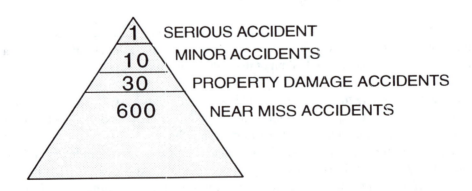

Causes of Accidents

As stated earlier, the term "accident" will be used to describe occupational injuries, illnesses, and unwanted events. There are four major categories of accident causation: people, equipment, materials, and environment. What are the real or specific causes of accidents? To a casual observer it might seem that if a worker was not exposed to any hazardous chemicals or energy and if all machinery and equipment were guarded and "safe," then there would be no accidents. Historically, this has been the focus of safety practitioners. That is, emphasis was placed on correcting unsafe conditions rather than unsafe acts.

In recent years the trend of decreasing incident rates (the number of accidents per man-year worked) has been levelling off. It is believed that these rates have decreased in large part because equipment, materials, and the environment have become less hazardous. It is also believed that major improvement in the rates has slowed because one of the four reasons for accidents—people—has not shown the same rate of improvement. Why are people the cause of accidents? Accidents are most frequently caused by workers committing unsafe acts. An unsafe act can negate the design of even the safest machine. Unsafe acts are not caused by careless or unlucky employees who do not work "safely." They are caused by lack of understanding by the worker and through lack of management control. Just as poor sales, poor quality, late deliveries, etc. are a direct responsibility of management and must be controlled, so must unsafe acts.

Management Control

Management control involves commitment to a safety policy, in addition to other functions such as planning, organizing, and leading. Some symptoms of a lack of management control are an inadequate safety program, inadequate worker knowledge, inadequate work standards, and failure to perform according to standards. Lack of commitment, poor employee training, ineffective group meetings, poor standards, and lack of inspections are the root causes of accidents and unsafe acts and hence a reflection on management rather than the worker.

Accidents usually result from several levels of causation. These can be described as underlying and immediate causes. These then can result in a hazardous contact which will result in an injury or property damage accident. Underlying causes or root causes exist because of lack of management control. Underlying causes can be classified as being either personal or job related. Examples of personal factors could be lack of knowledge or skill, poor motivation, and physical difficulties. Job related factors could include inadequate work standards, poor maintenance or design of equipment, equipment wear and tear, or abnormal use of equipment. The immediate cause of the accident is usually the most apparent cause; however, it is usually the symptom as opposed to the true or underlying cause. Discovery and correction of underlying causes is very important to accident prevention since if not corrected they will be repeated. These immediate causes are normally called unsafe acts and unsafe conditions.

Examples of Unsafe Acts (ANSI Z16.2):

1. Operating without authority.

2. Operating equipment at an unsafe speed or in an improper manner.

3. Using defective equipment and tools.

4. Modifying or removing guards, interlocks, and other safety devices.

5. Failure to use personal protective equipment.

6. Failure to follow lockout/tagout procedures while maintaining equipment.

7. Failure to follow standard procedures.

Examples of Unsafe Conditions:

1. Inadequate guarding.

2. Defective tools, equipment, and substances.

3. Use of equipment which is not properly grounded.

4. Improper storage of hazardous or flammable materials.

5. Poor housekeeping and congestion.

6. Noise, vibration, and radiation hazards.

7. Inadequate illumination.

8. Inadequate ventilation.

9. Hazardous chemical exposures from gases, vapors, dusts, mists, fumes, and liquids.

10. Inadequate warning (alarm) systems.

These hazardous conditions are symptoms of an underlying cause which arises from a lack of management control. When unsafe acts or conditions exist, then the final link can be made—hazardous contact—which then results in an injury, accident or property damage, or both. Accidents can be very expensive for both the company and the worker. These costs are discussed next.

WORKERS' COMPENSATION AND OTHER COSTS OF ACCIDENTS

As described earlier, the National Safety Council estimated that the total cost of occupational accidents in 1990 was $64 billion. It is estimated that roughly half of the costs are direct costs which includes lost wages, insurance, and medical costs. The other half are indirect costs and includes the money value of time lost by workers, uninsured property damage, fire losses, and other costs.

Medical and hospital costs, which are a direct consequence of many accidents, have been rising faster than the inflation rate. Reducing both accident frequency (how often) and severity (how severe) by use of management control techniques can reduce these costs. Some of the uninsured or hidden costs are property damage, production delays, and product or material damage. Other categories include, in addition to wages paid to the injured for lost time and overtime costs, extra supervision costs, decreased output of the worker on return, costs of training new workers, costs of back-up workers, and costs of safety investigations and studies.

Workers' Compensation

When a worker is hurt or dies from work-related injuries or diseases, in addition to all the emotional and family impact, there can also be substantial financial impacts due to loss of earnings and medical expenses. Workers' compensation is designed to be a no-fault insurance program which helps remedy the financial impact on the worker and the family. The program is designed to replace lost earnings, although typically a two-thirds replacement ratio exists in most states. Additionally workers' compensation pays most medical costs related to an accident or illness including vocational rehabilitation. In general, all full-time manufacturing jobs are covered by workers' compensation, but farmwork, domestic service, railroad and maritime workers are not included. The costs of workers' compensation are borne by the employer through private insurance, state programs, self-funded insurance, or other means.

Some authorities and experts suggest that coverage is somewhat limited in that not all injuries or occupational diseases are covered. Other controversy surrounds death benefits and permanent partial benefits as being too limited. The fact remains that even if coverage is considered inadequate by some standards, the coverage provided represents a significant and substantial business expense. For example, almost $18 billion was spent in cash and medical benefits in 1983 through the workers' compensation system. These payments have soared to nearly $20 billion for 1990 based on Social Security Administration estimates.

Workers' compensation can represent a major cost of being in business. These and other costs of accidents and property damage present a significant reason to institute an effective safety system.

PREVENTING ACCIDENTS

To prevent accidents and near-misses, action must start at the top and permeate the company, plant, or organization. Safety should not be something imposed on a business; it must be an integral part of the management system. The essential components must be in place to prevent or minimize unwanted incidents. These basic elements were discussed in Chapter 2. They include a company policy on safety, work rules, hazard identification, and training. Accident prevention must involve everyone in the plant and requires many separate systems to be in place. However, this is no different than managing any other function such as production or quality.

The two absolutely essential elements of accident prevention are management commitment and training. Management commitment and responsibility were discussed earlier in this chapter and in Chapter 2. The other significant element—training—covers all aspects of employees knowing their jobs, supervisors knowing how to manage, and most importantly, how to recognize and prevent unsafe acts.

Below is a discussion of some examples of conclusions as to the real causes of accidents from several accident investigations that may illustrate these points. The underlying accident causes (real reasons) are indicated in parentheses after each accident investigation statement of cause. Causes of the accident:

1. Didn't follow instructions (Training).

2. Blundered ahead when he didn't know how to do the job (Training).

3. By-passed a safety rule to save time (Training and Supervision).

4. Failed to wear required personal protective equipment (Training and Supervision).

5. Didn't think ahead to possible consequences (Training and Job Safety Analysis).

6. Used the wrong equipment for the job (Training).

7. Used the correct equipment improperly (Training).

8. Used equipment that needed repair or replacement (Management).

9. Failed to use equipment guards or protective devices (Management).

On the surface, this accident investigation composite would indicate that the employee was always at fault. Indeed the employee was since most of the causes were unsafe acts. Why were the unsafe acts committed? Not because the worker was unlucky or did not work "safely." It was because the worker was not trained properly and supervision and senior management were not fully involved. Management involvement would require, among other things, quality assessment of training programs and enforcement of safety rules. If safety is not recognized as a management responsibility, then it will not be perceived by employees as an important part of their job. In some cases behavior modification of all individuals involved may be required to ensure this emphasis.

Reducing Workers' Compensation and Accident Costs

For the past several years the manufacturing sector in general has had over 110 lost workdays per one hundred employees based on reports of the Bureau of Labor Statistics. This is over one lost day per employee per year. Reducing the number of lost workdays is critical to reducing the cost of accidents.

Every day that an employee is off work due to an occupational injury or illness increases workers' compensation costs. Even though most companies are insured by outside firms, this insurance is typically "rated." In other words, more claims demand higher premiums. Aside from reducing both the number and severity of accidents to reduce workers compensation costs, these steps are suggested:

1. Return injured people to work as soon as possible. "Light duty" is one effective way of accomplishing this. Consider a work strengthening program.

2. Study attitudes and absenteeism. There is a correlation between absenteeism and injured or ill employees.

3. Do not pay overtime for lost time accidents on a holiday, weekend, etc.

4. Make sure workers are classified in the proper rate class. Lower rate classes will result in lower premiums.

5. Consider paying some small claims directly. Premiums on insurance will not escalate as rapidly.

6. Deal aggressively with insurance agents and ask for discounts.

7. Check all claims aggressively for clerical and other errors.

8. Change insurance companies if all else fails.

RECORDKEEPING AND REGULATORY COMPLIANCE

Recordkeeping and recordkeeping systems are discussed in detail in Chapter 5. The required OSHA injury and illness records can be used to analyze departments, situations, employees, plants, etc. The last part of Chapter 5 discusses instituting a safety information system and what it entails. Trends and rates of accidents will be key information points. Comparing rates for an industry, or SIC code, can also be done by comparison with the Bureau of Labor Statistics' *Annual Survey of Occupational Injuries and Illnesses*. This report lists various incident rates, both fatal and nonfatal, and allows for comparisons of your rates with the same of like industries.

Accident data can also reveal departmental problems and in some cases actual causes. As we have learned, these "actual causes" such as slips, falls, lifting, laceration, etc. are only symptoms, but the historical data can still guide you toward prevention.

Keeping a safe and healthy workplace has been discussed as a very practical and necessary part of doing business, but there is one more important consideration. Employers have at least two major legal requirements or obligations under the Occupational Safety and Health Act:

1. To provide each employee with employment which is free from recognized hazards that are causing or likely to cause death or serious physical harm. This is known as the General Duty Clause, which is Section 5(a) of the Act.

2. To comply with safety and health standards promulgated under the Act.

Because of potential citations and monetary penalties involved, employers have a further incentive to be committed to a safe and healthful plant. Safety not only makes sense, it's good business.

CHAPTER
8

ACCIDENT
INVESTIGATIONS

Accident investigation is an important aspect of any effective safety and health program. Investigation of all accidents from the seemingly minor near-miss and property accident through the injury accident should be a company policy. This chapter describes why accidents should be investigated, how they should be investigated, how to prepare accident reports, taking corrective action, and finally, compiling accident statistics and what to do with them.

WHY INVESTIGATE ACCIDENTS?

All accidents should be investigated to reduce the likelihood of their being repeated. Many times a bad situation that causes a near-miss or minor accident is not corrected and results in a much more serious accident the second time. Recognizing this, the near-miss and minor property damage accident should be investigated just as the more serious accident needs to be investigated. As the familiar saying goes: we must learn from our mistakes or we are doomed to repeat them. Additionally, a proactive approach and concern for safety and health are shown by a management commitment to the investigation of all accidents.

Accidents are expensive. They result in direct costs due to downtime and loss of production, damage to equipment, replacement costs for injured workers, increases in workers' compensation costs, and medical costs. Indirect costs from accidents include the costs of the accident investigation and other administrative costs as well as the negative publicity associated with an accident. In order to improve the bottom line, it is important to reduce accidents and accident costs. Accident investigation is one of the key ways to accomplish this task.

What types of accidents are there? A listing of categories of accidents ranked by severity and importance from an accident investigation standpoint are as follows:

❖ Near-miss accidents for property

❖ Property damage accidents (without injury)

❖ Accidents requiring first aid or minor medical treatment

❖ Near-miss accidents with a serious injury potential

❖ Accidents resulting in lost time

❖ Accidents requiring hospitalization

❖ Accidents resulting in a fatality

REPORTING ACCIDENTS

In order to be able to establish a policy of investigating accidents, it is necessary that a policy for the reporting of accidents be established first. Any accident that requires more than minor first aid at the work site or that results in lost work time must be recorded on the OSHA 200 log and should be investigated (see Chapter 7 on accident reporting for more detail). Therefore, you will need to establish a policy and forms for recording accidents. Major accidents that result in property damage or serious injury will usually reach management quickly and receive prompt attention; however, the minor accidents and near-misses are often times not reported. An illustration of this point based on an actual series of events follows:

> The wood shop was given the task of cutting wedges from wooden blocks on a table saw every few days for a construction project at the plant. The wooden blocks exceeded the height capacity of the guard for the saw blade. The guard was removed for this particular job and a push stick made to allow the pieces to be cut. In doing the job, there were several near-miss accidents using the temporary push rod with fingers coming very close to the blades when the pieces would slip during the wedge cutting. After several days of this task as might be suspected, a worker amputated three fingers as the wood piece slipped and his hand went through the blade. Although removal of the guard was clearly a contributing factor, the near-miss accidents should have resulted in a reevaluation of the task and the design of the push rod.

While there is no "cure-all" solution to this reporting problem, there are ways to encourage the reporting of near-miss and minor accidents. First, it is important that personnel be encouraged to report all accidents including the near-miss accidents and accidents resulting in property damage. *Negative reinforcement by assigning blame will discourage reporting.* It is important to positively reinforce the reporting of accidents even if they appear to be the result of carelessness. Accidents should be reported through the immediate supervisor and move up the management structure to a level where there is someone who can evaluate trends among areas or departments. For some companies, the person in charge of the maintenance department keeps a log of property damage accidents that can be used to study trends and recognize reoccurring problems since they are the most likely persons to repair damage or modify machinery. For example, continual damage to structural supports in a warehouse probably means there is a need for forklift operator training or a different means of stockage. Finally, the person responsible for distributing first aid supplies (e.g., bandages) or the person administering first aid can keep a running log of minor accidents for review on a plant-wide or area-wide basis. For example, frequent minor cuts might suggest the need for gloves or a different means of handling materials.

ACCIDENT INVESTIGATION

Accidents should be investigated as soon as possible after their occurrence (except when it delays medical treatment or the person involved is distraught). The longer the period of delay in the investigation, the more likely the chance of witnesses not remembering the facts of the accident. Normally the immediate supervisor should be the one to do the initial investigation because of his or her knowledge of the employees, equipment, and work practices in their area. There are five main methods of gathering accident information: (1) interviewing the accident victim, (2) interviewing accident witnesses, (3) investigation of the accident scene, (4) re-enactment of the accident, and (5) reconstruction of accident. Each method is detailed below:

1. *Interviewing the accident victim.* Interviewing the employee that had the accident will probably be the most important means of getting the facts on the accident that occurred. It is extremely important that the interviewer not attempt to assign blame or create a negative feeling in the employee that had the accident. The interview should be conducted in a positive fashion otherwise the "facts" of the case may become very biased or unclear as the employee senses that the interviewer is "looking for someone to blame". It is suggested that the interviewer not complete the accident report until after the investigation is finished. It also suggested that the investigator take notes and review the accident facts as known with the accident victim. The employee should be asked if the facts that have been recorded are correct. *AGAIN, IT IS VERY IMPORTANT THAT YOU DO NOT ASSIGN BLAME DURING THE INTERVIEW.* If the accident facts do not appear to be clear or reasonable, consider a second interview with the employee after a reasonable time for the employee to think about the occurrence. Finally, the interviewer should ask the accident victim his or her opinions concerning measures that should be taken to prevent a recurrence of the accident.

2. *Interviewing accident witnesses.* Many accidents will require the interviewing of witnesses to the accident. The accuracy of a fatality report, for example, relies heavily on the ability to gather information from witnesses. Witnesses need not be limited to those persons actually at the scene of the accident when it occurred, but should include others who might know of the circumstances relating to the accident. Most of the principles for interviewing the witnesses are the same as those for the accident victim. They are:

   ❖ Interview the witnesses as promptly as possible after the accident.

   ❖ Interview the witnesses separately.

   ❖ Reassure the witnesses of the positive nature of the investigation and the desire to conduct the investigation to prevent a reoccurrence of the accident.

   ❖ Do not interrupt the witnesses unless specific clarification is needed and, again, do not assign blame.

❖ Ask the witnesses for their opinions on ways to prevent a recurrence of the accident.

3. *Investigation of the accident scene.* It is important for the accident scene to remain untouched until after the accident investigation. If this is not possible, photographs or video can be taken of the area to attempted to record physical conditions immediately after the accident. It is also recommended that the victim and witnesses be interviewed (separately) at the scene of the accident if possible. Areas of special attention when investigating the scene of the accident are the lighting, condition of the working and walking areas, machinery guards, etc.

4. *Re-enactment of the accident.* Under some circumstances, it may be desirable to re-enact the accident, especially if the causes for the accident are not clear. This can involve the employee or witnesses if appropriate. Obviously, it is very important that precautions be taken to prevent the accident from being repeated.

5. *Reconstruction of the accident.* If the accident victim cannot be interviewed and there are no witnesses, it will be necessary to attempt to reconstruct the accident based on the information available. This situation may require the assistance of expert help from an outside consultant. It may also involve a regulatory agency, such as OSHA, if there is a fatality.

Information obtained in the accident investigation should be used to develop plans and to change operations to prevent a more serious recurrence. This information is normally recorded on a form called the "Accident Report." An example form and instructions for its use are described in the next section of this chapter.

THE ACCIDENT REPORT

As stated previously, the findings from the accident investigation should be recorded on an accident report form. An example of a brief form that can be used by the employee's immediate supervisor is shown in Figure 8-1 on page 104. Instructions for completion of the form are presented below. You may wish to modify the form to fit your specific needs. Additionally, a serious injury or fatality will require a more detailed investigation than indicated by the one page supervisor's report form although the overall approach should be the same.

Completing the Accident Report Form

❖ Description of the injury or illness:

Provide a brief description of the injury or illness and the body part(s) involved. State briefly the facts of the accident including the location, time of occurrence, equipment involved, etc. Include your estimate of the lost work time caused by the accident. (For example, broken leg and cracked ribs. Estimate of three months off the job. Employee being treated at Mount Hope hospital by Dr. Snort.)

❖ Employee's remarks:

This section is intended for the employee to provide his or her description and analysis of the accident. This should include suggestions for eliminating the cause(s) of the accident. If the employee is unable to complete this section because of the accident, all other sections should be completed and the section left blank until the employee is able to provide this information.

❖ Unsafe acts/conditions causing the accident/illness:

This section should be completed by the supervisor listing all the factors that may have contributed to the accident/illness event.

❖ Corrective actions required:

Identify the specific actions that need to be completed and the estimated schedule for completion. The person responsible for the action should also be listed. Many times the corrective action will include formal or on-the-job training for the employee.

CORRECTIVE ACTIONS

The accident investigation will usually yield a cause or causes for the accident and measures that will reduce the likelihood of a recurrence. Accident reports that include such statements as "...caused by carelessness—accident prone employee—caused by a stupid employee mistake—not paying attention when he/she should have been..." should not be accepted since they indicate an incomplete investigation and a lack of understanding of the causes of accidents. While it is true that most accidents are caused by an unsafe employee act, it is equally true that measures can be taken to lessen the chance of these acts recurring. One good approach to the correction of conditions that lead to an accident is to make these the responsibility of a union/management safety and health committee or an employee/management safety and health committee. They could also be used for the actual accident investigations.

COMPILING AND USING ACCIDENT STATISTICS

One copy of the supervisor's accident report should be sent to the person responsible for completing the OSHA accident and injury logs. This person should compile a monthly or quarterly report, as appropriate, listing the frequency and types of accidents that have occurred. It is also recommended that these compiled reports include copies of the supervisors' accident reports and be sent to the most senior official in the company. This shows the concern of management for accident prevention and is good business since accidents are expensive. Additionally, supervisors will be more careful in their investigations if they know a senior company official will be reviewing their work. These should be evaluated for trends such as an increase or excess of eye injuries, cuts, falls, back injuries, etc. Upward trends may suggest a company-wide problem and possible solutions. For example, a trend of increased reports of foot injuries might suggest the need for protective footwear.

Figure 8-1
Supervisory Report of Occupational Illness or Injury
(To be completed immediately after the accident, even when there are no injuries)

Accident Date:	Department:	Supervisor:
Employee name:	S.S.#	Date of birth:
Service date:	Time on present job:	Job classification:
Briefly describe the accident/illness (include estimated lost work days), exact accident location, time of occurrance, equipment involved, etc.		
Employee's remarks.		
Unsafe acts/conditions causing the accident/illness.		
Corrective actions required (note specific actions and schedules).		
Supervisor's Signature:	Date:	

Organizations such as Bureau of Labor Statistics (BLS) and the National Safety Council (NSC) compile and publish accident statistics. You can use these statistics to compare your performance to others in your industry or similar industries. The BLS reports are compiled from data collected under mandatory reporting requirements for injuries and illness under OSHA (see Chapter 5 on recordkeeping) reporting requirements. The NSC statistics are generated from many sources beyond the work place and through voluntary reporting by member companies or organizations through local and national councils. The major statistical measures reported are incidence rates for injuries and illness and a lost workdays rate. The BLS incidence rate is the number of recordable injuries and illness per 100 full-time employee equivalents (100 workers $\times$ 40 hours per week $\times$ 5 days per week = 200,000 hours). It is calculated by the following formula:

$$\text{Incidence rate} = \frac{\text{number of recordable injuries or illnesses} \times 200,000 \text{ hrs}}{\text{total number of employee hours worked}}$$

The lost workdays rate is calculated in a similar manner:

$$\text{Lost workdays rate} \ = \ \frac{\text{number of lost workdays} \times 200{,}000 \text{ hrs}}{\text{total number of employee hours worked}}$$

Comparing your company performance to national averages and other published statistics can give you a sense of your success at accident prevention. For example, in 1990 the country experienced approximately 6.8 million job-related injuries and illnesses with an incidence rate of 8.8 for total cases and an incidence rate of 4.1 for lost workday cases. This means that a firm employing 100 full-time workers would have had about nine reportable accidents in a year of normal work with about four lost workdays. For comparison purposes, the shipbuilding and repairing industry (SIC code 3731) had a rate six times higher for total cases (46.2) while the legal profession (SIC code 81) had a rate almost 15 times lower (0.6) as might be expected of an office environment.

CHAPTER
9
JOB ANALYSIS

Job Analysis (JA) is an analytical tool which can improve a company's overall performance by identifying and correcting undesirable events which could result in accidents, illnesses, injuries, reduced quality, and production. It is an employer/employee participation program in which job activities are observed, divided into individual steps, discussed, and recorded with the intent to identify, eliminate, or control undesirable events.

JA effectively accomplishes this goal because it operates at a very basic level. It reviews each job and breaks it down into an orderly series of smaller tasks. After these tasks have been determined, the same routine of observation, discussion, and recording is repeated, this time focusing on events which could have a negative impact on each step in the task. Once potential undesirable events are recognized, the process is repeated for a third time and corrective actions are identified. Although quality control and production, as well as safety and health, are involved in a JA, this chapter will concentrate on the safety aspects of the JA. An example of a JA form is shown in Figure 9-1.

ADVANTAGES OF JOB ANALYSIS

To have an effective JA program there must be input from all levels of the organization. Top management must supply moral and financial support to eliminate or modify potential hazards that were identified in the JA. They must also make the implementation and effectiveness of each component of the program part of the individual job performance rating for all supervisory personnel involved. The first line supervisor and plant safety officer can use the JA as a tool during an accident investigation—not to find fault but to identify and correct unsafe conditions. First line supervisors are the key to the success of JA. These individuals must be convinced of the commitment of top management and they must convey this support to the employees if the program is to work. First line supervisors are important because of their

Figure 9-1
Job Analysis

JOB ANALYSIS	JOB BEING ANALYZED	PAGE ___ OF ___	JOB ANALYSIS BY
FACILITY LOCATION	OPERATOR'S JOB TITLE	JOB ANALYSIS NO.	
		ORIGINAL ANALYSIS DATE	REVIEWED BY
DEPARTMENT	FOREMAN / SUPERVISOR	LAST REVISION DATE	APPROVED BY
REQUIRED AND / OR RECOMMENDED PERSONAL PROTECTIVE EQUIPMENT:			___ SAFETY & HEALTH ___ QUALITY CONTROL ___ PRODUCTION

SEQUENCE OF JOB STEPS	POTENTIAL DETRIMENTAL EVENTS	RECOMMENDED PROCEDURES OR CONTROLS

understanding of the process, its potential hazards, and the need for corrective actions instituted at each step. This also provides the interaction with hourly employees necessary to complete the JA. When the employees see that their input has been important in the establishment of work practices and modifications to equipment, they will be more willing to follow and participate in the program and to make suggestions for modifications to JA's when necessary.

WHO PERFORMS THE JOB ANALYSIS?

The responsibility for the development of a JA lies with first line supervision. These individuals have firsthand knowledge of the processes and daily contact with employees performing the job. Initially, first line supervisors must receive training in hazard recognition and procedures necessary to perform a JA. This training will give them the knowledge necessary to explain the JA to employees, what it is expected to accomplish, how it is conducted, and what their part will be in the program.

CREATING A JOB ANALYSIS

One of the questions usually asked by first line supervisors is what to analyze. From a safety standpoint this can easily be determined by reviewing accident records. The most obvious candidates for a JA are those jobs which have involved fatalities. The next priority are those that have resulted in injuries or illnesses which have caused employees to lose time from work. These can be further ranked by their severity and frequency of occurrence. Third, are jobs where repeated accidents involving individuals did not result in lost time from work but did require medical attention. The next category are jobs where an incident has occurred which did not result in a serious accident but could have (a near-miss incident). A JA should be conducted for all new operations.

After the job has been identified for which the JA is to be performed, information in the top part of the form can be completed (i.e., Facility Location, Department, Job Being Analyzed, Operator's Title, Foreman or Supervisor's Name, Page Number, and Job Analysis By). Other portions of this section are for individuals performing reviews of the particular sheet and for approvals by specified department heads. The box on the form for personal protective equipment (PPE) should be filled in with equipment generally required within the department and any specifically required for job. See the example in Figure 9-2.

Before beginning the observation portion of the JA an important decision must be made. There may be a number of people doing the job on the same or different shifts. The person chosen for the observation should have a thorough knowledge of the operation and be willing to share knowledge with you to improve the final product. Remember, after each stage of the JA is completed, it is prudent to circulate the JA to other supervisors and employees with similar jobs for comment before sending it for review and approval. After you have chosen the individual for observation, the individual should be contacted and given an explanation of the purpose of the JA.

The bottom half of the form is divided into three sections which can be completed independently if necessary.

Figure 9-2
Job Analysis: Personal Protection Equipment

KNR COMPANY JOB ANALYSIS	JOB BEING ANALYZED BAND SAW BLADE REPLACEMENT	PAGE ___ OF ___ JOB ANALYSIS NO. ___	JOB ANALYSIS BY K. B. SMITH
FACILITY LOCATION CLEVELAND, OHIO	OPERATOR'S JOB TITLE SAW OPERATOR	ORIGINAL ANALYSIS DATE	REVIEWED BY D. C. JOHNSON
DEPARTMENT BILLET PREPARATION	FOREMAN / SUPERVISOR J. A. CLARK	LAST REVISION DATE	APPROVED BY
REQUIRED AND / OR RECOMMENDED PERSONAL PROTECTIVE EQUIPMENT:	SAFETY GLASSES, SAFETY SHOES, LEATHER GLOVES		___ SAFETY & HEALTH ___ QUALITY CONTROL ___ PRODUCTION

SEQUENCE OF JOB STEPS	POTENTIAL DETRIMENTAL EVENTS	RECOMMENDED PROCEDURES OR CONTROLS

The first column is titled Sequence of Job Steps, and in many instances is the most difficult to complete correctly. In this column the job is broken down into basic steps. The problem arises because the breakdown may be too detailed, for example, pick up hammer with right hand, hold nail on board with left hand, strike nail with hammer, etc. On the other hand, it can be too general, such as, build a box. Each step in the JA should accomplish some major task. The wording for each step should begin with a verb or action word such as "hammer nail into the board," or, "sand edge of board smooth." The wording for each step should also be concise and to the point. If you find yourself describing the step with more than one simple sentence it can probably be broken down into two steps. See the example in Figure 9-3.

The second column is titled Potential Detrimental Events. Events which could occur during that particular job step that might cause harm to the individual, environment, product, or production should be listed. This section should only be started after the basic job steps have been identified. Subjects that must be addressed in this column involve unsafe actions taken by employees and unsafe conditions due to machinery or environmental control deficiencies. A general list of these are outlined below.

1. caught in, under, or between

2. fall to same level

3. fall to different level

4. struck by flying object

5. struck by falling object

6. strike against a stationary object

7. strains, sprains, or pulls from pushing, pulling, bending, lifting, or twisting

8. cuts, lacerations, or contusions

9. struck by moving object

10. overexposure to dust, fumes, mists, vapor, or gas

11. skin irritation from oils, solvents, soaps

12. high noise levels

13. contact with energized equipment

14. ionizing radiation (X or Gamma Rays, Alpha, Beta Particles)

15. non-ionizing radiation (Illumination, Infra-Red, Ultra Violet, Lasers)

16. environmental extremes (hot or cold)

17. ergonomic stresses

18. fire and explosion hazards

Remember it is acceptable to have a number of potentially undesirable events associated with a single job step. See the example in Figure 9-4.

Figure 9-3
Job Analysis: Sequence of Job Steps

KNR COMPANY JOB ANALYSIS	JOB BEING ANALYZED BAND SAW BLADE REPLACEMENT	PAGE ___ OF ___ JOB ANALYSIS NO. ___	JOB ANALYSIS BY K. B. SMITH
FACILITY LOCATION CLEVELAND, OHIO	OPERATOR'S JOB TITLE SAW OPERATOR	ORIGINAL ANALYSIS DATE	REVIEWED BY D. C. JOHNSON
DEPARTMENT BILLET PREPARATION	FOREMAN / SUPERVISOR J. A. CLARK	LAST REVISION DATE	APPROVED BY
REQUIRED AND / OR RECOMMENDED PERSONAL PROTECTIVE EQUIPMENT:	SAFETY GLASSES, SAFETY SHOES, LEATHER GLOVES		___ SAFETY & HEALTH ___ QUALITY CONTROL ___ PRODUCTION

SEQUENCE OF JOB STEPS	POTENTIAL DETRIMENTAL EVENTS	RECOMMENDED PROCEDURES OR CONTROLS
1. SHUT SAW OFF		
2. GET NEW BLADE AND INSPECT FOR DAMAGE		
3. REMOVE LOCK BOLT AND OPEN HOUSING DOOR		
4. REMOVE NUT HOLDING SAW BLADE ON SHAFT		
5. REMOVE OLD BLADE		
6. INSTALL NEW BLADE		
7. ALIGN FLANGE AND TIGHTEN NUT		
8. CLOSE HOUSING DOOR AND INSERT LOCK BOLT		
9. REMOVE LOCK AND TAGS FROM CONTROL SWITCH & ELECTRICAL BOX		

Figure 9-4
Job Analysis: Potential Detrimental Events

KNR COMPANY JOB ANALYSIS	JOB BEING ANALYZED	PAGE OF	JOB ANALYSIS BY
	BAND SAW BLADE REPLACEMENT	JOB ANALYSIS NO.	K. B. SMITH
FACILITY LOCATION CLEVELAND, OHIO	OPERATOR'S JOB TITLE SAW OPERATOR	ORIGINAL ANALYSIS DATE	REVIEWED BY D. C. JOHNSON
DEPARTMENT BILLET PREPARATION	FOREMAN / SUPERVISOR J. A. CLARK	LAST REVISION DATE	APPROVED BY
REQUIRED AND \ OR RECOMMENDED PERSONAL PROTECTIVE EQUIPMENT:	SAFETY GLASSES, SAFETY SHOES, LEATHER GLOVES		SAFETY & HEALTH QUALITY CONTROL PRODUCTION

SEQUENCE OF JOB STEPS	POTENTIAL DETRIMENTAL EVENTS	RECOMMENDED PROCEDURES OR CONTROLS
1. SHUT SAW OFF	1. CUTS, LACERATIONS OR CONTUSIONS OF HANDS & FINGERS IF SAW IS ENERGIZED.	
2. GET NEW BLADE	2A. CUTS AND LACERATIONS TO HANDS & FINGERS 2B. STRUCK BY FLYING PIECES OF THE NEW SAW BLADE	
3. REMOVE LOCK BOLT AND OPEN HOUSING DOOR	3. PINCH POINT TO HANDS AND FINGERS FROM MACHINE HOUSING	
4. REMOVE NUT HOLDING SAW BLADE ON SHAFT	4. WRENCH SLIPPING OFF OF NUT, CUTS TO FINGERS AND HANDS	
5. REMOVE OLD BLADE	5. CUTS, LACERATIONS OF HANDS, FINGERS, OR ARMS	
6. INSTALL NEW BLADE	6. CUTS, LACERATIONS OF HANDS, FINGERS, OR ARMS	
7. ALIGN FLANGE AND TIGHTEN NUT	7A. WRENCH SLIPPING OFF OF NUT, CUTS TO FINGERS AND HANDS 7B. BUMP AGAINST SAW BLADE, CUTS AND LACERATIONS TO HANDS OR ARM	
8. CLOSE HOUSING DOOR AND INSERT LOCK BOLT	8. PINCH POINT TO HANDS AND FINGERS BETWEEN DOOR AND HOUSING	
9. REMOVE LOCK AND TAGS FROM CONTROL SWITCH & ELECTRICAL BOX	9. NO APPARENT HAZARD	

The third and final part of the JA is determining the types of corrective actions which can be taken to reduce or eliminate the undesirable event. This step should emphasize employee input to take advantage of their knowledge and increase their acceptance of the program. This is also the step where management support plays a major role. If the implementation of suggestions made by employees and accepted by the supervisor as valid are rejected, delayed, or ignored by top management, the program will lose credibility and fail. When a recommended corrective action is rejected or delayed, the reasons for the action must be explained to the parties involved and acceptable short- or long-term measures outlined in a reasonable amount of time. Remember the corrective action can be as simple as the installation of a guard or worker awareness training. It can also be as complex and expensive as installation of a local ventilation system or complete automation of the task. In these cases it may take a considerable period of time to implement the control and this should be explained to those involved along with a tentative timetable in which temporary control measures (such as ear plugs or respirators) will be used.

When completing this section the solutions must be specific and complete. It is unacceptable to use statements such as "stay alert" or "pay attention" as a corrective measure. Corrective measures should not only state what to do but how to do it. For example, assume the job step is identified as lifting a box onto a conveyor and the undesirable event is identified as a back sprain or strain. A corrective action would then be to use proper lifting procedures such as keeping the back straight, knees bent, and lifting with the legs.

When developing corrective actions for each particular undesirable event there are general questions about each job step which should be asked. They include:

1. Is the step necessary to complete the job?

2. Is there a better way to do the job or individual step? (Modify job procedure)

3. Can the physical conditions which create the hazard be eliminated or modified?

4. Are there mechanical or procedural changes which can be implemented to reduce the frequency with which the job must be done?

5. Can potentially hazardous materials be satisfactorily replaced with less hazardous ones or can environmental controls be instituted?

Before going through an extensive analysis of developing solutions for hazards of a specific task or step, you should ask the following question: Is this task or step necessary to complete the job? First determine the goal of the job. After this has been accomplished, you can determine if safer ways can be developed to reach the desired outcome.

If the task or step cannot be eliminated or modified, then mechanical, administrative, or environmental changes should be considered. Mechanical changes can be in the form of work-saving tools such as hoists, conveyors, powered hand tools, and guards. Administrative changes can be in the form of work schedules, required personal protective equipment, training, and work procedures. Environmental changes include items such as sound absorbing materials, ventilation controls, and isolation booths. See the example in Figure 9-5.

After completion of the first draft of the JA it should be sent to all affected individuals (e.g., Safety, QC, and Production). You should not be discouraged if the sheet goes through a number of reviews by both management and employee representatives before it is complete.

Once the review is completed, the person who performed the JA, the JA reviewer, and the designated persons in the department should either sign or initial the final copy. Each completed

Figure 9-5
Job Analysis: Recommended Procedures or Controls

KNR COMPANY JOB ANALYSIS	JOB BEING ANALYZED	PAGE 1 OF 1	JOB ANALYSIS BY
	BAND SAW BLADE REPLACEMENT	JOB ANALYSIS NO. 001	K. B. SMITH
	OPERATOR'S JOB TITLE	**ORIGINAL ANALYSIS DATE**	**REVIEWED BY**
	SAW OPERATOR	1/15/90	D. C. JOHNSON
	FOREMAN / SUPERVISOR	**LAST REVISION DATE**	**APPROVED BY**
	J. A. CLARK	1/15/92	
			SAFETY & HEALTH
	REQUIRED AND / OR RECOMMENDED		QUALITY CONTROL
	PERSONAL PROTECTIVE EQUIPMENT: SAFETY GLASSES, SAFETY SHOES, LEATHER GLOVES		PRODUCTION

SEQUENCE OF JOB STEPS	POTENTIAL DETRIMENTAL EVENTS	RECOMMENDED PROCEDURES OR CONTROLS
1. SHUT SAW OFF	1. CUTS, LACERATIONS, OR CONTUSIONS OF HANDS & FINGERS IF SAW IS ENERGIZED	1. USE PROPER LOCK OUT TAGOUT PROCEDURE
2. GET NEW BLADE	2A. CUTS AND LACERATIONS TO HANDS & FINGERS	2A. WEAR LEATHER GLOVES WHEN HANDLING SAW BLADES
	2B. STRUCK BY FLYING PIECES OF THE NEW SAW BLADE	2B. ALWAYS INSPECT NEW SAW BLADES FOR DAMAGE BEFORE INSTALLATION
3. REMOVE LOCK BOLT AND OPEN HOUSING DOOR	3. PINCH POINT TO HANDS AND FINGERS FROM MACHINE HOUSING	3. GRASP DOOR AT THE TOP TO AVOID PINCH POINT WITH MACHINE HOUSING
4. REMOVE NUT HOLDING SAW BLADE ON SHAFT	4. WRENCH SLIPPING OFF OF NUT, CUTS TO FINGERS AND HANDS	4. INSPECT NUT FOR WEAR, REPLACE IF NEEDED. USE CORRECT SIZE WRENCH FOR NUT
5. REMOVE OLD BLADE	5. CUTS, LACERATIONS OF HANDS, FINGERS, OR ARMS	5. WEAR LEATHER GLOVES AND LONG SLEEVES WHEN WORKING WITH OR NEAR SAW BLADES
6. INSTALL NEW BLADE	6. CUTS, LACERATIONS OF HANDS, FINGERS, OR ARMS	6. WEAR LEATHER GLOVES AND LONG SLEEVES WHEN WORKING WITH OR NEAR SAW BLADES
7. ALIGN FLANGE AND TIGHTEN NUT	7A. WRENCH SLIPPING OFF OF NUT, CUTS TO FINGERS AND HANDS	7A. INSPECT NUT FOR WEAR, REPLACE IF NEEDED. USE CORRECT SIZE
	7B. BUMP AGAINST SAW BLADE, CUTS AND LACERATIONS LACERATIONS TO HANDS OR ARM	7B. WEAR LEATHER GLOVES AND LONG SLEEVES WHEN WORKING WITH OR NEAR SAW BLADES
8. CLOSE HOUSING DOOR AND INSERT LOCK BOLT	8. PINCH POINT TO HANDS AND FINGERS BETWEEN DOOR AND HOUSING	8. GRASP DOOR AT THE TOP TO AVOID PINCH POINT WITH HOUSING
9. REMOVE LOCK AND TAGS FROM CONTROL SWITCH & ELECTRICAL BOX	9. NO APPARENT HAZARD	9. RETURN TAGS AND LOCKS TO THEIR PROPER STORAGE LOCATION

form should also be given a unique analysis number and kept on file in a central location. Copies of the form should be placed in books, one located in the supervisors office and the other near the location where the job is performed. These should always be accessible to the employee.

Updating the Job Analysis

Once the initial JA is complete, there are a number of events which can occur that will require the document to be updated. Some of these events are as follows:

1. A possible undesirable event or control which was not recognized before.

2. No one can identify all possible hazards within a company 100% of the time; therefore, when an accident occurs on a job for which a JA has been completed, it should be reviewed to see if it correctly identified the particular hazard, its effects, and appropriate corrective actions. If this is not the case, the JA should be updated immediately.

3. Modification to a process, equipment, or material being used may require the JA to be updated or replaced.

4. A new local, state, or federal regulation may require tighter control or elimination of a material used in a process.

The JA is a dynamic document and the program must be monitored by a central person or committee familiar with changes within the company. This person or committee must also be thoroughly familiar with regulatory requirements. Once the JA is completed and distributed it should be scheduled for review every twelve to eighteen months.

It has been proven that a well-organized and maintained JA program can have a very beneficial effect on accident prevention, improved production, and product quality. Emphasis for this program, as with any other program, must start at the top and be conveyed down the line to all employees.

CHAPTER
10
ERGONOMICS

This chapter is intended to provide an overview of practical ergonomic concepts. This begins with a brief overview of other disciplines associated with ergonomics, a listing of repetitive motion diseases, and physical variations in workers. This is followed by materials handling and a detailed section on manual lifting since back injuries are such a major problem for business. The chapter concludes with some general guidance on work station design.

DEFINING ERGONOMICS

The term ergonomics is derived from two root Greek words, *erg* (work) and *nomos* (study of). A working definition of ergonomics is the study of human characteristics for the design of the work environment. You may have also heard the terms "human factors" used to describe ergonomics. It is most often used by industrial engineers who design work stations that promote production efficiency. In this chapter, the term ergonomics will be used (as used by OSHA), to refer to the interrelationship of workers and their work stations.

Ergonomics is a science that uses many different disciplines to evaluate body stressors. These include:

❖ Anthropometry—The scientific measurement of the body to better design work stations considering the physical variables of employees.

❖ Biomechanics—The science of studying the mechanics of living organisms by combining the sciences of biology and mechanical engineering. This field of science enables the ergonomist to determine the stresses put on body parts from various types of loading in tasks requiring lifting, pushing, pulling, twisting, reaching, etc.

❖ Physiology—The science that deals with functions and body processes. Data from this field enables the ergonomist to estimate metabolic rates and needs of the cardiovascular system (gas exchanges and heart rates) for work tasks.

❖ Psychology—The study of behavioral responses that include those influenced by the work environment.

CUMULATIVE TRAUMA DISORDERS

When employees are over stressed from tasks such as lifting, twisting, straining, stretching, static loads and repetitive motions, they are likely to develop symptoms associated with a variety of injuries and illnesses. These often cause employees to miss work, result in performance restrictions, or may cause disability. These types of injuries and illnesses cause industry to lose billions of dollars annually.

Employees exposed to these stressors are susceptible to injury and illness from overexertion of their musculo-skeletal and nervous systems. Injuries from overexertion of musculo-skeletal, nervous system groups can be caused by a single exposure. These types of injuries are often attributed to one or more of the following conditions:

❖ Lifting too heavy a load

❖ Shifting of an unstable load (bagged material)

❖ Improper lifting technique

❖ Pushing or pulling too heavy a load

Illnesses from overexertion are generally classified as cumulative trauma disorders (CTD) or repetitive motion disorders (RMD). CTDs are fast becoming the leading cause of workers' compensation claims. Recent statistics indicate they may account for up to 60 percent of all OSHA reportable illnesses and injuries. In response to this alarming increase, OSHA has placed special emphasis on the control of CTDs. They have issued citations with penalties exceeding $1 million to companies with high incidence rates of CTDs (such as the meat packing industry). This emphasis has lead many industries to develop their own special programs to review work processes.

The following is a list of common diseases that can result from poor ergonomic design or conditions:

❖ *Carpal Tunnel Syndrome* (CTS) is generally caused by tasks with high repetitive wrist and finger movements and tasks that cause the wrist to be held at an angle. Symptoms include tingling, pain or numbness in the first three fingers and thumb. These symptoms are manifested when the median nerve is compressed and tendons become inflamed from overuse or unnatural positions causing the tendons to be scraped against the inside of the carpal tunnel in the wrist.

❖ *Raynaud's Syndrome* (White finger) is associated with severely vibrating tools. The symptoms are loss of color in the fingers (hence it is called White Finger disease), reduced blood flow, sensitivity to cold, numbness, and the inability to sense heat. Excessive vibration is also associated with the following conditions:

   1) Bone cysts of hands and fingers

   2) Muscle, bone, and joint disorders

   3) Tendinitis of fingers and wrists

   4) Carpal Tunnel Syndrome

   5) Dupuytren's contracture.

❖ **Tenosynovitis** is the inflammation of tendon sheaths of the wrist. Generally associated with extreme side-to-side wrist movements causing pain or soreness. Related diseases include **trigger finger** from excessive flexing of a finger other than the thumb and **DeQuervian's disorder** where the sheath of the long and short thumb abductor muscles narrow.

❖ **Tendinitis** is generally caused by excessive twisting of a muscle group resulting in sore and painful muscles. Tendinitis is most often associated with athletic injuries such as "tennis elbow" and "golf elbow." In industry it is generally associated with long-term tasks such as driving screws for a full shift.

A detailed discussion on the design of tools and work stations for the prevention of these cumulative trauma disorders is beyond the scope of this chapter. However, there are some common sense guidelines which can be provided. Avoid or eliminate prolonged exposure to:

❖ Repetitive, twisting hand movements

❖ Jobs that require prolonged bending of the wrists

❖ Tools and equipment with high levels of vibration (especially below 1000 cycles per second) unless protective equipment is used

❖ Jobs that require high levels of force exertion from awkward positions

❖ Jobs which place excessive pressure on parts of the hand, wrist, or arm

❖ Jobs which require awkward movements or extremes of movement

Well-designed work stations and tasks reduce body stresses that may cause injury and illness while usually increasing productivity. Improved worker morale is an additional benefit.

APPLICATION OF ERGONOMIC PRINCIPLES TO THE WORKPLACE

Physical Variations

Workers come in a variety of sizes and shapes, with varying degrees of physical and mental capabilities. It is impractical to purchase machines and equipment that meet the size specifications of each and every worker from the midget to the basketball player. Architects and design engineers are aware of the variability of people and try to design machines, equipment, and work stations to fit the "norm" or the average person. Examples of average body dimensions for U.S. Civilian males and females are given in Table 10-1. Even dealing with the average person you are faced with a size difference between the male and female population. This does not mean that you are faced with an impossible task to fit employee physical differences with the machines and equipment available to you. But it does mean that you need to design the work stations with this level of variation in mind.

Most design work uses the 95th percentile since all but 5 percent of the population should "fit" these values. For example, 95 percent of the U.S. population should have a knee height when sitting of no more than 23.4 inches. Therefore, desk or table clearance should be at least this height.

You must also consider the wide range of physical capability of individual employees. Physical capabilities differ based on stature, muscle tone, natural abilities, and other factors, as well as limitations imposed by injuries or illnesses and the process of aging. For example, under

Table 10-1
US Civilian Body Dimensions of Males and Females in Inches*
50th Percentile Values for Ages 20-60 Years

	Male	Female
Heights Standing		
Stature	68.3	63.2
Eye	63.9	58.6
Shoulder	56.2	51.6
Elbow	43.3	39.8
Knuckle	29.7	27.4
Heights Sitting		
Seat to head	35.7	33.5
Eye	30.9	28.9
Shoulder	23.4	21.9
Elbow (from seat)	9.6	9.2
Knee height	21.4	19.6
Thigh clearance	5.7	5.4
Depths/Reaches		
Chest depth	9.5	9.5
Elbow - finger tip	18.8	16.6
Forward reach	32.5	28.0
Buttock-knee (sit)	23.4	22.4
Buttock-back leg (sit)	19.5	18.9
Breaths		
Elbow to elbow	16.4	15.2
Hip breath (sit)	13.9	14.3
Foot Dimensions		
Foot length	10.6	9.5
Foot breadth	3.9	3.5
Weight (lbs.)	163.0	134.7

*Adapted from Table 13-C, *Fundamentals of Industrial Hygiene*, National Safety Council, Chicago, 1988.

the new Americans with Disability Act (ADA) accommodations must be made for those persons not meeting normal physical requirements. This usually means mechanical assistance.

MANUAL MATERIAL HANDLING TASKS

Manual material handling tasks can be defined as any task where material is lifted, lowered, held, carried, pushed or pulled by employees without the benefit of powered mechanical assistance. Tasks using simple machines such as push trucks, rope blocks, and chain falls should be considered manual material handling tasks.

Manual material handling tasks should be evaluated by the following applicable parameters:

❖ Weight of the load to be moved

❖ The dimensions of the load

❖ The starting and ending elevations of loads for lifting and lowering tasks

❖ The distance of travel from the start of a lift or lowering of a load to the completion of the task

❖ The stability of the load (loose bagged material vs. stable rigid material)

❖ The distance the load is moved on a horizontal plane

❖ The time a load is suspended by an employee

❖ Frequency of the task (repetitions per hour).

When making ergonomic evaluations or simply observing manual material tasks that are thought to have been causing injuries and illnesses or new tasks that could cause injuries or illnesses, you should consider the following potential solutions:

❖ Reduce the size or weight of the load

❖ Use powered mechanical handling equipment

❖ Provide self-leveling devices to reduce bending or reaching by employees

❖ Add additional employees to assist in the task

❖ Modify the task by:

  –altering the distance the load is to be lifted or lowered

  –altering the starting point and/or the finish point of manual lifts or lowering tasks

  –eliminating twisting of the torso during lifting and lowering tasks

  –avoidance of one-hand and side lifts

  –reducing the number of repetitions of the task

  –provide handles for movement of the load

  –increasing the diameter of wheel sizes for manual pushed/pulled vehicles

  –altering the handles on manually pushed/pulled vehicles to a level that enables employees to keep their backs straight

  –eliminating or reducing the distance a load must be carried.

MANUAL LIFTING

There were over 370,000 disabling work-related back injuries reported in 1991. It has been estimated that 80 percent of all Americans will suffer from lower back pain at some point in their working lives. The average cost is estimated to be about $7,500 per case. Back injuries also represent about 30 percent to 40 percent of all workers compensation costs. Of these impressive statistics, it is estimated that over half of all back injuries occurred as a result of manual lifting. Since this is such an important issue for all businesses, most of the remainder of this chapter will address manual lifting safety.

Evaluation of Manual Lifting Tasks

The National Institute of Occupational Safety and Health's (NIOSH) *Work Practices Guide for Manual Material Lifting* provides specific weight guidelines for ideal lifting tasks. NIOSH developed a relatively easily applied formula that allows you to evaluate lifting tasks to determine maximum load limits, and action limits (AL), for specific lifting tasks that over 99 percent of the male and 75 percent of the female population can be expected to lift without excessive risk of injury. The formula is also adaptable to calculate the load limits for the maximum permissible limit (MPL) for lifting tasks which less than 25 percent of the male and 1 percent of the female population can be expected to handle without an excessive risk of injury.

The NIOSH formulas are only applicable under the following conditions:

- ❖ Slow smooth lifts without sudden movements or shocks. The formula is not applicable for lowering, pushing, pulling and other similar task functions
- ❖ Unencumbered lifting posture
- ❖ Favorable physical environment (level, stable lifting surface, etc.)
- ❖ Two-handed symmetrical lifts, load directly in front of body and no twisting or turning lifts
- ❖ Good couplings, hands with handles on the load, and feet with shoes on flat floor surfaces
- ❖ Loads of moderate width, 30 inches or less
- ❖ Lifts that are performed by employees who are physically capable and accustomed to the task
- ❖ Where there is minimal combining of tasks (holding, carrying, pushing, etc.) in combination with the lifting task

For lifting tasks where the NIOSH formula result exceeds the AL, you should modify the task to reduce the risk of injury. It is difficult to find many lifting tasks that meet the criteria for using the NIOSH formula. When the task does not meet all the specified conditions for using the formula you may assume the risk of injury may exceed good safety practices even when the weight of the load is below the AL. Tasks that equal or exceed the NIOSH MPL should be modified in some manner to reduce the risk of employee injury.

The NIOSH formulas for calculating the action limit (AL) and maximum permissible limit (MPL) for a lifting task are:

$$AL = 90(6/H)(1-.01V-30)(.7+3/D)(1-F/F_{max})$$

$$MPL = 3 \times AL$$

NIOSH addresses four critical factors when calculating the AL for a lifting task. These factors and their influence on the lifting task follow:

1. *Horizontal hand location* (H) is the distance in front of the midpoint between the ankles at the origin of the lift to the load handles of the lift. If the load being lifted

does not have handles you would measure the distance from the midpoint between the ankles to the center line of the hands where they grip the load. The H dimension is restricted by body size and the length of the employee's arms, ranging from approximately 6 inches to 32 inches.

2. *Vertical location* (V) is the location of the load at the start of the lift. The average employee has a knuckle height of 30 inches as measured from the surface on which the employee is standing. This function has the following interesting principles:

   ❖ A lift starting at knuckle height will result in a factor of one.

   ❖ The closer the starting point of the lift is to the floor, or the greater the distance from knuckle height, with less weight an employee can be expected to lift safely.

   ❖ The further the starting point of the lift is above knuckle height, the less weight an employee can be expected to lift safely.

3. Distance of travel (D) is the distance the load is moved (vertically) from the start of the lift to the completion of the lift. The minimum and maximum travel distances for this function are 9.8 inches and 78.7 inches.

4. Frequency of task (F) is the number of times the task or lift is made per hour as compared to the maximum number of times the lift can be made per hour as shown in Table 10-2. If the lift is made less than once per 5 minutes then F = 0. The F factor becomes 0 when the number of lifts per minute equal the maximum lifts per hour.*

Table 10-2
NIOSH Maximum Lift Values per Minute

Period	V 30" Standing	V 30" Stooped
1 hour	18 lifts/min.	15 lifts/min.
8 hours	15 lifts/min.	12 lifts/min.

The following hypothetical lifting problem is an example of how to use the NIOSH formula:

Your storeroom foreman wants to add a fourth tier of storage racks that will require the storeroom attendant to place 20 pound machine components on the top shelf. The shelf is 60 inches above floor level. The components are delivered to stores on a four-wheeled hand truck with a bed 12 inches above the floor. The new task will require the attendant to make 2 lifts a minute in a one hour period to place the machine parts in storage from the height of the four-wheeled hand truck. The machine parts require the storeroom attendant to grip the

---

*   Distances should be in inches for calculating ALs in pounds or in centimeters for calculating ALs in kilograms.

parts at a distance of 20 inches from the midpoint of his ankles (H value). Using the NIOSH formula, should you permit the addition of the fourth tier?

Calculate the action limit:

$$AL = 90(6/H)(1-.01V-30)(.7+3/D)(1-F/F_{max})$$

AL = action limit in lbs.

D = 48 inches (60 inch shelf less 12 inch truck)

H = 20 inches   F = 2 lifts/min.

V = 12 inches F_{max} = 18 lifts per/min.

$AL = 90(6/20)(1-.01|12-30|)(.7+3/48)(1-2/18)$

$AL = 90(.30)(.82)(.76)(.89)$

$AL = 14.97 = 15$ lbs

Calculate the maximum permissible limit:

MPL = (3)(AL)

MPL = 3 x 15

MPL = 45 lbs.

The results of the analysis shows you that the proposed task has an action limit of 15 lbs. The weight of the machine parts (20 pounds) are above the AL and below the MPL (45 pounds) for the intended task as it currently exists. As a manager you will need to consider modifying work tasks that require employees to lift objects having weights that exceed the task AL or MPL. Alternative actions include:

<u>Weights between the AL and the MPL</u>

❖ Restrict the task to persons having the physical capabilities to handle the task without excess risk of injury.

❖ Train employees for safe task performance.

❖ Increase the AL by altering one or more of the task variables.

<u>Weights that exceed the MPL</u>

❖ Automate the job.

❖ Use mechanical material handling devices.

❖ Modify the task to increase the MPL value.

As a manager you must determine what alternative can best be implemented to alter the task to achieve an acceptable risk factor or action limit. Table 10-3 provides a matrix of alternatives for modification of the storeroom shelf stocking task.

Table 10-3
Storeroom Shelf Stocking Task

Alternative	Feasable (Y/N)
1. Reduce weight of machine parts	N
2. Shorten reach distance (H)	N
3. Shorten travel distance for lift (D);	Y
4. Reduce frequency of lifts per minute	Y/N

The weight of the machine parts cannot be altered because of critical design requirements. The size of the machine part and the employee's body size prevent reducing the distance from the center point of the ankles to the grasping points of the load. Therefore, this alternative is not feasible for this task.

Shortening the distance of travel for the loads provides two possibilities. First, modify the hand truck to raise the starting point of the lift or the (V) function. Secondly, lower the height of the shelf on which the machine parts are to be placed. Since you have already calculated the AL you have enough information to substitute a lighter object for storage on the top shelf and place the machine parts on a lower shelf or to eliminate the proposed top shelf if space is not critical.

A reduction in the frequency of the lifts is possible but may not be practical. You could either use two employees to stock the shelf or restrict the number of lifts per hour. The unavailability of additional labor or the blocking of the storeroom aisle could be reasons that would not permit use of these alternatives.

General Manual Lifting Guidelines

❖ The further the load is extended from the body the more stress on the employee.

❖ The ideal starting point for a lift is knuckle height. As the distance increases above or below knuckle height, the stress on the employee increases.

❖ Employee stress is increased directly with the distance a load is lifted.

❖ Employee stress increases directly with the frequency per minute that loads are lifted.

By keeping these principles in mind you should be able to make quick preliminary evaluations for manual lifting tasks. A simple reminder should also be given to employees to hold the load as close as practical to their body to reduce their risk of injury. You may not always be able to evaluate lifting tasks or determine corrective actions through observations. However, you should be able to determine if additional studies need to be initiated.

WORK STATION DESIGN

There are four fundamental ergonomic risk factors:

❖ Awkward Postures

❖ Forceful Exertion

❖ Mechanical Stress

❖ Repetitiveness

Many of the work station design criteria can be derived from the avoidance of the four simple risk factors. From these come some very simple guidelines:

❖ Place all work within reach.

❖ Eliminate tasks which require work above the shoulders or with arms outstretched.

❖ Permit the worker to stand on a flat surface with both feet on the ground.

❖ Provide a footrest for work that must be done while standing.

❖ For most work keep the work bench just below elbow height. For heavy work, lower it to 4 to 8 inches below elbow height. Adjustable benches are best.

❖ Try to position work so that the elbows are bent at about 90 degrees.

❖ Allow for rotation between jobs that use different muscle groups.

❖ Keep work off the floor by using stands.

❖ Keep all repetitive lifts between knuckle and shoulder height.

❖ Avoid jobs that require twisting, stretching, or leaning while carrying or manipulating a load.

❖ Avoid work stations or tasks that are in constricted spaces.

❖ Provide controls that are dimensioned to fit the hand and are easily located.

❖ Follow conventional practices for controls (e.g., down is off for power and clockwise increases speed).

❖ Place emergency controls within easy reach and color them red.

❖ Avoid foot controls unless force is needed.

❖ Avoid narrow viewing angles and minimize detail in control displays.

❖ Group related controls together.

❖ Provide work areas that are comfortable without extremes of heat or cold.

❖ Maintain clean and nonslippery walking and working surfaces.

❖ Provide lighting appropriate for the work task.

By following these simple guidelines, you will help your workforce to "work smarter, not harder."

CHAPTER
11
PLANT INSPECTIONS

It is clear that a well-designed inspection program for safety and health hazards can significantly reduce the economic losses and human suffering resulting from on-the-job injuries and illnesses. This chapter discusses the preparation for inspections, areas and items to be inspected, completion of the inspection report, and how to handle outside inspectors.

There are several methods for identifying in-plant safety and health hazards or, as referred to earlier, unsafe acts and conditions. These include examining accident and illness reports and informal conversations with your employees. However, the most common and most effective method of discovering unsafe conditions is to conduct periodic and structured inspections of your workplace. Observing operations within your plant is the only way you can determine the actual work conditions and existing or potential hazards.

Thorough plant inspections can take as little time as a few hours, depending on the inspection area and the complexity of the operations. Inspections can be done by a single person who is knowledgeable about the area (especially for small departments). However, it is best to utilize a team approach for plant-wide inspections or for those covering large areas or complex operations. A comprehensive inspection can be done with as few as two or three other knowledgeable employees. The importance of plant management's being visibly involved in part or all of the inspection and supporting the process should not be overlooked. This is a very good way to build confidence in your employees by demonstrating that you are genuinely concerned about their health and well-being.

Plant inspections should be conducted as fact-finding exercises rather than attempts to fix blame for hazards and to point fingers at responsible employees. The fact-finding approach to the inspection process makes the most sense and is the most effective. If employees are led to believe that these inspections are primarily designed to find faults and assign blame to the employee, then they are much less likely to be open and fully cooperative during the process.

Plant safety and health inspections are just one of several important monitoring functions conducted within your plant or business. They are no less important than the accounting, process design and control, inventory management, quality control, or other monitoring functions you perform on a routine basis. Inspections are designed to uncover, document, and correct existing or potential hazards in your workplace that have the capacity to cause accidents or illness. Inspections should be regarded as an important and necessary managerial tool, not as a frill or a gimmick for maintaining labor relations.

PREPARING FOR THE INSPECTION

In order to conduct effective inspections, you need to prepare for them. This includes gathering information and formulating a plan of attack prior to the inspection. If you do this in an organized manner prior to the initial inspection, the time required for the preparation and planning stages of all subsequent inspections can be significantly reduced.

The first step in the preparation process is determining the scope of your inspections. You should consider the following questions in determining this:

- ❖ What is the previous safety and workers' compensation history of your plant?
- ❖ Does your plant have high hazard areas?
- ❖ Are there areas of your plant where only minor hazards exist, such as office areas and seldom-used warehouses?
- ❖ What is the overall employee attitude about health and safety issues in the workplace?

Once you have answered these questions, you will have a better idea of the scope and focus of your inspections. It is probably a wise decision to spend extra time during the first inspection looking thoroughly at both high and low hazard areas of your plant so that you can document current conditions and can target areas of greatest concern for future inspections. For example, you may find that some areas of your plant need to be inspected only every three to six months, while others may require weekly inspection.

A good safety and health inspection plan requires at least the following elements:

1. Thorough knowledge of plant operations and production processes

2. Working knowledge of applicable federal, state, and local standards, regulations, codes, and recommendations

3. A logical approach to the inspection process such as starting with incoming raw material receiving, following materials through the production process, and ending with finished goods shipping

4. A way of recording and analyzing the inspection data so that it can be used to identify and correct deficiencies

In order to include these elements in your program, you should choose employees for the inspection team who know your plant operations well, such as your maintenance foreman and several key production supervisors. If your plant has a safety and health committee, you should include at least one representative from this committee. You should also include someone who is familiar with the applicable safety and health standards and regulations. Other chapters in this book should prove to be valuable references for familiarizing the team with these requirements. You may also find it useful to carry with you plant materials inventories and a reasonably detailed plant layout showing the locations of key machinery and operations.

One final element of the inspection preparation process is to assign someone to accurately record each potentially hazardous condition that is observed during the inspection. This list should include a description of the deficiency and what the team recommends to correct it. Sometimes this task can be accomplished more quickly by using a tape recorder or video camera (with audio) for initial verbal note-taking and later transcription.

Once you feel that you have assembled the necessary information and personnel, it is always a good idea to inform interested parties about what the team will be doing and what kinds of things they will be looking for. Surprise inspections may be used in the future if you feel they are warranted, but the initial efforts should be directed toward identifying hazards and informing employees and supervisors of ways to correct them.

CONDUCTING THE INSPECTION

How the inspection is conducted and what items the inspection team examines depends to a great extent on the size and type of facility and the complexity of your operations. However, there are a number of general inspection items common to most industrial facilities. These include:

1. Receiving, shipping, and warehousing areas

    ❖ Are these areas kept orderly with clear aisles and unobstructed access to exits?

    ❖ Are storage areas being used properly and in good repair?

    ❖ Are lighting and temperature levels adequate?

2. Processing and production areas, laboratories, etc.

    ❖ Are all machines with moving parts adequately guarded?

    ❖ Are hazardous chemicals being stored and handled properly?

3. Building and grounds

    ❖ Are floors, walls, ceilings, stairways, ramps, doors, etc. in good condition?

4. Housekeeping

    ❖ Are spills and leaks cleaned up promptly and completely?

    ❖ Is machinery cleaned properly after use?

    ❖ Are floors in aisles and work areas kept relatively clean and dry?

5. Electrical equipment and systems

- ❖ Are switch boxes, receptacles, fixtures, wiring, fuse boxes, etc. protected from physical damage and kept in good repair?
- ❖ Do all wiring installations comply with applicable OSHA and National Electrical Code requirements? Are motors and other high voltage equipment grounded properly?
- ❖ Are fixed machines supplied with permanent wiring?

6. Lighting and temperature levels

- ❖ Are work areas well lit and comfortable?
- ❖ Is specific task lighting, heating, or cooling provided when necessary?

7. Moving machinery

- ❖ Are all moving parts of machinery and points of operation guarded adequately?
- ❖ Are belts, pulleys, gears, flywheels, etc. maintained properly?
- ❖ Are machines completely deenergized during maintenance and repair operations?

8. Worker training and knowledge

- ❖ Have operators been properly trained on their machines?
- ❖ Are they using all protective equipment and clothing properly?
- ❖ Are they aware of the hazards from the machinery and chemicals they are using?

9. Hand and power tools

- ❖ Are these tools being used, stored, and maintained properly?

10. Hazardous chemicals

- ❖ Are containers of hazardous chemicals labelled properly?
- ❖ Are there any potential employee overexposures to these materials which need to be evaluated?
- ❖ Has effective hazard communication training been done?

11. Fire protection

❖ Are all areas of the plant adequately covered with fire extinguishers or sprinkler systems?

❖ Are these systems inspected on a regular basis?

❖ Are flammable and combustible materials stored properly?

❖ Is there an adequate emergency evacuation plan?

12. Maintenance programs

❖ Are all machines covered by a routine preventive maintenance program?

❖ Are maintenance records and logs kept properly?

❖ Is there an effective lockout/tagout program in place?

13. Personal protective equipment (PPE)

❖ Are all gloves, goggles, boots, hard hats, respirators, ear plugs, etc. being used and maintained properly?

❖ Has appropriate PPE been selected based on a hazard analysis?

❖ Is there an adequate written respirator program?

14. Physical hazards

❖ Have all physical hazards such as noise and radiation been assessed properly?

❖ Are there hazard control programs in place for these hazards?

15. Medical and first aid facilities

❖ Are there adequate medical and first aid supplies on hand?

❖ Are emergency showers, eye wash stations, etc. operating properly?

❖ Are trained medical or first aid personnel available on all shifts?

❖ Is there a procedure for transporting injured or sick employees to the nearest medical facility?

16. Walking and working surfaces

❖ Are they kept clear of obstructions?

❖ Are elevated surfaces and floor openings properly guarded?

17. Material handling equipment and vehicles

❖ Are chains, hoists, slings, ropes, cranes, etc. inspected regularly?

❖ Are employees trained in their proper use?

❖ Are forklift operators trained and licensed?

The conduct of the inspection should follow a predetermined route through your plant. For example, it may make the most sense to follow the same route as the materials that flow through your plant (i.e., from the receiving end to shipping). Be sure you do not skip areas such as chemical or physical testing labs, workshops, maintenance and janitorial areas, chemical storage buildings, office areas, etc., at least during your initial effort. Subsequent inspections can be targeted to high hazard areas.

OSHA has developed a number of self-inspection checklists for use by small businesses. These are reproduced in Appendix 11-A at the end of the chapter, and are fairly comprehensive, but not all-inclusive. You may find them useful as starting points for developing your own custom-tailored plant inspection checklists. They are reproduced from the 1990 edition of the *OSHA Handbook for Small Businesses*. This booklet is available at no cost through your local, area, or regional OSHA office.

AFTER THE INSPECTION

Once your inspection team has finished the actual plant walk-through, it should meet as soon as possible to discuss each of the possible deficiencies that were observed. This post-inspection meeting is a more appropriate time to discuss and debate the merits of each suggestion and to rank the seriousness of each deficiency. Some items may even require a second inspection or research into applicable standards and regulations. Those items that required immediate attention and were corrected during the inspection should still be mentioned and listed even though they have already been corrected.

Once your team has finished its deliberations, someone should be assigned to write a report of the inspection based on the field notes and follow-up discussions. This report should briefly list all deficiencies or potentially unsafe conditions and acts that were observed, the recommended corrective actions, and the status of each item. This status section should include the estimated completion date of items where a substantial amount of work will be required and the person or department responsible for correcting the deficiency. The report should then be posted in a prominent place in your workplace, such as in the lunch room or on your employee bulletin board, so that all your employees can read it.

Each inspection report should be used as a starting point for the next inspection. These reports can also be used to determine the frequency of reinspections. For example, if there were numerous deficiencies found in the electrical systems throughout the plant, more frequent inspections of these systems may be required until the number of deficiencies on each inspection is substantially reduced. Items marked as completed on one inspection report should not be removed from the inspection agenda until their status is verified during a subsequent inspection.

The frequency of reinspections depends on the number and severity of the deficiencies found during the initial inspection. During the early stages of your inspection program, you should probably conduct reinspections at least once every two weeks. You should soon get a feel for whether you need to increase or reduce this frequency. For example, if the inspections are uncovering a large number of new deficiencies each time, weekly inspections may be warranted for awhile. On the other hand, if there are relatively few new items, you may want to conduct monthly inspections.

Responsibility for the inspection process may eventually be turned over to another key employee, but it is important that you continue to demonstrate your strong support for it. Employees should understand that they are the most important element in maintaining a safe workplace.

Once you have conducted several inspections, begin to analyze the inspection reports by grouping the items into general areas and looking for trends. For example, if large numbers of problems are occurring involving forklifts, such as physical damage to the building, improperly stacked goods and supplies in the warehouse, etc., you may need to conduct some retraining for all forklift operators. Also, if numerous employees are observed not wearing protective clothing or equipment properly or at all, some retraining in how and why PPE should be used is probably indicated.

OBTAINING OUTSIDE HELP

You may find it necessary to obtain outside help to deal with some of the problems or deficiencies uncovered during your inspections. For example, designing effective machine guards can be a difficult task that may require the help of a consulting engineer. Another area where you may need outside expertise is chemical safety. If several of your employees who work with hazardous chemicals are complaining of similar adverse health effects, you may need to have a thorough professional evaluation of their exposures.

There are several sources of outside help which you should consider. Help from OSHA Area Offices and state consultation programs is available in many areas. Another possibility is to hire an outside safety and health consultant to evaluate your operations. The most important consideration in deciding on an outside consultant is the qualifications of the individual or firm you are considering. Look for credentials such as the professional engineer (PE) designation, certification as an industrial hygienist (CIH) or safety professional (CSP) or academic degrees from recognized institutions and experience. Listings of consultants can be found in the yellow pages of your telephone book or may be available through your local safety council. You may also contact the American Society of Safety Engineers (ASSE) or the American Industrial Hygiene Association (AIHA) for a listing of consultants in your area.

HANDLING GOVERNMENT INSPECTIONS

Chapter 1 discussed the development of a plan and an approach for handling government inspectors. However, it is worthwhile to repeat the essentials here.

Your facility may at some time be the subject of a government safety and health inspection. Typically these will be OSHA inspections because OSHA has primary responsibility for enforcing safety and health standards in the workplace. Other inspections may also be conducted by various state and local agencies.

Regardless of which agency is conducting the inspection, you should do several things as soon as the inspector arrives. These include:

❖   Ask the inspector to see his or her credentials

❖ Ask the inspector to explain the reason for his or her visit

❖ Explain briefly and concisely the nature of your workplace and operations stressing your commitment to providing a safe and healthful work environment for all your employees

❖ Ask the inspector when and how the inspection will be conducted.

Once you have completed this process, you may want to arrange for a representative to accompany the inspector through your workplace (but only if you are absolutely not available yourself) and to get any written records or documentation the inspector requests. If you feel that any of the inspector's requests are unreasonable or would unduly disrupt your operations, or if you would like to have your own expert present during the inspection, you may wish to seek the advice of legal counsel at this point.

Before allowing any government inspectors into your workplace, you should expect and have a right to receive a thorough explanation of why they are there, what they want to do or observe, and what your legal rights are during and after the inspection. You should determine ahead of time procedures for such things as taking photographs, collecting air samples, examining records, and conducting employee interviews.

After the inspection has been completed, you should have a closing conference with the inspector. This is the time for a frank discussion of the inspector's findings. If the inspector indicates that he or she has found deficiencies in your safety program or performance and that there may be citations or violations issued as a result, you should ask enough questions to be sure that you understand the situation. You should also ask for an explanation of your rights of appeal. If it is an OSHA inspection, you will be given a booklet explaining your rights and responsibilities after the inspection. Be sure to read and understand this important document.

There are several cardinal rules to follow during any government inspection. They are:

❖ Always be courteous to the inspector.

❖ Answer their questions as honestly and forthrightly as possible, but do not volunteer any additional information that is not requested.

❖ Be sure that you understand your rights throughout the inspection process.

❖ Do not intentionally antagonize the inspector or become argumentative during the inspection.

Remember that most inspectors do not enter your plant with a chip on their shoulders. They are simply there to try to make your plant a safer place to work.

APPENDIX 11-A

SELF-INSPECTION CHECK LISTS

These check lists are by no means all-inclusive. You should add to them or delete portions or items that do not apply to your operations. However, carefully consider each item as you come to it and then make your decision.

EMPLOYER POSTING

☐ Is the required OSHA workplace poster displayed in a prominent location where all employees are likely to see it?

☐ Are emergency telephone numbers posted where they can be readily found in case of emergency?

☐ Where employees may be exposed to any toxic substances or harmful physical agents, has appropriate information concerning employee access to medical and exposure records, and "Material Safety Data Sheets," etc., been posted or otherwise made readily available to affected employees?

☐ Are signs concerning "Exiting from buildings," room capacities, floor loading, exposures to x-ray, microwave, or other harmful radiation or substances posted where appropriate?

☐ Is the Summary of Occupational Illnesses and Injuries posted in the month of February?

RECORDKEEPING

☐ Are all occupational injury or illnesses, except minor injuries requiring only first aid, being recorded as required on the OSHA 200 log?

☐ Are employee medical records and records of employee exposure to hazardous substances or harmful physical agents up-to-date?

☐ Have arrangements been made to maintain required records for the legal period of time for each specific type record? (Some records must be maintained for at least 40 years.)

☐ Are operating permits and records up-to-date for such items as elevators, air pressure tanks, liquefied petroleum gas tanks, etc.?

SAFETY AND HEALTH PROGRAM

☐ Do you have an active safety and health program in operation?

☐ Is one person clearly responsible for the overall activities of the safety and health program?

☐ Do you have a safety committee or group made up of management and labor representatives that meet regularly and report in writing on its activities?

☐ Do you have a working procedure for handling in-house employee complaints regarding safety and health?

☐ Are you keeping your employees advised of the successful effort and accomplishments you and/or your safety committee have made in assuring they will have a workplace that is safe and healthful?

MEDICAL SERVICES AND FIRST AID

☐ Do you require each employee to have a pre-employment physical examination?

☐ Is there a hospital, clinic, or infirmary for medical care in proximity of your workplace?

☐ If medical and first aid facilities are not in proximity of your workplace, is at least one employee on each shift currently qualified to render first aid?

☐ Are medical personnel readily available for advice and consultation on matters of employees' health?

☐ Are emergency phone numbers posted?

☐ Are first aid kits easily accessible to each work area, with necessary supplies available, periodically inspected and replenished as needed?

☐ Have first aid kit supplies been approved by a physician, indicating that they are adequate for a particular area or operation?

☐ Are means provided for quick drenching or flushing of the eyes and body in areas where corrosive liquids or materials are handled?

FIRE PROTECTION

☐ Is your local fire department well acquainted with your facilities, its location and specific hazards?

☐ If you have a fire alarm system, is it certified as required?

☐ If you have a fire alarm system, is it tested at least annually?

☐ If you have interior stand pipes and valves, are they inspected regularly?

☐ If you have outside private fire hydrants, are they flushed at least once a year and on a routine preventive maintenance schedule?

☐ Are fire doors and shutters in good operating condition?

☐ Are fire doors and shutters unobstructed and protected against obstructions, including their counterweights?

☐ Are fire door and shutter fusable links in place?

☐ Are automatic sprinkler system water control valves, air and water pressure checked weekly/periodically as required?

☐ Is the maintenance of automatic sprinkler systems assigned to responsible persons or to a sprinkler contractor?

☐ Are sprinkler heads protected by metal guards, when exposed to physical damage?

☐ Is proper clearance maintained below sprinkler heads?

☐ Are portable fire extinguishers provided in adequate number and type?

☐ Are fire extinguishers mounted in readily accessible locations?

☐ Are fire extinguishers recharged regularly and noted on the inspection tag?

☐ Are employees periodically instructed in the use of extinguishers and fire protection procedures?

PERSONAL PROTECTIVE EQUIPMENT AND CLOTHING

☐ Are protective goggles or face shields provided and worn where there is any danger of flying particles or corrosive materials?

☐ Are approved safety glasses required to be worn at all times in areas where there is a risk of eye injuries such as punctures, abrasions, contusions or burns?

☐ Are employees who need corrective lenses (glasses or contacts) in working environments having harmful exposures, required to wear *only* approved safety glasses, protective goggles, or use other medically approved precautionary procedures.

☐ Are protective gloves, aprons, shields, or other means provided against cuts, corrosive liquids and chemicals?

☐ Are hard hats provided and worn where danger of falling objects exists?

☐ Are hard hats inspected periodically for damage to the shell and suspension system?

☐ Is appropriate foot protection required where there is the risk of foot injuries from hot, corrosive, poisonous substances, falling objects, crushing or penetrating actions?

☐ Are approved respirators provided for regular or emergency use where needed?

☐ Is all protective equipment maintained in a sanitary condition and ready for use?

☐ Do you have eye wash facilities and a quick Drench Shower within the work area where employees are exposed to injurious corrosive materials?

☐ Where special equipment is needed for electrical workers, is it available?

☐ Where lunches are eaten on the premises, are they eaten in areas where there is no exposure to toxic materials or other health hazards?

☐ Is protection against the effects of occupational noise exposure provided when sound levels exceed those of the OSHA noise standard?

☐ Are adequate work procedures, protective clothing and equipment provided and used when cleaning up spilled toxic or otherwise hazardous materials or liquids?

GENERAL WORK ENVIRONMENT

☐ Are all worksites clean and orderly?

☐ Are work surfaces kept dry or appropriate means taken to assure the surfaces are slip-resistant?

☐ Are all spilled materials or liquids cleaned up immediately?

☐ Is combustible scrap, debris and waste stored safely and removed from the worksite promptly?

☐ Are accumulations of combustible dust routinely removed from elevated surfaces including the overhead structure of buildings, etc.?

☐ Is combustible dust cleaned up with a vacuum system to prevent the dust going into suspension?

☐ Is metallic or conductive dust prevented from entering or accumulating on or around electrical enclosures or equipment?

☐ Are covered metal waste cans used for oily and paintsoaked waste?

☐ Are all oil and gas fired devices equipped with flame failure controls that will prevent flow of fuel if pilots or main burners are not working?

☐ Are paint spray booths, dip tanks, etc., cleaned regularly?

☐ Are the minimum number of toilets and washing facilities provided?

☐ Are all toilets and washing facilities clean and sanitary?

☐ Are all work areas adequately illuminated?

☐ Are pits and floor openings covered or otherwise guarded?

WALKWAYS _____

☐ Are aisles and passageways kept clear?

☐ Are aisles and walkways marked as appropriate?

☐ Are wet surfaces covered with non-slip materials?

☐ Are holes in the floor, sidewalk or other walking surface repaired properly, covered or otherwise made safe?

☐ Is there safe clearance for walking in aisles where motorized or mechanical handling equipment is operating?

☐ Are materials or equipment stored in such a way that sharp projectives will not interfere with the walkway?

☐ Are spilled materials cleaned up immediately?

☐ Are changes of direction or elevations readily identifiable?

☐ Are aisles or walkways that pass near moving or operating machinery, welding operations or similar operations arranged so employees will not be subjected to potential hazards?

☐ Is adequate headroom provided for the entire length of any aisle or walkway?

☐ Are standard guardrails provided wherever aisle or walkway surfaces are elevated more than 30 inches above any adjacent floor or the ground?

☐ Are bridges provided over conveyors and similar hazards?

FLOOR AND WALL OPENINGS _____

☐ Are floor openings guarded by a cover, a guardrail, or equivalent on all sides (except at entrance to stairways or ladders)?

☐ Are toeboards installed around the edges of permanent floor opening (where persons may pass below the opening)?

☐ Are skylight screens of such construction and mounting that they will withstand a load of at least 200 pounds?

☐ Is the glass in the windows, doors, glass walls, etc., which are subject to human impact, of sufficient thickness and type for the condition of use?

☐ Are grates or similar type covers over floor openings such as floor drains, of such design that foot traffic or rolling equipment will not be affected by the grate spacing?

☐ Are unused portions of service pits and pits not actually in use either covered or protected by guardrails or equivalent?

☐ Are manhole covers, trench covers and similar covers, plus their supports designed to carry a truck rear axle load of at least 20,000 pounds when located in roadways and subject to vehicle traffic?

☐ Are floor or wall openings in fire resistive construction provided with doors or covers compatible with the fire rating of the structure and provided with self closing feature when appropriate?

STAIRS AND STAIRWAYS _____

☐ Are standard stair rails or handrails on all stairways having four or more risers?

☐ Are all stairways at least 22 inches wide?

☐ Do stairs have at least a 6'6'' overhead clearance?

☐ Do stairs angle no more than 50 and no less than 30 degrees?

☐ Are stairs of hollow-pan type treads and landings filled to noising level with solid material?

☐ Are step risers on stairs uniform from top to bottom, with no riser spacing greater than 7½ inches?

☐ Are steps on stairs and stairways designed or provided with a surface that renders them slip resistant?

☐ Are stairway handrails located between 30 and 34 inches above the leading edge of stair treads?

☐ Do stairway handrails have at least 1½ inches of clearance between the handrails and the wall or surface they are mounted on?

☐ Are stairway handrails capable of withstanding a load of 200 pounds, applied in any direction?

☐ Where stairs or stairways exit directly into any area where vehicles may be operated, are adequate barriers and warnings provided to prevent employees stepping into the path of traffic?

☐ Do stairway landings have a dimension measured in the direction of travel, at least equal to the width of the stairway?

☐ Is the vertical distance between stairway landings limited to 12 feet or less?

ELEVATED SURFACES _____

☐ Are signs posted, when appropriate, showing the elevated surface load capacity?

☐ Are surfaces elevated more than 30 inches above the floor or ground provided with standard guardrails?

☐ Are all elevated surfaces (beneath which people or machinery could be exposed to falling objects) provided with standard 4-inch toeboards?

☐ Is a permanent means of access and egress provided to elevated storage and work surfaces?

☐ Is required headroom provided where necessary?

☐ Is material on elevated surfaces piled, stacked or racked in a manner to prevent it from tipping, falling, collapsing, rolling or spreading?

☐ Are dock boards or bridge plates used when transferring materials between docks and trucks or rail cars?

EXITING OR EGRESS

☐ Are all exits marked with an exit sign and illuminated by a reliable light source?

☐ Are the directions to exits, when not immediately apparent, marked with visible signs?

☐ Are doors, passageways or stairways, that are neither exits nor access to exits and which could be mistaken for exits, appropriately marked "NOT AN EXIT," "TO BASEMENT," "STOREROOM," etc.?

☐ Are exit signs provided with the word "EXIT" in lettering at least 5 inches high and the stroke of the lettering at least ½-inch wide?

☐ Are exit doors side-hinged?

☐ Are all exits kept free of obstructions?

☐ Are at least two means of egress provided from elevated platforms, pits or rooms where the absence of a second exit would increase the risk of injury from hot, poisonous, corrosive, suffocating, flammable, or explosive substances?

☐ Are there sufficient exits to permit prompt escape in case of emergency?

☐ Are special precautions taken to protect employees during construction and repair operations?

☐ Is the number of exits from each floor of a building and the number of exits from the building itself, appropriate for the building occupancy load?

☐ Are exit stairways which are required to be separated from other parts of a building, enclosed by at least 2-hour fire-resistive construction in buildings more than four stories in height, and not less than 1-hour fire-resistive constructive elsewhere?

☐ Where ramps are used as part of required exiting from a building, is the ramp slope limited to 1 ft. vertical and 12 ft. horizontal?

☐ Where exiting will be through frameless glass doors, glass exit doors, storm doors, etc., are the doors fully tempered and meet the safety requirements for human impact?

EXIT DOORS

☐ Are doors which are required to serve as exits designed and constructed so that the way of exit travel is obvious and direct?

☐ Are windows which could be mistaken for exit doors, made inaccessible by means of barriers or railings?

☐ Are exit doors openable from the direction of exit travel without the use of a key or any special knowledge or effort when the building is occupied?

☐ Is a revolving, sliding or overhead door prohibited from serving as a required exit door?

☐ Where panic hardware is installed on a required exit door, will it allow the door to open by applying a force of 15 pounds or less in the direction of the exit traffic?

☐ Are doors on cold storage rooms provided with an inside release mechanism which will release the latch and open the door even if it's padlocked or otherwise locked on the outside?

☐ Where exit doors open directly onto any street, alley or other area where vehicles may be operated, are adequate barriers and warnings provided to prevent employees stepping into the path of traffic?

☐ Are doors that swing in both directions and are located between rooms where there is frequent traffic, provided with viewing panels in each door?

PORTABLE LADDERS

☐ Are all ladders maintained in good condition, joints between steps and side rails tight, all hardware and fittings securely attached and moveable parts operating freely without binding or undue play?

☐ Are non-slip safety feet provided on each ladder?

☐ Are non-slip safety feet provided on each metal or rung ladder?

☐ Are ladder rungs and steps free of grease and oil?

☐ Is it prohibited to place a ladder in front of doors opening toward the ladder except when the door is blocked open, locked or guarded?

☐ Is it prohibited to place ladders on boxes, barrels, or other unstable bases to obtain additional height?

☐ Are employees instructed to face the ladder when ascending or descending?

☐ Are employees prohibited from using ladders that are broken, missing steps, rungs, or cleats, broken side rails or other faulty equipment?

☐ Are employees instructed not to use the top step of ordinary stepladders as a step?

☐ When portable rung ladders are used to gain access to elevated platforms, roofs, etc., does the ladder always extend at least 3 feet above the elevated surface?

☐ Is it required that when portable rung or cleat type ladders are used, the base is so placed that slipping will not occur, or it is lashed or otherwise held in place?

☐ Are portable metal ladders legibly marked with signs reading "CAUTION" - Do Not Use Around Electrical Equipment" or equivalent wording?

☐ Are employees prohibited from using ladders as guys, braces, skids, gin poles, or for other than their intended purposes?

☐ Are employees instructed to only adjust extension ladders while standing at a base (not while standing on the ladder or from a position above the ladder)?

☐ Are metal ladders inspected for damage?

☐ Are the rungs of ladders uniformly spaced at 12 inches, center to center?

HAND TOOLS AND EQUIPMENT

☐ Are all tools and equipment (both company and employee-owned) used by employees at their workplace in good condition?

☐ Are hand tools such as chisels, punches, etc. which develop mushroomed heads during use, reconditioned or replaced as necessary?

☐ Are broken or fractured handles on hammers, axes and similar equipment replaced promptly?

☐ Are worn or bent wrenches replaced regularly?

☐ Are appropriate handles used on files and similar tools?

☐ Are employees made aware of the hazards caused by faulty or improperly used hand tools?

☐ Are appropriate safety glasses, face shields, etc. used while using hand tools or equipment which might produce flying materials or be subject to breakage?

☐ Are jacks checked periodically to assure they are in good operating condition?

☐ Are tool handles wedged tightly in the head of all tools?

☐ Are tool cutting edges kept sharp so the tool will move smoothly without binding or skipping?

☐ Are tools stored in dry, secure location where they won't be tampered with?

☐ Is eye and face protection used when driving hardened or tempered spuds or nails?

PORTABLE (POWER OPERATED) TOOLS AND EQUIPMENT

☐ Are grinders, saws and similar equipment provided with appropriate safety guards?

☐ Are power tools used with the correct shield, guard, or attachment, recommended by the manufacturer?

☐ Are portable circular saws equipped with guards above and below the base shoe?

☐ Are circular saw guards checked to assure they are not wedged up, thus leaving the lower portion of the blade unguarded?

☐ Are rotating or moving parts of equipment guarded to prevent physical contact?

☐ Are all cord-connected, electrically-operated tools and equipment effectively grounded or of the approved double insulated type?

☐ Are effective guards in place over belts, pulleys, chains, sprockets, on equipment such as concrete mixers, air compressors, etc.?

☐ Are portable fans provided with full guards or screens having openings ½ inch or less?

☐ Is hoisting equipment available and used for lifting heavy objects, and are hoist ratings and characteristics appropriate for the task?

☐ Are ground-fault circuit interrupters provided on all temporary electrical 15 and 20 ampere circuits, used during periods of construction?

☐ Are pneumatic and hydraulic hoses on power-operated tools checked regularly for deterioration or damage?

ABRASIVE WHEEL EQUIPMENT-GRINDERS

☐ Is the work rest used and kept adjusted to within ⅛ inch of the wheel?

☐ Is the adjustable tongue on the top side of the grinder used and kept adjusted to within ¼ inch of the wheel?

☐ Do side guards cover the spindle, nut, and flange and 75 percent of the wheel diameter?

☐ Are bench and pedestal grinders permanently mounted?

☐ Are goggles or face shields always worn when grinding?

☐ Is the maximum RPM rating of each abrasive wheel compatible with the RPM rating of the grinder motor?

☐ Are fixed or permanently mounted grinders connected to their electrical supply system with metallic conduit or other permanent wiring method?

☐ Does each grinder have an individual on and off control switch?

☐ Is each electrically operated grinder effectively grounded?

☐ Before new abrasive wheels are mounted, are they visually inspected and ring tested?

☐ Are dust collectors and powered exhausts provided on grinders used in operations that produce large amounts of dust?

☐ Are splash guards mounted on grinders that use coolant to prevent the coolant reaching employees?

☐ Is cleanliness maintained around grinders?

POWDER ACTUATED TOOLS

☐ Are employees who operate powder-actuated tools trained in their use and carry a valid operators card?

☐ Is each powder-actuated tool stored in its own locked container when not being used?

☐ Is a sign at least 7 inches by 10 inches with bold face type reading ''POWDER-ACTUATED TOOL IN USE'' conspicuously posted when the tool is being used?

☐ Are powder-actuated tools left unloaded until they are actually ready to be used?

☐ Are powder-actuated tools inspected for obstructions or defects each day before use?

☐ Do powder-actuated tool operators have and use appropriate personal protective equipment such as hard hats, safety goggles, safety shoes and ear protectors?

MACHINE GUARDING

☐ Is there a training program to instruct employees on safe methods of machine operation?

☐ Is there adequate supervision to ensure that employees are following safe machine operating procedures?

☐ Is there a regular program of safety inspection of machinery and equipment?

☐ Is all machinery and equipment kept clean and properly maintained?

☐ Is sufficient clearance provided around and between machines to allow for safe operations, set up and servicing, material handling and waste removal?

☐ Is equipment and machinery securely placed and anchored, when necessary to prevent tipping or other movement that could result in personal injury?

☐ Is there a power shut-off switch within reach of the operator's position at each machine?

☐ Can electric power to each machine be locked out for maintenance, repair, or security?

☐ Are the noncurrent-carrying metal parts of electrically operated machines bonded and grounded?

☐ Are foot-operated-switches guarded or arranged to prevent accidental actuation by personnel or falling objects?

☐ Are manually operated valves and switches controlling the operation of equipment and machines clearly identified and readily accessible?

☐ Are all emergency stop buttons colored red?

☐ Are all pulleys and belts that are within 7 feet of the floor or working level properly guarded?

☐ Are all moving chains and gears properly guarded?

☐ Are splash guards mounted on machines that use coolant to prevent the coolant from reaching employees?

☐ Are methods provided to protect the operator and other employees in the machine area from hazards created at the point of operation, ingoing nip points, rotating parts, flying chips, and sparks?

☐ Are machinery guards secure and so arranged that they do not offer a hazard in their use?

☐ If special handtools are used for placing and removing material, do they protect the operator's hands?

☐ Are revolving drums, barrels, and containers required to be guarded by an enclosure that is interlocked with the drive mechanism, so that revolution cannot occur unless the guard enclosures is in place, so guarded?

☐ Do arbors and mandrels have firm and secure bearings and are they free from play?

☐ Are provisions made to prevent machines from automatically starting when power is restored after a power failure or shutdown?

☐ Are machines constructed so as to be free from excessive vibration when the largest size tool is mounted and run at full speed?

☐ If machinery is cleaned with compressed air, is air pressure controlled and personal protective equipment or other safeguards utilized to protect operators and other workers from eye and body injury?

☐ Are fan blades protected with a guard having openings no larger than ½ inch, when operating within 7 feet of the floor?

☐ Are saws used for ripping, equipped with anti-kick back devices and spreaders?

☐ Are radial arm saws so arranged that the cutting head will gently return to the back of the table when released?

LOCKOUT BLOCKOUT PROCEDURES

☐ Is all machinery or equipment capable of movement, required to be de-energized or disengaged and blocked or locked-out during cleaning, servicing, adjusting or setting up operations, whenever required?

☐ Where the power disconnecting means for equipment does not also disconnect the electrical control circuit:

   Are the appropriate electrical enclosures identified?

   Is means provided to assure the control circuit can also be disconnected and locked-out?

☐ Is the locking-out of control circuits in lieu of locking-out main power disconnects prohibited?

☐ Are all equipment control valve handles provided with a means for locking-out?

☐ Does the lock-out procedure require that stored energy (mechanical, hydraulic, air, etc.) be released or blocked before equipment is locked-out for repairs?

☐ Are appropriate employees provided with individually keyed personal safety locks?

☐ Are employees required to keep personal control of their key(s) while they have safety locks in use?

☐ Is it required that only the employee exposed to the hazard, place or remove the safety lock?

☐ Is it required that employees check the safety of the lock-out by attempting a start up after making sure no one is exposed?

☐ Are employees instructed to always push the control circuit stop button prior to re-energizing the main power switch?

☐ Is there a means provided to identify any or all employees who are working on locked-out equipment by their locks or accompanying tags?

☐ Are a sufficient number of accident preventive signs or tags and safety padlocks provided for any reasonably foreseeable repair emergency?

☐ When machine operations, configuration or size requires the operator to leave his or her control station to install tools or perform other operations, and that part of the machine could move if accidentally activitated, is such element required to be separately locked or blocked out?

☐ In the event that equipment or lines cannot be shut down, locked-out and tagged, is a safe job procedure established and rigidly followed?

WELDING, CUTTING AND BRAZING

☐ Are only authorized and trained personnel permitted to use welding, cutting or brazing equipment?

☐ Does each operator have a copy of the appropriate operating instructions and are they directed to follow them?

☐ Are compressed gas cylinders regularly examined for obvious signs of defects, deep rusting, or leakage?

☐ Is care used in handling and storage of cylinders, safety valves, relief valves, etc., to prevent damage?

☐ Are precautions taken to prevent the mixture of air or oxygen with flammable gases, except at a burner or in a standard torch?

☐ Are only approved apparatus (torches, regulators, pressure-reducing valves, acetylene generators, manifolds) used?

☐ Are cylinders kept away from sources of heat?

☐ Are the cylinders kept away from elevators, stairs, or gangways?

☐ Is it prohibited to use cylinders as rollers or supports?

☐ Are empty cylinders appropriately marked and their valves closed?

☐ Are signs reading: DANGER—NO SMOKING, MATCHES, OR OPENLIGHTS, or the equivalent, posted?

☐ Are cylinders, cylinder valves, couplings, regulators, hoses, and apparatus kept free of oily or greasy substances?

☐ Is care taken not to drop or strike cylinders?

☐ Unless secured on special trucks, are regulators removed and valve-protection caps put in place before moving cylinders?

☐ Do cylinders without fixed and wheels have keys, handles, or non-adjustable wrenches on stem valves when in service?

☐ Are liquefied gases stored and shipped valve-end up with valve covers in place?

☐ Are provisions made to never crack a fuel-gas cylinder valve near sources of ignition?

☐ Before a regulator is removed, is the valve closed and gas released from the regulator?

☐ Is red used to identify the acetylene (and other fuel-gas) hose, green for oxygen hose, and black for inert gas and air hose?

☐ Are pressure-reducing regulators used only for the gas and pressures for which they are intended?

☐ Is open circuit (No Load) voltage of arc welding and cutting machines as low as possible and not in excess of the recommended limits?

☐ Under wet conditions, are automatic controls for reducing no load voltage used?

☐ Is grounding of the machine frame and safety ground connections of portable machines checked periodically?

☐ Are electrodes removed from the holders when not in use?

☐ Is it required that electric power to the welder be shut off when no one is in attendance?

☐ Is suitable fire extinguishing equipment available for immediate use?

☐ Is the welder forbidden to coil or loop welding electrode cable around his body?

☐ Are wet machines thoroughly dried and tested before being used?

☐ Are work and electrode lead cables frequently inspected for wear and damage, and relaced when needed?

☐ Do means for connecting cable lengths have adequate insulation?

☐ When the object to be welded cannot be moved and fire hazards cannot be removed, are shields used to confine heat, sparks, and slag?

☐ Are fire watchers assigned when welding or cutting is performed in locations where a serious fire might develop?

☐ Are combustible floors kept wet, covered by damp sand, or protected by fire-resistant shields?

☐ When floors are wet down, are personnel protected from possible electrical shock?

☐ When welding is done on metal walls, are precautions taken to protect combustibles on the other side?

☐ Before hot work is begun, are used drums, barrels, tanks, and other containers so thoroughly cleaned that no substances remain that could explode, ignite, or produce toxic vapors?

☐ Is it required that eye protection helmets, hand shields and goggles meet appropriate standards?

☐ Are employees exposed to the hazards created by welding, cutting, or brazing operations protected with personal protective equipment and clothing?

☐ Is a check made for adequate ventilation in and where welding or cutting is performed?

☐ When working in confined places, are environmental monitoring tests taken and means provided for quick removal of welders in case of an emergency?

COMPRESSORS AND COMPRESSED AIR

☐ Are compressors equipped with pressure relief valves, and pressure gauges?

☐ Are compressor air intakes installed and equipped so as to ensure that only clean uncomtaminated air enters the compressor?

☐ Are air filters installed on the compressor intake?

☐ Are compressors operated and lubricated in accordance with the manufacturer's recommendations?

☐ Are safety devices on compressed air systems checked frequently?

☐ Before any repair work is done on the pressure system of a compressor, is the pressure bled off and the system locked-out?

☐ Are signs posted to warn of the automatic starting feature of the compressors?

☐ Is the belt drive system totally enclosed to provide protection for the front, back, top, and sides?

☐ Is it strictly prohibited to direct compressed air towards a person?

☐ Are employees prohibited from using highly compressed air for cleaning purposes?

☐ If compressed air is used for cleaning off clothing, is the pressure reduced to less than 10 psi?

☐ When using compressed air for cleaning, do employees wear protective chip guarding and personal protective equipment?

☐ Are safety chains or other suitable locking devices used at couplings of high pressure hose lines where a connection failure would create a hazard?

☐ Before compressed air is used to empty containers of liquid, is the safe working pressure of the container checked?

☐ When compressed air is used with abrasive blast cleaning equipment, is the operating valve a type that must be held open manually?

☐ When compressed air is used to inflate auto ties, is a clip-on chuck and an inline regulator preset to 40 psi required?

☐ Is it prohibited to use compressed air to clean up or move combustible dust if such action could cause the dust to be suspended in the air and cause a fire or explosion hazard?

COMPRESSORS AIR RECEIVERS

☐ Is every receiver equipped with a pressure guage and with one or more automatic, spring-loaded safety valves?

☐ Is the total relieving capacity of the safety valve capable of preventing pressure in the receiver from exceeding the maximum allowable working pressure of the receiver by more than 10 percent?

☐ Is every air receiver provided with a drain pipe and valve at the lowest point for the removal of accumulated oil and water?

☐ Are compressed air receivers periodically drained of moisture and oil?

☐ Are all safety valves tested frequently and at regular intervals to determine whether they are in good operating condition?

☐ Is there a current operating permit used by the Division of Occupational Safety and Health?

☐ Is the inlet of air receivers and piping systems kept free of accumulated oil and carbonaceous materials?

COMPRESSED GAS CYLINDERS

☐ Are cylinders with a water weight capacity over 30 pounds, equipped with means for connecting a valve protector device, or with a collar or recess to protect the valve?

☐ Are cylinders legibly marked to clearly identify the gas contained?

☐ Are compressed gas cylinders stored in areas which are protected from external heat sources such as flame impingement, intense radiant heat, electric arcs, or high temperature lines?

☐ Are cylinders located or stored in areas where they will not be damaged by passing or falling objects or subjects to tampering by unauthorized persons?

☐ Are cylinders stored or transported in a manner to prevent them creating a hazard by tipping, falling or rolling?

☐ Are cylinders containing liquefied fuel gas, stored or transported in a position so that the safety relief device is always in direct contact with the vapor space in the cylinder?

☐ Are valve protectors always placed on cylinders when the cylinders are not in use or connected for use?

☐ Are all valves closed off before a cylinder is moved, when the cylinder is empty, and at the completion of each job?

☐ Are low pressure fuel-gas cylinders checked periodically for corrosion, general distortion, cracks, or any other defect that might indicate a weakness or render it unfit for service?

☐ Does the periodic check of low pressure fuel-gas cylinders include a close inspection of the cylinders' bottom?

HOIST AND AUXILLIARY EQUIPMENT

☐ Is each overhead electric hoist equipped with a limit device to stop the hook travel at its highest and lowest point of safe travel?

☐ Will each hoist automatically stop and hold any load up to 125 percent of its rated load, if its actuating force is removed?

☐ Is the rated load of each hoist legibly marked and visible to the operator?

☐ Are stops provided at the safe limits of travel for trolley hoist?

☐ Are the controls of hoist plainly marked to indicate the direction of travel or motion?

☐ Is each cage-controlled hoist equiped with an effective warning device?

☐ Are close-fitting guards or other suitable devices installed on hoist to assure hoist ropes will be maintained in the sheave groves?

☐ Are all hoist chains or ropes of sufficient length to handle the full range of movement of the application while still maintaining two full wraps on the drum at all times?

☐ Are nip points or contact points between hoist ropes and sheaves which are permanently located within seven feet of the floor, ground or working platform, guarded?

☐ Is it prohibited to use chains or rope slings that are kinked or twisted?

☐ Is it prohibited to use the hoist rope or chain wrapped around the load as a substitute, for a sling?

☐ Is the operator instructed to avoid carrying loads over people?

INDUSTRIAL TRUCKS—FORKLIFTS

☐ Are only employees who have been trained in the proper use of hoists allowed to operate them?

☐ Are only trained personnel allowed to operate industrial trucks?

☐ Is substantial overhead protective equipment provided on high lift rider equipment?

☐ Are the required lift truck operating rules posted and enforced?

☐ Is directional lighting provided on each industrial truck that operates in an area with less than 2 foot candles per square foot of general lighting?

☐ Does each industrial truck have a warning horn, whistle, gong, or other device which can be clearly heard above the normal noise in the areas where operated?

☐ Are the brakes on each industrial truck capable of bringing the vehicle to a complete and safe stop when fully loaded?

☐ Will the industrial trucks' parking brake effectively prevent the vehicle from moving when unattended?

☐ Are industrial trucks operating in areas where flammable gases or vapors, or combustible dust or ignitable fibers may be present in the atmosphere, approved for such locations?

☐ Are motorized hand and hand/rider trucks so designed that the brakes are applied, and power to the drive motor shuts off when the operator releases his or her grip on the device that controls the travel?

☐ Are industrial trucks with internal combustion engine, operated in buildings or enclosed areas, carefully checked to ensure such operations do not cause harmful concentration of dangerous gases or fumes?

SPRAYING OPERATIONS

☐ Is adequate ventilation assured before spray operations are started?

☐ Is mechanical ventilation provided when spraying operations is done in enclosed areas?

☐ When mechanical ventilation is provided during spraying operations, is it so arranged that it will not circulate the contaminated air?

☐ Is the spray area free of hot surfaces?

☐ Is the spray area at least 20 feet from flames, sparks, operating electrical motors and other ignition sources?

☐ Are portable lamps used to illuminate spray areas suitable for use in a hazardous location?

☐ Is approved respiratory equipment provided and used when appropriate during spraying operations?

☐ Do solvents used for cleaning have a flash point to 100°F or more?

☐ Are fire control sprinkler heads kept clean?

☐ Are "NO SMOKING" signs posted in spray areas, paint rooms, paint booths, and paint storage areas?

☐ Is the spray area kept clean of combustible residue?

☐ Are spray booths constructed of metal, masonry, or other substantial noncombustible material?

☐ Are spray booth floors and baffles noncombustible and easily cleaned?

☐ Is infrared drying apparatus kept out of the spray area during spraying operations?

☐ Is the spray booth completely ventilated before using the drying apparatus?

☐ Is the electric drying apparatus properly grounded?

☐ Are lighting fixtures for spray booths located outside of the booth and the interior lighted through sealed clear panels?

☐ Are the electric motors for exhaust fans placed outside booths or ducts?

☐ Are belts and pulleys inside the booth fully enclosed?

☐ Do ducts have access doors to allow cleaning?

☐ Do all drying spaces have adequate ventilation?

ENTERING CONFINED SPACES

☐ Are confined spaces thoroughly emptied of any corrosive or hazardous substances, such as acids or caustics, before entry?

☐ Are all lines to a confined space, containing inert, toxic, flammable, or corrosive materials valved off and blanked or disconnected and separated before entry?

☐ Is it required that all impellers, agitators, or other moving equipment inside confined spaces be locked-out if they present a hazard?

☐ Is either natural or mechanical ventilation provided prior to confined space entry?

☐ Are appropriate atmospheric tests performed to check for Oxygen deficiency, toxic substances and explosive concentrations in the confined space before entry?

☐ Is adequate illumination provided for the work to be performed in the confined space?

☐ Is the atmosphere inside the confined space frequently tested or continuously monitored during conduct of work?

☐ Is there an assigned safety standby employee outside of the confined space, when required, whose sole responsibility is to watch the work in progress, sound an alarm if necessary, and render assistance?

☐ Is the standby employee appropriately trained and equipped to handle an emergency?

☐ Is the standby employee or other employees prohibited from entering the confined space without lifelines and respiratory equipment if there is any question as to the cause of an emergency?

☐ Is approved respiratory equipment required if the atmosphere inside the confined space cannot be made acceptable?

☐ Is all portable electrical equipment used inside confined spaces either grounded and insulated, or equipped with ground fault protection?

☐ Before gas welding or burning is started in a confined space, are hoses checked for leaks, compressed gas bottles forbidden inside of the confined space, torches lightly only outside of the confined area and the confined area tested for an explosive atmosphere each time before a lighted torch is to be taken into the confined space?

☐ If employees will be using oxygen-consuming equipment such as salamanders, torches, furnaces, etc., in a confined space, is sufficient air provided to assure combustion without reducing the oxygen concentration of the atmosphere below 19.5 percent by volume?

☐ Whenever combustion-type equipment is used in a confined space, are provisions made to ensure the exhaust gases are vented outside of the enclosure?

☐ Is each confined space checked for decaying vegetation or animal matter which may produce methane?

☐ Is the confined space checked for possible industrial waste which could contain toxic properties?

☐ If the confined space is below the ground and near areas where motor vehicles will be operating, is it possible for vehicle exhaust or carbon monoxide to enter the space?

ENVIRONMENTAL CONTROLS

☐ Are all work areas properly illuminated?

☐ Are employees instructed in proper first air and other emergency procedures?

☐ Are hazardous substances identified which may cause harm by inhalation, ingestion, skin absorption or contact?

☐ Are employees aware of the hazards involved with the various chemicals they may be exposed to in their work environment, such as ammonia, chlorine, epoxies, caustics, etc.?

☐ Is employee exposure to chemicals in the workplace kept within acceptable levels?

☐ Can a less harmful method or produce be used?

☐ Is the work area's ventilation system appropriate for the work being performed?

☐ Are spray painting operations done in spray rooms or booths equipped with an appropriate exhaust system?

☐ Is employee exposure to welding fumes controlled by ventilation, use of respirators, exposure time, or other means?

☐ Are welders and other workers nearby provided with flash shields during welding operations?

☐ If forklifts and other vehicles are used in buildings or other enclosed areas, are the carbon monoxide levels kept below maximum acceptable concentration?

☐ Has there been a determination that noise levels in the facilities are within acceptable levels?

☐ Are steps being taken to use engineering controls to reduce excessive noise levels?

☐ Are proper precautions being taken when handling asbestos and other fibrous materials?

☐ Are caution labels and signs used to warn of asbestos?

☐ Are wet methods used, when practicable, to prevent the emisison of airborne asbestos fibers, silica dust and similar hazardous materials?

☐ Is vacuuming with appropriate equipment used whenever possible rather than blowing or sweeping dust?

☐ Are grinders, saws, and other machines that produce respirable dusts vented to an industrial collector or central exhaust system?

☐ Are all local exhaust ventilation systems designed and operating properly such as air flow and volume necessary for the application, ducts not plugged or belts slipping?

☐ Is personal protective equipment provided, used and maintained wherever required?

☐ Are there written standard operating procedures for the selection and use of respirators where needed?

☐ Are restrooms and washrooms kept clean and sanitary?

☐ Is all water provided for drinking, washing, and cooking potable?

☐ Are all outlets for water not suitable for drinking clearly identified?

☐ Are employees' physical capacities assessed before being assigned to jobs requiring heavy work?

☐ Are employees instructed in the proper manner of lifting heavy objects?

☐ Where heat is a problem, have all fixed work areas been provided with spot cooling or air conditioning?

☐ Are employees screened before assignment to areas of high heat to determine if their health condition might make them more· susceptible to having an adverse reaction?

☐ Are employees working on streets and roadways where they are exposed to the hazards of traffic, required to wear bright colored (traffic orange) warning vests?

☐ Are exhaust stacks and air intakes so located that contaminated air will not be recirculated within a building or other enclosed area?

☐ Is equipment producing ultra-violet radiation properly shielded?

FLAMMABLE AND COMBUSTIBLE MATERIALS

☐ Are combustible scrap, debris and waste materials (oily rags, etc.) stored in covered metal receptacles and removed from the worksite promptly?

☐ Is proper storage practiced to minimize the risk of fire including spontaneous combustion?

☐ Are approved containers and tanks used for the storage and handling of flammable and combustible liquids?

☐ Are all connections on drums and combustible liquid piping, vapor and liquid tight?

☐ Are all flammable liquids kept in closed containers when not in use (e.g. parts cleaning tanks, pans, etc.)?

☐ Are bulk drums of flammable liquids grounded and bonded to containers during dispensing?

☐ Do storage rooms for flammable and combustible liquids have explosion-proof lights?

☐ Do storage rooms for flammable and combustible liquids have mechanical or gravity ventilation?

☐ Is liquidified petroleum gas stored, handled, and used in accordance with safe practices and standards?

☐ Are no smoking signs posted on liquified petroleum gas tanks?

☐ Are liquified petroleum storage tands guarded to prevent damage from vehicles?

☐ Are all solvent wastes, and flammable liquids kept in fire-resistant, covered containers until they are removed from the worksite?

☐ Is vacuuming used whenever possible rather than blowing or sweeping combustible dust?

☐ Are firm separators placed between containers of combustibles or flammables, when stacked one upon another, to assure their support and stability?

☐ Are fuel gas cylinders and oxygen cylinders separated by distance, fire resistant barriers, etc. while in storage?

☐ Are fire extinguishers selected and provided for the types of materials in areas where they are to be used?

Class A   Ordinary combustible material fires.

Class B   Flammable liquid, gas or grease fires.

Class C   Energized-electrical equipment fires.

☐ Are appropriate fire extinguishers mounted within 75 feet of outside areas containing flammable liquids, and within 10 feet of any inside storage area for such materials?

☐ Are extinguishers free from obstructions or blockage?

☐ Are all extinguishers serviced, maintained and tagged at intervals not to exceed one year?

☐ Are all extinguishers fully charged and in their designated places?

☐ Where sprinkler systems are permanently installed, are the nozzle heads so directed or arranged that water will not be sprayed into operating electrical switch boards and equipment?

☐ Are "NO SMOKING" signs posted where appropriate in areas where flammable or combustible materials are used or stored?

☐ Are safety cans used for dispensing flammable or combustible liquids at a point of use?

☐ Are all spills of flammable or combustible liquids cleaned up promptly?

☐ Are storage tanks adequately vented to prevent the development of excessive vacuum or pressure as a result of filling, emptying, or atmosphere temperature changes?

☐ Are storage tanks equipped with emergency venting that will relieve excessive internal pressure caused by fire exposure?

☐ Are "NO SMOKING" rules enforced in areas involving storage and use of hazardous materials?

HAZARDOUS CHEMICAL EXPOSURE _____

☐ Are employees trained in the safe handling practices of hazardous chemicals such as acids, caustics, etc.?

☐ Are employees aware of the potential hazards involving various chemicals stored or used in the workplace such as acids, bases, caustics, epoxies, phenols, etc.?

☐ Is employee exposure to chemicals kept within acceptable levels?

☐ Are eye wash fountains and safety showers provided in areas where corrosive chemicals are handled?

☐ Are all containers, such as vats, storage tanks, etc., labeled as to their contents, e.g., "CAUSTICS"?

☐ Are all employees required to use personal protective clothing and equipment when handling chemicals (gloves, eye protection, respirators, etc.)?

☐ Are flammable or toxic chemicals kept in closed containers when not in use?

☐ Are chemical piping systems clearly marked as to their content?

☐ Where corrosive liquids are frequently handled in open containers or drawn from storage vessels or pipe lines, is adequate means readily available for neutralizing or disposing of spills or overflows properly and safely?

☐ Have standard operating procedures been established and are they being followed when cleaning up chemical spills?

☐ Where needed for emergency use, are respirators stored in a convenient, clean, and sanitary location?

☐ Are respirators intended for emergency use adequate for the various uses for which they may be needed?

☐ Are employees prohibited from eating in areas where hazardous chemicals are present?

☐ Is personal protective equipment provided, used and maintained whenever necessary?

☐ Are there written standard operating procedures for the selection and use of respirators where needed?

☐ If you have a respirator protection program, are your employees instructed on the correct usage and limitations of the respirators? Are the respirators NIOSH approved for this particular application? Are they regularly inspected and cleaned, sanitized and maintained?

☐ If hazardous substances are used in your processes, do you have a medical or biological monitoring system in operation?

☐ Are you familiar with the Threshold Limit Values or Permissible Exposure Limits of airborne contaminants and physical agents used in your workplace?

☐ Have control procedures been instituted for hazardous materials, where appropriate, such as respirators, ventilation systems, handling practices, etc.?

☐ Whenever possible are hazardous substances handled in properly designed and exhausted booths or similar locations?

☐ Do you use general dilution or local exhaust ventilation systems to control dusts, vapors, gases, fumes, smoke, solvents or mists which may be generated in your workplace?

☐ Is ventilation equipment provided for removal of contaminants from such operations as: Production grinding, buffing, spray painting, and/or vapor degreasing, and is it operating properly?

☐ Do employees complain about dizziness, headaches, nausea, irritation, or other factors of discomfort when they use solvents or other chemicals?

☐ Is there a dermatitis problem? Do employees complain about dryness, irritation, or sensitization of the skin?

☐ Have you considered the use of an industrial hygienist or environmental health specialist to evaluate your operation?

☐ If internal combustion engines are used, is carbon monoxide kept within acceptable levels?

☐ Is vacuuming used, rather than blowing or sweeping dusts whenever possible for clean-up?

☐ Are materials which give off toxic asphyxiant, suffocating or anesthetic fumes, stored in remote or isolated locations when not in use?

HAZARDOUS SUBSTANCES COMMUNICATION _____

☐ Is there a list of hazardous substances used in your workplace?

☐ Is there a written hazard communication program dealing with Material Safety Data Sheets (MSDS), labeling, and employee training?

☐ Is each container for a hazardous substance (i.e., vats, bottles, storage tanks, etc.) labeled with product identity and a hazard warning (communication of the specific health hazards and physical hazards)?

☐ Is there a Material Safety Data Sheet readily available for each hazardous substance used?

☐ Is there an employee training program for hazardous substances?

Does this program include:

☐ (1) An explanation of what an MSDS is and how to use and obtain one.

☐ (2) MSDA contents for each hazardous substance or class of substances.

☐ (3) Explanation of "Right to Know."

☐ (4) Identification of where an employee can see the employers written hazard communication program and where hazardous substances are present in their work areas.

☐ (5) The physical and health hazards of substances in the work area, and specific protective measures to be used.

☐ (6) Details of the hazard communication program, including how to use the labeling system and MSDS's.

ELECTRICAL ───────────

☐ Do you specify compliance with OSHA for all contract electrical work?

☐ Are all employees required to report as soon as practicable any obvious hazard to life or property observed in connection with electrical equipment or lines?

☐ Are employees instructed to make preliminary inspections and/or appropriate tests to determine what conditions exist before starting work on electrical equipment or lines?

☐ When electrical equipment or lines are to be serviced, maintained or adjusted, are necessary switches opened, locked-out and tagged whenever possible?

☐ Are portable electrical tools and equipment grounded or of the double insulated type?

☐ Are electrical appliances such as vacuum cleaners, polishers, vending machines, etc., grounded?

☐ Do extension cords being used have a grounding conductor?

☐ Are multiple plug adaptors prohibited?

☐ Are ground-fault circuit interrupters installed on each temporary 15 or 20 ampere, 120 volt AC circuit at locations where construction, demolition, modifications, alterations or excavations are being performed?

☐ Are all temporary circuits protected by suitable disconnecting switches or plug connectors at the junction with permanent wiring?

☐ Do you have electrical installations in hazardous dust or vapor areas? If so, do they meet the National Electrical Code (NEC) for hazardous locations?

☐ Is exposed wiring and cords with frayed or deteriorated insulation repaired or replaced promptly?

☐ Are flexible cords and cables free of splices or taps?

☐ Are clamps or other securing means provided on flexible cords or cables at plugs, receptacles, tools, equipment, etc., and is the cord jacket securely held in place?

☐ Are all cord, cable and raceway connections intact and secure?

☐ In wet or damp locations, are electrical tools and equipment appropriate for the use or location or otherwise protected?

☐ Is the location of electrical power lines and cables (overhead, underground, underfloor, other side of walls, etc.) determined before digging, drilling or similar work is begun?

☐ Are metal measuring tapes, ropes, handlines or similar devices with metallic thread woven into the fabric prohibited where they could come in contact with energized parts of equipment or circuit conductors?

☐ Is the use of metal ladders prohibited in areas where the ladder or the person using the ladder could come in contact with energized parts of equipment, fixtures or circuit conductors?

☐ Are all disconnecting switches and circuit breakers labeled to indicate their use or equipment served?

☐ Are disconnecting means always opened before fuses are replaced?

☐ Do all interior wiring systems include provisions for grounding metal parts of electrical raceways, equipment and encloures?

☐ Are all electrical raceways and enclosures securely fastened in place?

☐ Are all energized parts of electrical circuits and equipment guarded against accidental contact by approved cabinets or enclosures?

☐ Is sufficient access and working space provided and maintained about all electrical equipment to permit ready and safe operations and maintenance?

☐ Are all unused openings (including conduit knockouts) in electrical enclosures and fittings closed with appropriate covers, plugs or plates?

☐ Are electrical enclosures such as switches, receptacles, junction boxes, etc., provided with tight-fitting covers or plates?

☐ Are disconnecting switches for electrical motors in excess of two horsepower, capable of opening the circuit when the motor is in a stalled condition, without exploding? (Switches must be horsepower rated equal to or in excess of the motor hp rating.)?

☐ Is low voltage protection provided in the control device of motors driving machines or equipment which could cause probable injury from inadvertent starting?

☐ Is each motor disconnecting switch or circuit breaker located within sight of the motor control device?

☐ Is each motor located within sight of its controller or the controller disconnecting means capable of being locked in the open position or is a separate disconnecting means installed in the circuit within sight of the motor?

☐ Is the controller for each motor in excess of two horsepower, rated in horsepower equal to or in excess of the rating of the motor it serves?

☐ Are employees who regularly work on or around energized electrical equipment or lines instructed in the cardio-pulmonary resuscitation (CPR) methods?

☐ Are employees prohibited from working alone on energized lines or equipment over 600 volts?

NOISE

☐ Are there areas in the workplace where continuous noise levels exceed 85dBA?

☐ Is there an ongoing preventive health program to educate employees in: safe levels of noise, exposures; effects of noise on their health; and the use of personal protection?

☐ Have work areas where noise levels make voice communication between employees difficult been identified and posted?

☐ Are noise levels being measured using a sound level meter or an octave band analyzer and records being kept?

☐ Have engineering controls been used to reduce excessive noise levels? Where engineering controls are determined to not be feasible, are administrative controls (i.e. worker rotation) being used to minimize individual employee exposure to noise?

☐ Is approved hearing protective equipment (noise attenuating devices) available to every employee working in noisy areas?

☐ Have you tried isolating noisy machinery from the rest of your operation?

☐ If you use ear protectors, are employees properly fitted and instructed in their use?

☐ Are employees in high noise areas given periodic audiometric testing to ensure that you have an effective hearing protection system?

FUELING

☐ Is it prohibited to fuel an internal combustion engine with a flammable liquid while the engine is running?

☐ Are fueling operations done in such a manner that likelihood of spillage will be minimal?

☐ When spillage occurs during fueling operations, is the spilled fuel washed away completely, evaporated, or other measures taken to control vapors before restarting the engine?

☐ Are fuel tank caps replaced and secured before starting the engine?

☐ In fueling operations, is there always metal contact between the container and the fuel tank?

☐ Are fueling hoses of a type designed to handle the specific type of fuel?

☐ Is it prohibited to handle or transfer gasoline in open containers?

☐ Are open lights, open flames, or sparking, or arcing equipment prohibited near fueling or transfer of fuel operations?

☐ Is smoking prohibited in the vicinity of fueling operations?

☐ Are fueling operators prohibited in building or other enclosed areas that are not specifically ventilated for this purpose?

☐ Where fueling or transfer of fuel is done through a gravity flow system, are the nozzles of the self-closing type?

IDENTIFICATION OF PIPING SYSTEMS

☐ When nonpotable water is piped through a facility, are outlets or taps posted to alert employees that it is unsafe and not to be used for drinking, washing or other personal use?

☐ When hazardous substances are transported through above ground piping, is each pipeline identified at points where confusion could introduce hazards to employees?

☐ When pipelines are identified by color painting, are all visible parts of the line so identified?

☐ When pipelines are identified by color painted bands or tapes, are the bands or tapes located at reasonable intervals and at each outlet, valve or connection?

☐ When pipelines are identified by color, is the color code posted at all locations where confusion could introduce hazards to employees?

☐ When the contents of pipelines are identified by name or name abbreviation, is the information readily visible on the pipe near each valve or outlet?

☐ When pipelines carrying hazardous substances are identified by tags, are the tags constructed of durable materials, the message carried clearly and permanently distinguishable and are tags installed at each valve or outlet?

☐ When pipelines are heated by electricity, steam or other external source, are suitable warning signs or tags placed at unions, valves, or other serviceable parts of the system?

MATERIAL HANDLING

☐ Is there safe clearance for equipment through aisles and doorways?

☐ Are aisleways designated, permanently marked, and kept clear to allow unhindered passage?

☐ Are motorized vehicles and mechanized equipment inspected daily or prior to use?

☐ Are vehicles shut off and brakes set prior to loading or unloading?

☐ Are containers of combustibles or flammables, when stacked while being moved, always separated by dunnage sufficient to provide stability?

☐ Are dock boards (bridge plates) used when loading or unloading operations are taking place between vehicles and docks?

☐ Are trucks and trailers secured from movement during loading and unloading operations?

☐ Are dock plates and loading ramps constructed and maintained with sufficient strength to support imposed loading?

☐ Are hand trucks maintained in safe operating condition?

☐ Are chutes equipped with sideboards of sufficient height to prevent the materials being handled from falling off?

☐ Are chutes and gravity roller sections firmly placed or secured to prevent displacement?

☐ At the delivery end of the rollers or chutes, are provisions made to brake the movement of the handled materials?

☐ Are pallets usually inspected before being loaded or moved?

☐ Are hooks with safety latches or other arrangements used when hoisting materials so that slings or load attachments won't accidentally slip off the hoist hooks?

☐ Are securing chains, ropes, chockers or slings adequate for the job to be performed?

☐ When hoisting material or equipment, are provisions made to assure no one will be passing under the suspended loads?

☐ Are material safety data sheets available to employees handling hazardous substances?

TRANSPORTING EMPLOYEES AND MATERIALS

☐ Do employees who operate vehicles on public thoroughfares have valid operator's licenses?

☐ When seven or more employees are regularly transported in a van, bus or truck, is the operator's license appropriate for the class of vehicle being driven?

☐ Is each van, bus or truck used regularly to transport employees, equipped with an adequate number of seats?

☐ When employees are transported by truck, are provisions provided to prevent their falling from the vehicle?

☐ Are vehicles used to transport employees equipped with lamps, brakes, horns, mirrors, windshields and turn signals in good repair?

☐ Are transport vehicles provided with handrails, steps, stirrups or similar devices, so placed and arranged that employees can safely mount or dismount?

☐ Are employee transport vehicles equipped at all times with at least two reflective type flares?

☐ Is a full charged fire extinguisher, in good condition, with at least 4 B:C rating maintained in each employee transport vehicle?

☐ When cutting tools or tools with sharp edges are carried in passenger compartments of employee transport vehicles, are they placed in closed boxes or containers which are secured in place?

☐ Are employees prohibited from riding on top of any load which can shift, topple, or otherwise become unstable?

CONTROL OF HARMFUL SUBSTANCES BY VENTILATION

☐ Is the volume and velocity of air in each exhaust system sufficient to gather the dusts, fumes, mists, vapors or gases to be controlled, and to convey them to a suitable point of disposal?

☐ Are exhaust inlets, ducts and plenums designed, constructed, and supported to prevent collapse or failure of any part of the system?

☐ Are clean-out ports or doors provided at intervals not to exceed 12 feet in all horizontal runs of exhaust ducts?

☐ Where two or more different type of operations are being controlled through the same exhaust system, will the combination of substances being controlled, constitute a fire, explosion or chemical reaction hazard in the duct?

☐ Is adequate makeup air provided to areas where exhaust systems are operating?

☐ Is the source point for makeup air located so that only clean, fresh air, which is free of contaminates, will enter the work environment?

☐ Where two or more ventilation systems are serving a work area, is their operation such that one will not offset the functions of the other?

SANITIZING EQUIPMENT AND CLOTHING

☐ Is personal protective clothing or equipment that employees are required to wear or use, of a type capable of being cleaned easily and disinfected?

☐ Are employees prohibited from interchanging personal protective clothing or equipment, unless it has been properly cleaned?

☐ Are machines and equipment, which process, handle or apply materials that could be injurious to employees, cleaned and/or decontaminated before being overhauled or placed in storage?

☐ Are employees prohibited from smoking or eating in any area where contaminates that could be injurious if ingested are present?

☐ When employees are required to change from street clothing into protective clothing, is a clean change room with separate storage facility for street and protective clothing provided?

☐ Are employees required to shower and wash their hair as soon as possible after a known contact has occurred with a carcinogen?

☐ When equipment, materials, or other items are taken into or removed from a carcinogen regulated area, is it done in a manner that will contaminate non-regulated areas or the external environment?

TIRE INFLATION

☐ Where tires are mounted and/or inflated on drop center wheels, is a safe practice procedure posted and enforced?

☐ Where tires are mounted and/or inflated on wheels with split rims and/or retainer rings, is a safe practice procedure posted and enforced?

☐ Does each tire inflation hose have a clip-on chuck with at least 24 inches of hose between the chuck and an in-line hand valve and gauge?

☐ Does the tire inflation control valve automatically shutoff the air flow when the valve is released?

☐ Is a tire restraining device such as a cage, rack or other effective means used while inflating tires mounted on split rims, or rims using retainer rings?

☐ Are employees strictly forbidden from taking a position directly over or in front of a tire while it's being inflated?

CHAPTER
12
PERSONAL PROTECTIVE EQUIPMENT

The primary objective of any health and safety program is worker protection. It is the responsibility of management to carry out this objective. Part of this responsibility includes protecting workers from exposure to hazardous materials and hazardous situations that arise in the workplace. It is best for management to try to eliminate these hazardous exposures through changes in workplace design or engineering controls. When hazardous workplace exposures cannot be controlled by these measures, personal protective equipment (PPE) becomes necessary. When looking at hazardous workplace exposures, keep in mind that government regulations consider PPE the last alternative in worker protection because it does not eliminate the hazards. PPE only provides a barrier between the worker and the hazard. If PPE must be used as a control alternative, a positive attitude and strong commitment by management is required.

HOW TO IMPLEMENT A PERSONAL PROTECTIVE EQUIPMENT PROGRAM

For successful administration of a PPE program, complete the following tasks:

1. Write a policy on usage of PPE and communicate it to the employees.

2. Select the proper type of PPE based on the hazards.

3. Implement an effective and thorough training program.

4. Familiarize the employees with correct usage and maintenance of the PPE.

5. Enforce the usage of the PPE using a positive approach.

Policy

The written company policy is the key to success for any PPE program. A well-written company policy should include the following items:

❖ A written statement about management's commitment to workplace safety and health

❖ Statement about why PPE is being used

❖ Circumstances where PPE will be used

❖ Exceptions and limitations of PPE

Selection of Proper Equipment

The first step in selecting PPE is to analyze the job operation and determine what hazards are present. Then choose the PPE that provides the best degree of protection for that operation. Worker acceptance plays a major factor in use versus nonuse of PPE. One way to increase worker acceptance of PPE is to include them in the selection process. By including workers in the selection process, you will gain feedback on issues such as comfort, fit, and aesthetic appeal of various types of PPE. This worker involvement in the selection process will increase user acceptance and the likelihood the PPE will be worn.

Training

Development of an effective training program is necessary to help assure compliance with PPE requirements. Stress the need to wear the selected PPE. A successful training program would include the following:

❖ Describe what hazardous conditions are present in the workplace.

❖ Explain what can or cannot be done about hazardous conditions.

❖ Explain why a certain type of PPE has been selected.

❖ Discuss the capabilities and limitations of the PPE.

❖ Demonstrate how to fit, adjust, and use the PPE properly.

❖ Practice using the PPE correctly.

❖ Explain the written company policy and enforcement procedures for PPE.

❖ Discuss how to deal with emergency situations.

❖ Discuss how PPE will be maintained, cleaned, repaired, and replaced when necessary.

Maintenance and Use

All PPE needs to be maintained in proper working order so the equipment will be effective in controlling hazardous exposures. To effectively maintain PPE, complete the following:

❖ Perform periodic inspections of the PPE and maintain a record of the inspections.

❖ Follow inspection procedures recommended by the manufacturer when applicable.

❖ Repair any defective equipment using replacement parts supplied by the manufacturer, or replace the equipment altogether.

Enforcement

Enforcement of PPE use is necessary to assure adequate worker protection during hazardous job operations. The methods of enforcement differ in various industries. Many companies use some form of progressive disciplinary action. Other companies use a more positive approach, such as safety incentives for wearing PPE or recognition clubs for employees that have escaped injury by wearing PPE. The specific type of enforcement program your company chooses depends on worker attitude and acceptance of the PPE. By combining a positive disciplinary approach with thorough training, a very successful program can be attained.

TYPES OF PERSONAL PROTECTIVE EQUIPMENT

Six types of PPE will be discussed in detail in the following paragraphs:

1. Head Protection

2. Eye and Face Protection

3. Hearing Protection

4. Respiratory Protection

5. Foot Protection

6. Chemical Protective Clothing

HEAD PROTECTION

Head protection is needed for protection against the following hazards:

❖ Falling or flying objects
❖ Bumping the head against objects
❖ Getting hair caught in machinery

There are several industries where head protection is currently mandated, including tree trimming, construction work, shipbuilding, logging, mining, overhead line construction or maintenance, and basic metal or chemical production. It is important to choose the head protection based on the hazards presented during job operations. The types of head protection available include:

❖ Helmets
❖ Bump Caps
❖ Hair Nets and Caps

Helmets. There are a wide variety of helmets or hard hats available. They are usually made of molded thermoplastic or aluminum. Most can accommodate other protective equipment, such as hearing and eye protection. The American National Standards Institute (ANSI Z89.1-1986) defines three types of helmets:

Class A—Helmets designed to protect the head from the force of impact of falling objects and from electric shock during contact with exposed low voltage conductors.

Class B—Helmets are intended to protect from force of impact of falling objects and from electric shock during contact with exposed high voltage conductors.

Class C—Helmets are intended to protect the head from the force of impact of falling objects.

The ANSI standard requires identification inside the helmet shell that lists the manufacturer's name and helmet class. ANSI standards are designated by year, and the helmet identification should be no more than five years old. The Occupational Safety and Health Administration (OSHA) standard for head protection (29 CFR 1910.135) requires that helmets meet the requirements and specifications in ANSI Z89.1-1986.

Headbands and suspension webs inside the helmet should have 1 inch clearance between the helmet shell and the suspension. This suspension system distributes the force of the blow to the helmet over a large area of the skull and helps absorb the energy of the blow, thus preventing injury. Keep in mind that helmets do not protect against neck injuries.

Helmets should be inspected before each use. Items to inspect for include the following items:

❖ Cracks
❖ Signs of impact and rough treatment
❖ Excessive wear
❖ Loose or torn cradle straps
❖ Broken sewing lines
❖ Loose rivets
❖ Any other defects

These defects can affect the protective characteristics and performance of the helmets. Extreme temperatures, sunlight, and certain solvents can shorten the life of thermoplastic helmets. Helmets that exhibit chalking, cracking, less surface gloss, or any other damage should be replaced.

An important helmet maintenance function is cleaning and disinfecting. Clean the helmets with detergent and disinfectant solutions recommended by the manufacturer. Do not use organic solvents to clean plastic helmets, unless the manufacturer recommends a specific type of solvent. Once helmets are cleaned, they should be stored out of direct sunlight and away from extreme heat and cold.

Bump Caps.   Bump caps are not helmets (hard hats). They are lightweight and protect users from bumping their heads on stationary objects. They do not provide protection from falling objects. There are no standards covering the use of bump caps. Use bump caps if the severity of the potential injury is limited but not if ANSI-approved helmets are required. Bump caps can accommodate eye and hearing protection attached directly to the cap.

Hair Nets and Caps.   Hair nets or caps are needed to prevent long hair or beards from getting caught in chains, belts, or other moving equipment. They also may be used to prevent hair from falling into food processing operations and causing contamination. Hair protection can be paper, fabric, net caps, or covers. The important aspect in using this protection is that the hair is completely covered. If contact with sparks or hot metal is possible, then flame resistant materials are needed. No standards have been accepted for hair nets or caps, but they should be durable and capable of being laundered and disinfected. There should be several head sizes available or they should be adjustable to fit all wearers.

EYE AND FACE PROTECTION

Protection of the eyes and face from injury is an important aspect of any health and safety program. This type of PPE has the widest use and the widest range of styles, models, and types. The type of eye and face protection needed depends upon the job operation being conducted, materials involved, and the severity of the eye and/or face injury that could result. Remember that the severity of the eye and/or face injury will increase if the hazardous materials are hot or can react chemically with skin. The OSHA standard for eye and face protection (29 CFR 1910.133) requires that protection be used if there is a reasonable chance injury will occur. Eye and/or face protection needs to be provided if machines or operations present the following hazards:

❖ Flying objects and particles
❖ Airborne dusts
❖ Glare
❖ Splashing liquids
❖ Ultraviolet radiation
❖ Combinations of these hazards

The most common types of eye and face protection include:

❖ Spectacles
❖ Spectacles with side shields
❖ Goggles
❖ Faceshields
❖ Welding Helmets

To select the proper eye and face protection based on the hazards of the work operation, refer to Table 12-1.

Spectacles. Spectacles are designed to protect against frontal impact from particles and flying objects generated during grinding, hammering, or other operations that could generate high speed objects. The OSHA standards for eye and face protection specify that all spectacle lenses and frames shall meet performance standards listed in ANSI standard Z87.1-1968. Lenses must meet certain impact resistance criteria to be designated as safety glasses. Typical lenses are constructed of plastic or hardened glass.

Spectacles with Side Shields. Spectacles with side shields should be used in situations involving the possibility of falling or flying particles entering the eye from the side. Side shields

Table 12-1
Eye and Face Protection Selection Guide (OSHA Table E-1*)

1. GOGGLES, flexible fitting, regular ventilation
2. GOGGLES, flexible fitting, hooded ventilation
3. GOGGLES, cushioned fitting, rigid body
4.** SPECTACLES, metal frame, with side shields
5.** SPECTACLES, plastic frame, with side shields
6.** SPECTACLES, metal-plastic frame, with side shields

7. WELDING GOGGLES, eyecup type, tinted lenses
8. WELDING GOGGLES, coverspec type, tinted lenses
9. WELDING GOGGLES, coverspec type, tinted plate lens
10. FACE SHIELD (Available with plastic or mesh window)
11. WELDING HELMETS

OPERATION	HAZARDS	RECOMMENDED PROTECTORS
Acetylene-burning Acetylene-cutting Acetylene-welding	Sparks, harmful rays, molten metal, flying particles	7, 8, 9
Chemical Handling	Splash, acid burns, fumes	2, 10 (For severe exposure add 10 over 2)
Chipping	Flying particles	1, 3, 4, 5, 6
Electric (arc) welding	Sparks, intense (UV) rays, molten metal	9, 11 (11 in combination with 4, 5, 6 in tinted lenses)
Furnace operations	Glare, heat, molten metal	7, 8, 9 (For severe exposure add 10)
Grinding-light	Flying particles	1, 3, 4, 5, 6, 10
Grinding-heavy	Flying particles	1, 3 (for severe exposure add 10)
Laboratory	Chemical splash, glass breakage	2 (10 when in combination with 4, 5, 6)
Machining	Flying particles	1, 3, 4, 5, 6, 10
Molten metals	Heat, glare, sparks, splash	7, 8 (10 in combination with 4, 5, 6 in tinted lenses)
Spot welding	Flying particles, sparks	1, 3, 4, 5, 6, 10

*29 CFR 1926.102 (b) (1).
**Non-side shield spectacles are available for limited hazard use requiring only frontal protection.

can be solid, perforated, or wire mesh depending on the size of the particles. Grinding operations are an example of where spectacles with side shields would be needed.

Goggles. Goggles protect the eyes from flying particles, liquid splashes, molten metal, heat, and glare. There are several types of goggles available depending on the type of activity being conducted. Some activities require the protection of tight-fitting eye cups, other activities may require a different type of goggle. Some goggles are equipped with ventilation openings to prevent fogging. However, the ventilation openings should be suitable for the hazards present because some particles and liquids can pass through these openings. In general, for activities that potentially may involve liquid splashes of irritating or corrosive materials, a faceshield should be used since it protects the face. Goggles can be used to protect prescription glasses that do not provide adequate protection for the hazards presented. Goggles can also be used for protection against UV and optical radiation, provided they have the correct shading for the type of welding and cutting being conducted. To select the proper shading number for welding and cutting activities refer to Table 12-2.

Faceshields. For activities such as pouring of liquids or working with molten metals, a faceshield should be used. A faceshield has a large transparent panel made of plastic or wire mesh that extends over the front and sides of the face. Plastic faceshields are used for protection against irritating and corrosive liquids. Wire mesh faceshields are used for protection against splashes of molten metals. Faceshields are not recommended by ANSI Z87.1 as basic eye protection against impact. For impact protection, faceshields must be used in combination with safety glasses or goggles.

Welding Helmets. Welding helmets protect the face and eyes from damage due to ultraviolet radiation, sparks, and molten metals during electric arc welding. The helmet is equipped with a shaded window for radiation protection for the eyes. For more information on the hazards associated with welding operations refer to Chapter 22.

Contact Lenses

Over the last several years there has been considerable debate over the use of contacts versus corrective safety glasses. Before establishing a contact lens policy, it would be prudent to contact a reputable manufacturer, optometrist, regulatory agency, or other appropriate resource for current information on the subject of contact lenses in the workplace environment. Recent accident data and research studies suggest that contact lens wearers do not appear to have problems when their eyes are properly protected in the workplace. According to guidelines published by the National Society to Prevent Blindness, individuals may be allowed to wear contacts except for situations in which there exists significant risks of eye injury. It maintains that contact lens wearers who have experienced long-term success with contacts will be able to judge for themselves whether or not they will be able to wear contact lenses in their occupational environment. However, employees must conform to the policies and directives of management regarding contact use. Contact lens use should be restricted when the work environment involves exposure to chemicals, vapors, splashes, radiant or intense heat, molten metals, or high concentrations of particulate. Contact lens use in the work environment should be made on a case-by-case basis in conjunction with any future guidelines established by NIOSH or OSHA.

The National Society to Prevent Blindness makes the following recommendations regarding contact lens use in the occupational environment:

1. Occupational safety eyewear meeting or exceeding ANSI Z87.1 standards should be worn at all times by individuals in designated areas.

2. Employees and visitors should be advised of defined areas where contacts are allowed.

3. At work stations where contacts are allowed, the type of eye protection required should be specified.

Table 12-2
Filter Lens Shade Numbers for Protection Against Radiant Energy (OSHA Table E-2*)

Welding Operation	Shade No.**
Shielded metal-arc welding: 1/16, 3/32, 1/8 or 5/32 in. electrode	10
Gas-shielded arc welding (nonferrous): 1/16, 3/32, 1/8 or 5/32 in. electrode	11
Gas-shielded arc welding (ferrous): 1/16, 3/32, 1/8 or 5/32 in. electrode	12
Shielded metal-arc welding: 3/32, 7/32 or 1/4 in. electrodes	12
5/16 or 3/8 in. electrodes	14
Atomic hydrogen welding	10 - 14
Carbon arc welding	14
Soldering	2
Torch brazing	3 or 4
Light cutting, up to 1 in.	3 or 4
Medium cutting, 1 to 6 in.	4 or 5
Heavy cutting, 6 in. and greater	5 or 6
Gas welding (light) up to 1/8 in.	4 or 5
Gas welding (medium) 1/8 to 1/2 in.	5 or 6
Gas welding (heavy) 1/2 in. and greater	6 or 8

* 29 CFR 1926.102(a) (5)
**Note: Shades more dense than those listed may be used to suit the individual's needs.

4. A specific written management policy on contact lens use should be developed with employee consultation and involvement.

5. Restrictions on contact lens wear do not apply to office or secretarial employees.

6. A directory should be developed which lists all employees who wear contacts. This list should be maintained in the plant medical facility for easy access by trained first-aid personnel. Foreman and supervisors should be informed of individual employees wearing contact lenses.

7. Medical and first-aid personnel should be trained in the proper procedures and equipment for removing both hard and soft contact lenses from conscious and unconscious workers.

8. Employees should be required to keep a spare pair of contacts and/or a pair of up-to-date prescription spectacles in their possession. This action will allow the employees to perform their job functions, should they damage or lose a lens while working.

9. Employees who wear contact lenses should be instructed to remove contacts immediately if redness of the eye, blurring of vision, or pain in the eye associated with contact lens use occurs.

Comfort and Fit

Eye and face protection needs to be fitted properly to be comfortable and effective. Any corrective type of eyewear should be fitted by a qualified ophthalmologic professional. Employees can be taught to fit, adjust, and properly maintain their specific piece of eye and face protection. If the particular piece of face and eye protection does not fit, is not comfortable, or employees are not satisfied with its appearance, that piece of equipment may not be utilized and will not provide adequate protection.

A common problem interfering with the fit and usage of eye and face protection is lens fogging. Lens fogging causes safety concerns as well as comfort problems, and will decrease employee usage of the equipment. If lens fogging is a serious problem, use sprays or wipes containing anti-fogging agents. These sprays or wipes are available from a variety of manufacturers.

Maintenance

All eye and face protection needs to be maintained in proper working order. The lenses of spectacles, goggles, and faceshields need to be inspected and cleaned periodically so a clear field of vision is maintained. If lenses become cracked or significantly scratched they should be replaced. Follow the manufacturer's cleaning, inspection, and maintenance recommendations where applicable.

HEARING PROTECTION

Noise-induced hearing loss has long been recognized by health and safety professionals. The hearing loss can be temporary or permanent depending on the noise level and length of exposure. For cases in which engineering or administrative controls are not effective in reducing employee exposures below permissible levels, hearing protection should be provided. Hearing protection should be selected based on noise exposure levels and the noise reduction (attenuation) capabilities of the protection. An aid in the selection of hearing protection is a regulation that requires all protectors to have a label indicating their Noise Reduction Rating (NRR). This number provides an estimate of the effectiveness and can generally be subtracted from the dBA value of the workplace noise. This value indicates the approximate noise level being received in the worker's ear. Several factors must be considered when using the NRR ratings:

- ❖ The NRR is derived in an ideal laboratory test.
- ❖ Wearing the device on the job is not the ideal laboratory situation.
- ❖ Noise frequencies and levels will vary hence the noise reduction may vary.

Because of the above factors the actual noise reduction received may be fifty percent or less than the published NRR value. OSHA uses the NRR value less 7 dBA and divides this value by 2 [(NRR-7)/2] as the maximum noise reduction achieved by employees using hearing protection at their job.

Several types of hearing protectors are available:

- ❖ Circumaural (Muffs)
- ❖ Aural Insert (Plugs)
- ❖ Canal Caps (Superaural)
- ❖ Combinations

Circumaural. These are also referred to as cups or earmuffs. They cover the external ear and provide an acoustic barrier. Muffs are more effective for high frequency noise. Attenuation properties of earmuffs are affected by size, shape, seal material, shell mass, and type of suspension. Muffs are available with liquid, grease-filled, plastic, and foam cushions. The liquid or grease-filled give better noise attenuation than the foam and plastic, but are subject to perforation. Attenuation is also affected because temple bars of safety spectacles interfere with proper fit. To obtain proper fit employees should use flat cable temple pieces (i.e. straps). Some styles of earmuffs can be attached directly to safety helmets.

Aural Inserts. These are commonly called earplugs. They are generally inexpensive but have limited service life. There are three categories of earplugs: (1) formable, (2) custom molded, and (3) molded.

Formable plugs have properties and features that allow them to adjust to the user. These are typically made of foam rubber that is placed in the external ear canal where they expand to the proper fit. They are widely used in many industrial situations. Employees need to be instructed on proper fitting so they will be effective in controlling noise exposure.

The custom molded plugs are molded to the ear canal of the user. A prepared mixture is placed in the employee's outer ear with a small portion of it in the ear canal. The material sets and takes the shape of the employee's ear and external ear canal. These should only be fitted by trained personnel. This type of plug is not widely used in industry since it can be quite expensive.

Molded (or premolded) inserts are usually made of soft silicone rubber or plastic. These plugs must fit snugly in the ear to be effective. However, this may cause discomfort because of the irregular shape of the ear. If the molded inserts are uncomfortable and do not fit, another type of protection should be utilized.

Superaural. These are also called canal caps. Their effectiveness depends on sealing the external edge of the ear canal in order to achieve sound reduction. The caps are made of soft rubber-like material attached to a spring band or head suspension.

Combinations. For severe noise environments such as airports, a combination of muffs and plugs will provide the greatest protection and sound attenuation.

Maintenance

Earmuffs should be inspected and cleaned periodically to maintain them in proper working order. If damaged equipment is found, replace or repair it following the manufacturer's recommendations. Before issuing hearing protection it is important to:

1. Have the employee's ear canals examined by a physician, because certain diseases may prevent the use of hearing protection.

2. Teach the employees the proper insertion techniques.

3. Teach proper sanitation and inspection procedures.

RESPIRATORY PROTECTION

Employees may need protection from airborne contaminants which are health hazards. Proper protection from airborne contaminants is the responsibility of management. When airborne hazards cannot be eliminated by other methods of control, proper selection and use of respiratory protection is part of that responsibility. Remember that respiratory protection is more than simply distributing respirators to workers who are exposed to airborne hazards. Effective protection demands that a well-planned program be implemented, including medical evaluation, evaluation of the airborne hazard, proper selection of respirators, fit testing, regular maintenance, and employee training. Some firms supply respirators to employees without first establishing a comprehensive respirator program, relying on the excuse that the respirators are not really required because contaminant concentrations in the work environment do not exceed permissible exposure limits (PELs). This is a serious mistake since allowing use of a respirator implies there is a hazard. A well-written program includes provisions for:

- ❖ written standard operating procedures
- ❖ evaluation of workplace hazards
- ❖ respirators selected on the basis of hazards present
- ❖ types of respirators to be used
- ❖ instructions, training, and fit testing of the user
- ❖ cleaning and disinfection
- ❖ proper storage

❖ maintenance and inspection

❖ evaluation of the respirator program

❖ medical evaluation of user

❖ use of certified equipment

The inclusion of the above items in your written program will also meet the OSHA requirements for an acceptable respirator program.

Written Standard Operating Procedures

The OSHA respiratory protection standard (29 CFR 1910.134) requires the establishment of a written respirator program, including standard operating procedures (SOPs) governing the selection and use of respirators. However, the standard does not provide guidance on preparation of these procedures and does not differentiate between small and large users. Some guidance concerning the general content of the procedures is provided by NIOSH and should help large or small users formulate their own procedures. The written SOPs should contain all information needed to maintain an effective respirator program to meet the user's individual requirements. SOPs should be written so they are useful to those directly involved in the respirator program. It is not necessary that the SOPs be written for the wearer, although in a very small program it may be desirable to direct the content to the wearer.

The SOPs should contain all the information needed to ensure proper respiratory protection of a specific group of workers against a specific hazard or several different hazards. The hazard(s) should be assessed thoroughly, otherwise the written procedures will have only limited validity. The point is that all information needed to establish a respirator program should be in writing.

The exact format of SOPs may vary widely. The large user with many workers wearing respirators and with several respiratory hazards to consider, may formulate separate SOPs for the selection and use of respirators for each hazard. For the small user, with a few workers to protect from limited hazards, a simplified document may be adequate. However, it must cover the same subjects. In general, the complexity of the SOPs increases as respirator use increases. The SOPs also become more extensive as the toxicity of the respiratory hazards increases, demanding better and more reliable protection. It is better to be overly detailed in developing the written SOPs than not detailed enough. A sample program for respirators is included in Appendix 12-A at the end of the chapter.

Hazard Evaluation

Before respirators are used in the workplace it is necessary to conduct a hazard evaluation. Once the types of airborne hazards have been identified, then a walk-through evaluation of the plant is necessary to identify employee groups or processes, or worker environments where respirators are needed. This will require some knowledge of the types of hazardous atmospheres where respiratory protection is needed. The evaluation may also involve the use of instruments to determine the concentrations of airborne contaminants so the correct respirators can be selected. The air monitoring should be done by qualified individuals. If your facility does not have qualified personnel, outside consultation may be required.

Airborne hazards are divided into the following classifications:

❖ Gaseous contaminants

❖ Particulate contaminants

❖ A combination of contaminants

❖ Oxygen-deficient atmospheres

Gaseous Contaminants. Gaseous contaminants include gases and vapors. Gases are substances that are similar to air in their ability to diffuse or spread freely throughout the area. Examples include carbon dioxide, nitrogen, carbon monoxide, hydrogen, and helium. Vapors are the gaseous state of substances that are normally liquids or solids at room temperature. Vapors are formed when solids or liquids evaporate. As an example, boiling water releases steam which is water vapor. Other examples of liquids that may form vapors include gasoline, paint thinners, and degreaser solvents.

Particulate Contaminants. Particulate contaminants are composed of tiny solid particles or liquid droplets of a substance. Many of these particles float in air for extended periods of time and may be inhaled. Particulates include dusts, fumes, and mists.

Dusts are created when solid materials are broken into fine particles that float in air before settling due to gravity. Dusts are produced by operations such as drilling, sanding, blasting, grinding, milling, and crushing.

Fumes are created when solid materials (e.g., metals) vaporize under extreme heat such as in the case of arc welding. The metal vapors quickly cool and condense into extremely fine particles, with a particle size of 1 micrometer or less in diameter. Fumes come from operations such as welding, smelting, and pouring of molten metals. It is a common misconception to call odorous vapors from liquids and solids "fumes." Fumes require use of a filter cartridge respirator while vapors require a sorbent cartridge respirator.

Mists are particles formed from liquids by atomization and condensation processes. For example, mists can be created by spraying operations, plating operations, mixing, and cleaning operations. Because mists are made up of discrete particles, a filter respirator can be used even though it may be a liquid. Some mists require both filtration and the use of a sorbent.

Combination Contaminants. Combinations of contaminants frequently occur together in many work operations. For example, paint spraying operations produce both paint mist (particulate) and solvent vapors (gaseous).

Oxygen-Deficient Atmospheres. Oxygen-deficient atmospheres result when there is an insufficient amount of oxygen within a work area to sustain human life. Oxygen-deficient atmospheres are typically found in confined spaces or poorly ventilated areas. The Occupational Safety and Health Administration (OSHA) considers any area containing less than 19.5% oxygen to be oxygen-deficient. These atmospheres are also classified as immediately dangerous to life and health (IDLH). IDLH is described as a very hazardous atmosphere where exposure can: (1) cause serious injury or death in a short period of time, or (2) cause serious delayed effects. Very high concentrations of certain gases, vapors, and particulates can also create IDLH atmospheres.

Respirator Selection

The type of respirator selected will be based on information obtained during the hazard evaluation. The types of hazard will dictate the choice of device or at least greatly narrow the field of choice. Devices are approved for maximum concentrations of classes of substances, and unapproved devices must not be used. Sometimes the excuse is given that no approved

Table 12-3
Guide for Selection of Respiratory Protection (OSHA Table E-4*)

Hazard**	Respirator
Oxygen Deficiency	Positive pressure self-contained breathing apparatus (SCBA)
	Positive pressure combination air-line with auxilliary SCBA
	or air-storage receiver with alarm
Gas and vapor Contaminants	
Immediately Dangerous to Life	Positve pressure SCBA
and Health (IDLH)	Gas mask for contaminant (ESCAPE ONLY)
	Self-rescue mouthpiece respirator (ESCAPE ONLY)
	Positive pressure combination air-line with auxilliary SCBA
	or an air-storage receiver with alarm
Not IDLH	Chemical cartridge respirator
	Gas mask
	Hose mask with or without blower
	Air-line respirator
Particulate Contaminants	
IDLH	Positive pressure SCBA
	Gas mask for contaminant (ESCAPE ONLY)
	Positve pressure combination air-line with auxilliary SCBA
	or air-storage receiver with alarm
Not IDLH	Dust, fume, mist respirator
	Air-line respirator
	Hose mask with or without blower
Combination Gas, Vapor, Particulate Contaminants	
IDLH	Positive pressure SCBA
	Gas mask for contaminant (ESCAPE ONLY)
	Positive pressure combination air-line with auxilliary SCBA
	or air-storage receiver with alarm
Not IDLH	Chemical cartridge with appropriate filter
	Gas mask with appropriate filter
	Hose mask with or without blower
	Air-line respirator

*29 CFR 1926.103
**Note: For the purposes of this table, IDLH is defined as a condition that either poses an immediate threat to life and health
or an immediate threat of severe exposure to contaminants that are likely to have adverse delayed effects on health.

device exists for a particular toxic substance, but this excuse is a poor one and dangerous because the worker must be protected from the hazardous atmosphere. To select the correct type of respiratory protection based on the hazards, see Table 12-3.

Types of Respirators

There are two broad categories of respiratory equipment available which are:

❖ Air-purifying respirators

❖ Air-supplying respirators

Air-Purifying Respirators.   These respirators are designed to remove certain gaseous and particulate contaminants from the atmosphere. These respirators include particulate respirators, gas and vapor respirators, and gas masks. The purifying elements can be attached to a full facepiece, which covers the eyes, nose, and mouth or to a half facepiece, which covers the nose and mouth. Some air-purifying respirators are disposable.

Particulate respirators are also known as mechanical filter respirators. Depending on their design, they filter out dusts, fumes, or mists by passing the contaminated air through a filter or pad. Filter respirators are not effective for gases, vapors, or oxygen-deficiency. The respirator (disposable variety) or other filter cartridges should be changed when they become clogged, or when it becomes difficult to breathe through them.

Gas and vapor respirators are also called chemical cartridge respirators. They remove gaseous contaminants by passing contaminated air through an adsorbent material, such as charcoal, which traps the contaminant. Cartridges should be matched with the contaminants present in the work environment. Table 12-4 lists the color coding of canisters and cartridges for different types of contaminants. Cartridges and canisters also have labels that state the type and concentrations of gaseous contaminants for which they have approval. These cartridges should only be used for gaseous contaminants that have adequate warning properties of smell or irritation. Cartridges should be changed if the user develops irritation or smells the characteristic odor of the material. Cartridges should not be used after their expiration date.

Gas masks are designed with canisters for gases, vapors and particulates. The canisters are typically too large and too heavy to hang directly from the chin. The canisters are suspended by a harness connected to the face mask by a corrugated flexible breathing tube. The gas mask must still be used in atmospheres containing sufficient oxygen. Gas masks are sometimes used as escape respirators for very high concentrations of gases, vapors, and particulates.

Major limitations of chemical cartridge respirators are:

1. Should not be used for protection against gaseous contaminants which are extremely toxic in small concentrations.

2. Should not be used for exposure to materials which cannot be detected by odor below a level that is hazardous.

3. Should not be used against gaseous contaminants in concentrations which are highly irritating to the eyes without sufficient eye protection.

4. Cannot be used for protection against gaseous contaminants which are not effectively stopped by the adsorbent used, regardless of concentration.

5. Can **never** be used in oxygen-deficient atmospheres.

Air-supplying respirators.   Air-supplying respirators provide a supply of breathable air different from the workplace air. These respirators include the following types: (1) supplied-air respirators and (2) self-contained breathing apparatus (SCBA).

Supplied-air respirators receive air through an air line or air hose. The air may be supplied by a compressor or cylinder. The method of delivery of air to the user results in three different possible modes of operation: continuous flow, demand flow, and pressure demand flow.

Table 12-4
Color Codes for Cartridges and Gas Mask Canisters

Atmospheric Contaminants to be Protected Against	Color Assigned
Acid gases	White
Organic vapors	Black
Ammonia gas	Green
Carbon monoxide gas	Blue
Acid gases and organic vapors	Yellow
Acid gases, ammonia, and organic vapors	Brown
Acid gases, ammonia, carbon monoxide, and organic vapors	Red
Other vapors and gases not listed above	Olive
Radioactive materials (except tritium and noble gases) and asbestos	Purple (Magenta)
Dusts, fumes, and mists (other than radioactive materials)	Orange

Notes:
1. A purple stripe shall be used to identify radioactive materials in combination with any othe vapor or gas.
2. An orange stripe shall be used to identify dusts, fumes, and mists in combinations with any other vapor or gas.
3. Where labels only are colored to conform with this table, the canister or cartridge body shall be gray or a metal canister or cartridge body may be left in its natural metallic color.
4. The user shall refer to the wording of the label to determine the type and degree of protection the canister or cartridge will afford

In the continuous flow mode, the air line respirator receives fresh air without any action on the part of the user; that is, the flow is forced by the apparatus. One of the advantages of continuous flow mode is that it permits use of a loose fitting hood. The positive pressure differential between the inside and outside of the hood maintains the air flow outward, preventing toxic agent entry. The continuous flow mode needs an unlimited supply of air, so a special compressor is used instead of bottled air.

In the demand flow mode, air does not flow until a valve opens, caused by a negative pressure created when the user inhales. Exhalation, in turn, closes the valve. This mode has the advantage of using less air, so it is feasible with bottled air. However, the disadvantage is the need for a tight-fitting facepiece. Since inhalation causes a negative pressure differential to develop, a poorly fitted, leaky facepiece may draw in toxic contaminants. In fact, if the facepiece is too leaky, the inhalation valve will fail to open, making the use of the facepiece hazardous. For this reason the demand flow mode is being replaced by the pressure demand mode.

The pressure demand mode has features of both continuous flow and demand flow modes. A continuous flow with a positive pressure differential is maintained. The positive pressure is maintained by a preset exhalation valve. Despite its advantages, the pressure demand flow mode still requires a good fitting mask; use by a person with a beard is not acceptable.

Supplied-air respirators are used if air-purifying respirators are not adequate and the atmosphere is not IDLH. The maximum hose length for supplied-air respirators is 300 feet. The air used in supplied-air respirators must be ASTM Grade D or better. Grade D air meets certain criteria established by the Compressed Gas Association (Air Specification G-7.1). Supplied-air systems can be used with various masks, hoods, and suits depending on the type of work operations involved.

SCBA respirators provide a transportable supply of breathing air to the user. Users carry a compressed airtank or cylinder on a backpack. These units have limited air supply and use times. They consist of either closed-circuit or open-circuit units. The closed-circuit units recycle the exhaled breath, restoring oxygen levels. These units are smaller and lighter than open-circuit devices. However, they require more extensive maintenance and training than do the open-circuit units. Open-circuit devices discharge exhaled air directly to the atmosphere. Most modern SCBA units are open-circuit devices.

Combination self-contained and supplied-air respirators have both an air line and a small cylinder of compressed air. The air line permits long-term work use, and the small cylinder provides emergency air (usually a 5 minute supply) for the worker if the air line system fails. This type of system can be used in IDLH atmospheres.

Training

Training is an important part of the respiratory protection program. Each wearer should be trained initially in the use of respirators and should be retrained periodically. Training helps insure that respirators are used in the correct manner, so adequate protection is achieved.

To effectively train employees, complete the following:

1. Explain the reason for respirators, explain why other methods of control are not being used, explain other efforts being used to reduce the hazards.

2. Explain the respirator selection process, including the identification and evaluation of the hazards.

3. Demonstrate proper fitting, donning, wearing, and removing of the respirator.

4. Explain the limitations, capabilities, and operation of the respirator.

5. Demonstrate proper maintenance and storage procedures.

6. Allow the employee to wear the respirator in a safe atmosphere to permit them to become familiar with its characteristics.

7. Correctly fit the respirator using qualitative or quantitative fit test procedures. (29 CFR 1910.134)

8. Teach the user how to recognize and cope with emergency situations.

9. Instruct the user in any special uses of the respirator.

10. Explain the regulations governing the use of respirators.

The instructors for respirator training should be qualified individuals such as industrial hygienists, safety professionals, or the respirator manufacturer's representative.

Fit Testing

An important element of an effective respirator program is fit testing or leak testing. Facial characteristics vary significantly, and fit tests are necessary to determine which particular model best fits a worker. Even the best respirators on the market will not fit all workers. Other respirators may need to be obtained for special facial shapes.

One facial shape that is impossible to fit is the bearded one. A beard should be prohibited for any worker who must wear a close fitting respirator for his safety and health. In fact, OSHA requires that respirators are not to be worn if facial hair interferes with the face-to-facepiece seal. The argument against facial hair on jobs requiring respirators is well-documented.

Respirator fit testing can be conducted quantitatively using various protocols. Quantitative fit testing assures the best fitting respirator is used. Respirators can also be fitted using qualitative fit testing protocols. These protocols use either isoamyl acetate (banana oil) or an irritant smoke to test the fit of the respirator. An explanation of fit testing protocols can be found in the OSHA regulations.

Once fit-tested for a particular make and model of respirator, every respirator of the same make and model is acceptable for that user unless there has been a change in facial features.

Cleaning and Disinfection

A respirator should be cleaned and sanitized after each use. The actual cleaning can be done in several different ways:

1. Respiratory protective equipment which is made of rubber should be disassembled and washed with dishwashing detergent in warm water using a soft brush, thoroughly rinsed to remove any detergent residue, and then air dried in a clean place. Be careful not to damage the respirator during the cleaning process. This procedure is typically used for a small respirator program or if each worker cleans his/her own respirator.

2. For large respirator programs, a standard domestic clothes washer or dishwasher can be used if a rack is installed to keep the respirators in a fixed position.

The detergents used should contain some type of biocide to disinfect the respirator. Detergents can be obtained from the respirator manufacturer. Organic solvents should not be used to clean respirators because they may deteriorate the elastomeric facepiece. Be sure all detergent and disinfectant residues are removed from the respirator or skin irritation and dermatitis may result. After washing allow the respirators to air dry. If high heat is used it may damage the elastomers of the facepiece.

Storage

Improper storage of respirators can cause damage to the respirator and reduce the protection supplied. OSHA requires that respirators be stored to protect against dust, sunlight, heat, extreme cold, excessive moisture, and damaging chemicals. Respirators should also be protected against mechanical damage. Leaving a respirator unprotected, as on a workbench, or in a tool cabinet or tool box among heavy wrenches, is not proper storage. It is strongly recommended that cleaned respirators be placed in a reusable plastic bag until reissue. They should be stored in a clean, dry location away from direct sunlight. They should be stored in a single layer with the facepiece and exhalation valve in a more or less normal position to prevent distortion.

Maintenance and Inspection

The maintenance of respirators should be made an integral part of the overall respirator program. Manufacturers' instructions for cleaning, inspection, and maintenance of respirators should be followed to ensure that the respirator continues to function properly. Wearing poorly maintained or malfunctioning respirators may be more dangerous than not wearing a respirator at all. The worker wearing a defective device may falsely assume that protection is being provided. Emergency escape and rescue devices are particularly vulnerable to inadequate inspection and maintenance, although they generally are used infrequently, and then in the most hazardous and demanding circumstances. The consequences of wearing a defective emergency escape and rescue device are lethal.

The OSHA standards for respiratory protection (29 CFR 1910.134) strongly emphasize the importance of an adequate maintenance program, but permit its being tailored to the type of plant, working conditions, and hazards involved. A proper maintenance program ensures that the worker's respirator remains as effective as when it was new.

Probably the most important part of the respirator maintenance program is frequent inspection. If conscientiously performed, inspections will identify damage or malfunctioning respirators before they are used. The OSHA requirements outline two primary types of inspection: inspection while the respirator is in use and inspection while it is being cleaned. In a small operation, where workers maintain their own respirators, the two inspections become essentially one and the same. In a large facility, the inspections may differ.

OSHA requires inspection of all respirators before and after use. It also requires that those not used routinely, such as emergency escape and rescue devices, be inspected after each use and at least monthly. NIOSH recommends that all stored SCBA be inspected weekly, because of the hazard that undetected loss of breathing air for emergency SCBA will present to the wearer. In one case, the respirator is inspected both before and after each use, in the other case, only after use. However, it is highly unlikely that anyone needing a respirator in a hurry, as during an emergency, is going to inspect it. In fact, it may be dangerous to take time to do so.

Inspection procedures differ depending on the type of respirator involved, and whether the inspection is to be conducted at work during use, or during routine cleaning.

OSHA standards require that respirator inspections include:

1. a check of tightness of the connections

2. a check of the facepiece, valves, connecting tube, canisters

3. a check of the regulator and warning devices on SCBA for proper functioning

Use the following NIOSH checklist to conduct respirator inspections:

Inspect disposable respirators for:

❖ holes in filter

❖ stretching and deterioration of straps

❖ deterioration of metal nose clips

Check air-purifying respirators for:

1. The facepiece

    ❖ excessive dirt

    ❖ cracks, tears, holes, or distortion from improper storage

    ❖ inflexibility (stretch and massage to restore flexibility)

    ❖ cracked, scratched, or loose-fitting lenses in full facepieces.

2. The headstraps

    ❖ breaks or tears

    ❖ loss of elasticity

    ❖ broken or malfunctioning buckles or attachments

    ❖ excessively worn serrations on the harness that might allow the facepiece to slip

3. The inhalation and exhalation valves

    ❖ detergent residue, dust particles, or dirt on the valve or valve seat

    ❖ cracks, tears, or distortion in the valve material or valve seat

    ❖ missing or defective valve cover

    ❖ improper installation of the valve in the valve body.

4. The filter elements

    ❖ incorrect cartridge, canister, or filter for the hazards present

    ❖ incorrect installation, loose connections, missing or worn gaskets, or cross-threading in the holder

    ❖ worn threads (both filter threads and facepiece threads)

    ❖ cracks or dents in the filter, cartridge, or canister housing

    ❖ deterioration of gas mask canister harness

    ❖ service life indicator, expiration date, or end-of-service date

    ❖ evidence of prior use of sorbent cartridge or canister, indicated by absence of sealing material, tape, foil, etc., over inlet.

5. The gas mask

❖ cracks or holes

❖ missing or loose hose clamps

❖ service-life indicator on the canister.

Inspect air-supplying respirators for:

1. If the device has a tight-fitting facepiece, use the procedures outlined above for air-purifying respirators, except those pertaining to the air-purifying elements.

2. If the device is a hood, helmet, blouse, or full suit, inspect for the following:

   ❖ rips or torn seams

   ❖ examine the protective headgear, if required, for general condition with emphasis on the suspension inside the headgear

   ❖ cracks or breaks in the faceshield

   ❖ protective screen to see that it is intact and fits correctly over the faceshield, abrasive blasting hoods, and blouses.

3. The air supply system

   ❖ breathing air quality (Grade D or better)

   ❖ proper operation of carbon monoxide alarms or high-temperature alarms

   ❖ breaks or kinks in air supply hoses and end fitting attachments

   ❖ tightness of connections

   ❖ proper setting of regulators and valves

   ❖ correct operation of air purifying elements

On SCBA units, determine that the high pressure cylinder of compressed air is sufficiently charged for the intended use, preferably fully charged (mandatory on emergency devices).

On closed-circuit SCBA, be sure that a fresh canister of CO_2 sorbent is installed before use, or in accordance with manufacturer's instructions. On open-circuit SCBA, recharge the cylinder if less than 80% of the useful service time remains. However, it is preferred that an open-circuit SCBA be fully charged before use.

When air-purifying or atmosphere-supplying devices are used nonroutinely, all of the above procedures should be followed after each use. OSHA requires a monthly inspection record for emergency devices that are inspected. The inspection record needs to contain the inspection date and findings from the inspection. Typically, this is accomplished by use of an inspection tag such as those used on fire extinguishers.

If defects are found during the inspection, two remedies are available. If the defect is minor, repair and/or adjustment may be made on the spot. If the defect is major, the device should be removed from service until it can be repaired or replaced. Under no circumstances should a device that is known to be defective be used or stored for future use.

Respirator Program Evaluation

The OSHA standard (29 CFR 1910.134) requires regular inspection and evaluation of the respirator program to determine its continued effectiveness in protecting employees. Periodic air monitoring is also required to determine if the workers are adequately protected. The overall program should be evaluated at least annually, and the written SOPs modified if necessary. A sample respiratory program evaluation checklist is included in Appendix 12-A at the end of the chapter.

Frequent inspection of respirator use will determine whether correct respirators are being used and worn properly. Examination of respirators in storage will indicate how well they are being maintained. Wearers should be consulted about their acceptance of respirators, including discomfort, resistance to breathing, fatigue, interference with vision and communication, restriction of movement and interference with job performance, and confidence in the respirator's effectiveness.

The results of the periodic inspections should be reviewed, studied, and analyzed to determine the effectiveness of the respirator program. Evidence of excessive exposure to hazards should be followed up to determine why inadequate protection was provided, and action should be taken to remedy the problem. The results of the program evaluation should be presented in a written report that lists plans to correct faults and target dates for corrective actions.

Medical Surveillance

The OSHA respiratory protection standard requires that no employee be assigned to a task requiring the use of a respirator unless it has been determined that the person is able to perform under such conditions. In addition, once a determination is made as to physical ability to wear a respirator and perform the work task, a review of the employee's health status must be made. A physician with knowledge of pulmonary disease and respiratory protection practices should determine what medical factors are pertinent, which tests will be performed and ultimately whether or not and employee may wear a respirator. A variety of physical and psychological problems may prevent a person from wearing a respirator and thus working in a contaminated area. Some of these problems include:

- ❖ Diabetes, insipidus or mellitus
- ❖ Epilepsy, Grand mal or petit mal
- ❖ Alcoholism
- ❖ Use of certain medications
- ❖ Punctured ear drum
- ❖ Skin sensitivities
- ❖ Impaired or nonexistent sense of smell
- ❖ Emphysema
- ❖ Chronic pulmonary obstructive disease
- ❖ Bronchial asthma
- ❖ X-ray evidence of pneumoconiosis
- ❖ Evidence of reduced pulmonary function
- ❖ Coronary artery disease or cerebral blood disease

❖ Severe or progressive hypertension

❖ Pernicious anemia

❖ Pneumomediastinum gap

❖ Communication of sinus through upper jaw to oral cavity

❖ Experiences breathing difficulty when wearing a respirator

❖ Experiences claustrophobia when wearing a respirator

❖ Any condition that the physician determines to place the employee at added physical risk

This process sounds costly and time consuming, but there is a shortcut that can expedite the personnel screening process, eliminating a large majority of the problem cases. The shortcut is to use a questionnaire which can identify obvious problems before employment. The employee questionnaire can be a source of protection from future liability in case physical problems are present. The questionnaire should be developed with the assistance of a physician. If physical problems are identified by the preliminary questionnaire, the employee should be examined by the physician for a final determination.

Respirator Certification

The National Institute of Occupational Safety and Health (NIOSH) under authorization of the Federal Mine Safety and Health Act (MSHA) of 1977 and the OSHAct of 1970, provides testing, approval, and certification of respiratory protection. This assures the availability of safe respiratory devices. The use of the terms "approved" and "certified" reflect these federal regulations.

There will be an approval number (TC number) for most respiratory protective equipment. Most respirators, cartridges, or other parts of respiratory protection are approved as a unit. The parts from one manufacturer may not be interchanged with parts from another manufacturer when the specific combination was subjected to approval testing. Interchanging parts may nullify the approval and could make the respiratory device defective. If in doubt about a specific respirator's approval status contact the manufacturer. The manufacturers will be aware of the approval status of their equipment. Remember to use only approved devices for your respiratory protection program.

PROTECTIVE FOOTWEAR

Protective footwear should be used if there is a possibility of damage to the foot from falling objects. Other hazards where foot protection may be needed are slipping, stepping on protruding nails, contact with hot materials, wet materials and contact with chemicals.

Analysis of accidents by the Bureau of Labor Statistics shows that in seventy-five percent of foot injuries employees were not wearing safety shoes.

Several types of protective footwear are available:

❖ Safety Shoes

❖ Metatarsal or Instep Guards

❖ Steel Insoles
❖ Rubber or Plastic Boots
❖ Electrical Hazard Shoes
❖ Foundry Shoes
❖ Conductive Soles and Nonsparking Shoes
❖ Nonconductive Shoes

Safety Shoes. Standard safety shoes have steel toes that meet testing requirements found in ANSI Z41-1983. The ANSI standard requires that safety toes meet minimum requirements for impact and compression. The OSHA standard for foot protection (29 CFR 1910.136) requires that safety toe shoes meet the requirements of the ANSI standard. Steel, reinforced plastic, and hard rubber are used for safety toes, with the choice depending on the level of protection desired and the shoe design. The test requirements are the same for men's and woman's shoes. Safety shoes are available in many styles of work or dress. Comfort is important to safety shoes hence the correct type and size are important factors when purchasing them.

Metatarsal or Instep Guards. To protect the upper foot area from impacts, metal guards that extend over the foot rather than just over the toes should be worn. These metatarsal or instep guards are made of heavy-gauge, flanged, and corrugated sheet metal. There are currently no standards developed for these items.

Steel Insoles. Another optional feature of safety shoes is puncture-resistant soles. These are used in construction work or other locations where there is a danger of stepping on sharp objects that can penetrate the soles of standard safety shoes.

Rubber or Plastic Boots. These are used when working in wet or muddy processes or where exposure to chemicals is likely. In wet or muddy areas, such as dairies and in food processing, the boots should provide good traction and waterproof properties. For chemical exposures, the type of boot material is an important consideration. Some boot materials will degrade when exposed to certain chemicals. Therefore, choose the boot that will provide the best protection for the chemicals being used. These boots can be ankle high or can extend over the entire thigh. In some situations waist high boots are used.

Electrical Hazard Shoes. This protective shoe is intended to minimize the hazard resulting from contact with electrical currents where the path of the current would be from point of contact to the ground. There are several general types available. Contact your local safety supplier for more information on this type of protective footwear. Keep in mind that if these become damp or badly worn they cannot be depended on for protection.

Foundry Shoes. These shoes are referred to as "congress" or gaiter shoes. They are used in plants to protect against splashes of molten metal. These shoes do not have any fasteners, so they can be easily and quickly removed in an emergency. Some foundries and steel mills have reported that serious burns have occurred to workers who were unable to remove ordinary work shoes during an emergency. During molten metal operations the tops of the shoes should be covered by the trouser leg, spats, or leggings to keep out molten metal from a splash.

Conductive Soles and Nonsparking Shoes. Safety shoes, boots, and rubber overshoes may be obtained with a conductive sole to allow a drain off of static charges, and with nonferrous metal parts to reduce the possibility of friction sparks. This type of shoe is used if there is a danger of fire or explosion from flammable gases and vapors. These shoes are used in refinery and munition facilities.

Nonconductive Shoes. These shoes are used with high voltage electrical equipment. They are electrically insulated to prevent electrical shock and flow of current through the shoes.

PROTECTIVE CLOTHING

In industrial environments there may be exposure to fire, extreme heat, cold, molten metal, corrosive chemicals, body impact, cuts from materials handled, and other specialized hazards. Specialized protective equipment may be needed to protect against these hazards. Special protective clothing includes the following:

❖ Gloves
❖ Thermal Protective Clothing
❖ Chemical Protective Clothing

Gloves. Gloves are used to protect against cuts, bruises, and abrasions on most jobs where heavy, sharp, or rough materials are handled. The materials used in the gloves depends on the materials being handled. For most light work, a cotton or canvas glove may be used. For rough or abrasive work, a leather glove or leather glove reinforced with metal stitching is required. Leather reinforced with metal stitching provides protection from sharp edged tools, as in butchering or similar occupations. Metal mesh or highly cut-resistant plastic or Kevlar gloves are also available. Gloves should not be used while working on moving machinery where they may be caught such as drills, saws, grinders, or other rotating and moving equipment. Gloves for protection against solvents, skin irritants, and corrosive materials are discussed in the chemical protective clothing section.

In addition to gloves, there are mittens, pads, thumb guards, finger cots, wrist and forearm protectors, elbow guards, sleeves, and capes, all of varying materials and lengths. These can be selected where more or less protection is needed depending on the job.

Thermal Protective Clothing. Thermal protective clothing is used if there is potential contact with heat and hot metals. These materials can be made of leather, wool, asbestos substitutes, or aluminized materials. Each type of material has its own protective characteristics and limitations. To select specific protection for high temperature applications, contact a local safety supplier.

Chemical Protective Clothing. Chemical protective clothing is designed to protect the skin against gaseous, liquid, and particulate chemical hazards. Chemical protective clothing is fabricated into various types of clothing depending on the hazards involved. It includes gloves, boots, aprons, and full body protection. Examples of materials used to make chemical protective clothing include natural rubber, synthetic rubber, neoprene, vinyl, polypropylene and polyethylene films, and fabrics coated with these materials.

Although a wide variety of chemical protective products are available, research has shown that chemicals can pass through (permeate) these devices and come in direct contact with the skin and potentially be absorbed into the blood. This has led to considerable research to find materials and construction best suited for the specific chemical challenges. Testing of protective materials is conducted using the American Society of Testing Materials (ASTM) F739-85, Standard Test Method for Resistance of Protective Clothing to Permeation by Liquids and Gases. This laboratory method establishes breakthrough times and steady-state permeation rates for specific chemical and protective clothing combinations. The breakthrough time is the time taken for a chemical to pass through the protective material. The permeation rate is the rate or speed of movement of the chemical through the protective material once it has broken through. Changes in physical properties of the material resulting from the contact with the chemical is referred to as degradation.

When selecting chemical protective clothing careful consideration must be given to test results. Protective clothing with long breakthrough times, low permeation rates, and no degradation should be chosen for the chemical you are using in your facility. Manufacturers of protective clothing can be contacted for permeation data on their protective materials. Some important points to keep in mind when selecting chemical protective clothing are:

❖ All chemicals will permeate a protective barrier given enough time.

❖ Permeation can take place without any visible indication.

❖ A material that protects against one chemical may not protect against a different chemical.

❖ No single protective material is an absolute barrier against all chemicals.

❖ Protective gloves and clothing may look the same. Therefore, choose the correct material for the job you are doing.

❖ Do not depend on color and appearance when choosing protective materials.

❖ Once a chemical has permeated a protective material, it will continue to pass through the material.

❖ If the protective material is contaminated by breakthrough, it must be decontaminated or disposed of before it is used again. The manufacturer's decontamination procedures should be followed where applicable.

❖ The best way to select the proper chemical protective clothing is to test the material against the chemical(s) you are using.

Appendix 12-A

SAMPLE RESPIRATOR PROGRAM AND

EVALUATION CHECKLIST*

* From Bollinger, Nancy J. and Shultz, Robert H., *NIOSH Guide to Industrial Respiratory Protection*, NIOSH Publication No. 87-116, Cincinnati, Ohio, 1987.

The following is a sample respirator program:

A B C COMPANY

RESPIRATOR PROGRAM

Purpose:

The purpose of this operating procedure is to ensure the protection of all employees from respiratory hazards, through proper use of respirators. Respirators are to be used only where engineering control of respirator hazards is not feasible, while engineering controls are being installed, or in emergencies.

Responsibility

The company Safety Officer is _____
_____. He/she is solely responsible for all facets of this program and has full authority to make necessary decisions to ensure success of this program. This authority includes hiring personnel and equipment purchases necessary to implement and operate the program. The Safety Officer will develop written detailed instructions covering each of the basic elements in this program, and is the sole person authorized to amend these instructions.

The ABC Company has expressly authorized the Safety officer to halt any operation of the company where there is danger of serious personal injury. This policy includes respiratory hazards.

Program Elements

1. The Safety Officer will develop detailed written standard operating procedures governing the selection and use of respirators, using the NIOSH Respirator Decision Logic as a guideline. Outside consultation, manufacturer's assistance, and other recognized authorities will be consulted if there is any doubt regarding proper selection and use. These detailed procedures will be included as appendices to this respirator program. Only the Safety Officer may amend these procedures.

2. Respirators will be selected on the basis of hazards to which the worker is exposed. All selections will be made by the Safety Officer. Only MSHA/NIOSH-certified respirators will be selected and used.

3. The user will be instructed and trained in the proper use of respirators and their limitations. Both supervisors and workers will be so instructed by the Safety Officer. Training should provide the employee an opportunity to handle the respirator, have it fitted properly, test its facepiece-to-face seal, wear it in normal air for a long familiarity period, and finally to wear it in a test atmosphere. Every respirator wearer will receive fitting instructions, including demonstrations and practice in how the respirator should be worn, how to adjust it, and how

to determine if it fits properly.

Respirators should not be worn when conditions prevent a good face seal. Such conditions may be a growth of beard, sideburns, a skull cap that projects under the facepiece, or temple pieces on glasses. No employees of A B C, who are required to wear respirators, may wear beards. Also the absence of one or both dentures can seriously affect the fit of a facepiece. The worker's diligence in observing these factors will be evaluated by periodic checks. To assure proper protection, the facepiece fit will be checked by the wearer each time the wearer puts on the respirator. This will be done by following the manufacturer's facepiece-fitting instructions.

4. Where practicable, the respirators will be assigned to individual workers for their exclusive use.

5. Respirators will be regularly cleaned and disinfected. Those issued for the exclusive use of one worker will be cleaned after each day's use, or more often if necessary. Those used by more than one worker will be thoroughly cleaned and disinfected after each use. The Safety Officer will establish a respirator cleaning and maintenance facility and develop detailed written cleaning instructions.

6. The central respirator cleaning and maintenance facility will store respirators in a clean and sanitary location.

7. Respirators used routinely will be inspected during cleaning. Worn or deteriorated parts will be replaced. Respirators for emergency use such as self-contained devices will be thoroughly inspected at least once a month and after each use. Inspection for SCBA breathing gas pressure will be performed weekly.

8. Appropriate surveillance of work area conditions and degree of employee exposure or stress will be maintained.

9. There will be regular inspection and evaluation to determine the continued effectiveness of the program. The Safety Officer will make frequent inspections of all areas where respirators are used to ensure compliance with the respiratory protection programs.

10. Persons will not be assigned to tasks requiring use of respirators unless it has been determined that they are physically able to perform the work and use the equipment. The ABC Company physician will determine what health and physical conditions are pertinent. The respirator user's medical status will be reviewed annually.

11. Certified respirators will be used.

John Doe
President, ABC Company

The following is a sample respirator program evaluation checklist:

Respirator Program Evaluation Checklist

In general, the respirator program should be evaluated for each job or at least annually, with program adjustments, as appropriate, made to reflect the evaluation results. Program function can be separated into administration and operation.

A. Program Administration

_____ (1) Is there a written policy which acknowledges employer responsibility for providing a safe and healthful workplace, and assigns program responsibility, accountability, and authority?

_____ (2) Is program responsibility vested in one individual who is knowledgeable and who can coordinate all aspects of the program at the jobsite?

_____ (3) Can feasible engineering controls or work practices eliminate the need for respirators?

_____ (4) Are there written procedures/statements covering the various aspects of the respirator program, including:

_____ designation of an administrator;
_____ respirator selection;
_____ purchase of MSHA/NIOSH certified equipment;
_____ medical aspects of respirator usage;
_____ issuance of equipment;
_____ fitting;
_____ training;
_____ maintenance, storage, and repair;
_____ inspection;
_____ use under special condition; and
_____ work area surveillance?

B. Program Operation

(1) Respiratory protective equipment selection

_____ Are work area conditions and worker exposures properly surveyed?

_____ Are respirators selected on the basis of hazards to which the worker is exposed?

_____ Are selections made by individuals knowledgeable of proper selection procedures?

_____ (2) Are only certified respirators purchased and used; do they provide adequate protection for the specific hazard and concentration of the contaminant?

_____ (3) Has a medical evaluation of the prospective user been made to determine physical and psychological ability to wear the selected respiratory protective equipment?

_____ (4) Where practical, have respirators been issued to the users for their exclusive use, and are there records covering issuance?

(5) Respiratory protective equipment fitting

_____ Are the users given the opportunity to try on several respirators to determine whether the respirator they will subsequently be wearing is the best fitting one?

_____ Is the fit tested at appropriate intervals?

_____ Are those users who require corrective lenses properly fitted?

_____ Are users prohibited from wearing contact lenses when using respirators?

_____ Is the facepiece-to-face seal tested in a test atmosphere?

_____ Are workers prohibited from wearing respirators in contaminated work areas when they have facial hair or other characteristics may cause faceseal leakage?

(6) Respirator use in the work area

_____ Are respirators being worn correctly (i.e., head covering over respirator straps)?

_____ Are workers keeping respirators on all the time while in the work area?

(7) Maintenance of respiratory protective equipment

Cleaning and Disinfecting

_____ Are respirators cleaned and disinfected after each use when different people use the same device, or as frequently as necessary for devices issued to individual users?

_____ Are proper methods of cleaning and disinfecting utilized?

Storage

_____ Are respirators stored in a manner so as to protect them from dust, sunlight, heat, excessive cold or moisture, or damaging chemicals?

_____ Are respirators stored properly in a storage facility so as to prevent them from deforming?

_____ Is storage in lockers and tool boxes permitted only if the respirator is in a carrying case or carton?

Inspection

_____ Are respirators inspected before and after each use and during cleaning?

_____ Are qualified individuals/users instructed in inspection techniques?

_____ Is respiratory protective equipment designated as "emergency use" inspected at least monthly (in addition to after each use)?

_____ Are SCBA incorporating breathing gas containers inspected weekly for breathing gas pressure?

_____ Is a record kept of the inspection of "emergency use" respiratory protective equipment?

Repair

_____ Are replacement parts used in repair those of the manufacturer of the respirator?

_____ Are repairs made by manufacturers or manufacturer-trained individuals?

(8) Special use conditions

_____   Is a procedure developed for respiratory protective equipment usage in atmospheres immediately dangerous to life or health?

_____   Is a procedure developed for equipment usage for entry into confined spaces?

(9) Training

_____   Are users trained in proper respirator use, cleaning, and inspection?

_____   Are users trained in the basis for selection of respirators?

_____   Are users evaluated, using competency-based evaluation, before and after training?

SAMPLE RESPIRATOR INSPECTION RECORD

1. TYPE_____ 2. NO. _____

3. DEFECTS FOUND:

    A.   Facepiece_____

    B.   Inhalation Valve_____

    C.   Exhalation Valve Assembly_____

    D.   Headbands_____

    E.   Cartridge Holder_____

    F.   Cartridge/Canister _____

    G.   Filter _____

    H.   Harness Assembly_____

    I.   Hose Assembly _____

    J.   Speaking Diaphragm _____

    K.   Gaskets_____

    L.   Connections_____

    M.   Other Defects_____

_____

_____

_____

_____

_____

_____

CHAPTER
13
EMERGENCY AND DISASTER PLANNING

This chapter begins by reviewing types of naturally occurring emergencies and other disasters requiring planning. This is followed by a discussion of the elements necessary to develop an emergency plan. The chapter closes with a discussion of what should be done after the emergency is over.

Recent events, such as the bomb explosion in the World Trade Center, have reinforced the critical need for preplanning for emergencies whether they be man-made or natural in origin. Effective preplanning and emergency procedures can save lives and business property. Most states and OSHA require written emergency plans. OSHA requires a written emergency evacuation plan for fire for most employers. OSHA also requires a separate written emergency plan for businesses that fall within the scope of their Hazardous Waste Operations and Emergency Response (HAZWOPER) standard.

Emergencies can include:

❖ criminal activities such as sabotage, riots, and extortion (bomb threats)

❖ natural disasters such as tornados, hurricanes, earthquakes, blizzards, floods, dust storms, and heavy thunderstorms

❖ accidents such as fire, explosion, hazardous materials spills, vehicle crashes, and major mechanical or structural failures

TYPES OF EMERGENCIES REQUIRING PLANNING

The first step in the process is to determine the types of emergencies that are most likely to occur. For natural disasters, geography plays a major role in determining risk. For example, it is very unlikely that a snow storm would require the evacuation of employees in Miami,

Florida. However, this is a location that is subject to hurricanes. Kansas City, in the middle of the country, is not affected by hurricanes but is known for a tornado season.

One peculiar aspect of planning for weather related emergencies is that most emergency planning centers on the safe evacuation of employees. For some weather related emergencies, employees may be required to stay in the building rather than to evacuate if the construction of the building is appropriate for shelter.

The potential for criminal activities and civil disobedience to be a major risk is very much dependent on the nature of the business and its location (e.g., rural versus urban and inner city versus suburban). For some businesses, the risk of these types of emergencies may seem very minimal. Nevertheless, most businesses should plan on a generic basis for emergencies that include bomb threats, riots, vandalism, or other criminal acts.

The last category of emergencies includes those of an accidental nature, such as fires and explosions, hazardous materials spills, and other similar emergencies. Again, the likelihood of these disasters is very much dependent on the nature of the business. For example, a paint manufacturer will have a much greater potential for fire and explosion than a soft drink bottler. Nevertheless, all businesses have some potential for a fire and thus the need for an employee evacuation plan.

PLANNING FOR EMERGENCIES

The OSHA standards specify the following minimum requirements for emergency plans in response to fire:

❖ Emergency escape procedures and emergency escape route assignments
❖ Procedures to be followed by employees who remain to operate critical plant operations before they evacuate
❖ Procedures to account for all employees after emergency evacuation has been completed
❖ Rescue and medical duties for those employees who are to perform them
❖ The preferred means of reporting fires and other emergencies
❖ Names or regular job titles of persons or departments who can be contacted for further information or explanation of duties under the plan

These elements required by OSHA are also generally applicable to all emergency planning. There are separate and similar requirements for emergencies from spills of hazardous materials under their HAZWOPER standard. Under HAZWOPER, a chemical spill represents an emergency if it poses an immediate threat to the health and safety of workers in the vicinity of the spill or an immediate threat to the environment. This does not include minor spills or normal maintenance activities. For example, spillage of 500 gallons of the chemical Toluene from a cracked bulk storage tank represents both a health and fire hazard while spillage of 5 gallons of paint containing Toluene does not represent an emergency, only a housekeeping problem. The HAZWOPER standard requires that all facilities with the potential for spills of hazardous materials or hazardous wastes have a written emergency plan. The written plan must include:

❖ pre-emergency planning

❖ personnel roles, lines of authority, training, and communication
❖ safe distances and places of refuge
❖ site security and control
❖ evacuation routes and procedures
❖ decontamination
❖ emergency medical treatment and facilities
❖ emergency alerting and response procedures

As has been noted, there are many common elements between the emergency plans for hazardous materials incidents and the emergency plans for fire. Key aspects of these are discussed in the following paragraphs.

Preplanning

Preplanning includes an assessment of the type and extent of emergencies that are likely to occur. This will usually result in the development of plans for assistance from outside agencies such as the fire service, police, and area hospitals or clinics.

This support may be evidenced by written agreements or other forms of cooperation and preplanning. One excellent technique is the involvement of emergency services in training drills. This allows for familiarization with your situation, personnel, and capabilities.

Chain of Command

The establishment of personnel roles and lines of authority are very important elements in emergency planning. Confusion is a major hazard during emergency evacuations or responses. To avoid a "Keystone Cops" event, it is very important to establish a chain of command during your emergency planning. This includes designation of a person in charge. This person would normally have responsibility to:

❖ Assess the situation to determine if there is a need to initiate emergency procedures.
❖ Direct all efforts in the area to include evacuations of personnel and steps necessary to minimize property damage.
❖ Direct the shutdown of plant operations as necessary.
❖ Obtain the services of outside help such as fire and medical (e.g., ambulance).

The designated plant emergency coordinator would then assign responsibilities to others as needed. However, it is important to have a leader who is clearly in charge rather than a group effort. This general scheme follows the Incident Command System (ICS) which is used by the fire service and military to handle emergencies.

Communications

Communications can easily become disrupted during an emergency due to lack of power, storm damage, fire damage, etc. Therefore, it is important to have alternative means available such as hand held two way radios, cellular phones, bull horns, etc. One other aspect that is also important is a list of key personnel and their home telephone or beeper numbers and agencies to contact in case of emergency. This would probably include the following:

❖ Critical plant personnel (e.g., Plant Manager, maintenance, etc.)
❖ Fire
❖ Police
❖ Ambulance
❖ Clinics and hospitals
❖ Utilities
❖ Clean up contractors

Shut-Down Procedures

Written shut-down procedures should be included in the emergency plan. These may be necessary for powered equipment, utilities, boilers, reactor vessels, etc., to reduce the probability of secondary damage in the event of an evacuation. These procedures should also be posted or readily available and personnel preassigned these roles.

Evacuation Routes and Places of Refuge

It is very important that all employees know their routes of evacuation within the plant or area in the event of an emergency. These evacuation routes are best learned through announced and unannounced drills. Additionally, these plans should be posted by area. This should be done using floor or area maps that are easily understood. Routing for fire evacuation must be through lower risk areas (less chance of fire) to an outside exit. Employees should also be provided with alternate routes of egress in case the primary routes are blocked. For some emergencies, off-site routing and places of refuge may be necessary (e.g., large chemical spill or possibility of explosion). In some cases, this can mean distances of over a mile (e.g., rail car derailment of butane).

Site Security and Control

For many emergencies, site security is important to prevent inadvertent or unauthorized access to plant property. Off-limits areas should be established at a safe distance as soon as possible. This is best done using preprinted hazard tape, ropes, signs, or other visible barriers or means of warning. Critical business records may also need to be removed and protected. This must be done in a fashion that does not jeopardize personnel. Obviously, it is best to have critical records duplicated and stored off-site in a secure area.

Emergency Teams

Emergency teams may be established for response to fires, chemical spills, and other emergencies. If they are used, the regulations require that they have specific training to qualify them for their assigned roles. The specifics of the training requirements are contained within the regulations.

FOLLOW-UP ACTIONS AFTER EMERGENCIES

The major follow-up action after an emergency is over is to determine what was done well and where improvements might be made. Questions that should be asked include:

❖ How effectively was the emergency assessed?

❖ How effectively was the information passed on to other personnel?

❖ Was someone in charge and were they effective?

❖ Was outside assistance called in a timely fashion and were they effectively directed to the areas of need?

❖ Did the alarm system work as designed?

❖ Did the evacuation scheme work as planned?

❖ Were shut-down procedures followed in a timely and effective manner?

❖ Were communications adequately maintained?

❖ If a response team was used, did they perform efficiently?

❖ Was there adequate emergency equipment available?

❖ Did equipment function properly?

❖ Was the media (always a hazard) handled properly?

In general, determine what changes should be made to allow for a better response the next time. There is an old axiom which states that you learn best from your mistakes. In the case of emergencies, it is important that most of the learning process come from drills since lives are at stake. The key is to preplan rather than to react.

PART THREE

CONTROL OF HAZARDS IN MANUFACTURING PLANTS

CHAPTER
14
WALKING AND WORKING SURFACES

Walking and working surfaces include floors, stairs, ladders, scaffolding, and all other surfaces that are used by employees in the performance of their jobs. Because these surfaces are so common, they are quite often neglected when considering the safety of the employee. Trips and falls represent one of the largest causes of injuries in the workplace.

An important factor in preventing accidents and injuries is to maintain all walking and working surfaces in a clear and uncluttered condition. Wet working surfaces present a slipping hazard and surfaces should be kept as dry as possible. Where wet processes or conditions are involved, appropriate drainage should be provided. The use of elevated work stations with drainage will also help eliminate contact with wet and potentially slippery surfaces.

In this chapter the OSHA required characteristics and properties of the following walking/working surfaces will be discussed:

- ❖ Stairs and Stairways
- ❖ Runways and Ramps
- ❖ Aisleways and Walkways
- ❖ Exit/Egress Requirements
- ❖ Floor and Wall Openings
- ❖ Ladders and Scaffolds

STAIRS AND STAIRWAYS

Stairs and stairways are discussed in this section except those intended for fire exit purposes, temporary devices used during construction, and articulated stairways as can be found on floating roof tanks and dock facilities. In general, spiral stairways are not allowed by codes except where it is not practical to provide a regular fixed stairway.

Considerable leeway is allowed in the construction of fixed stairways. The angle of ascent may vary between 30 and 50 degrees. A table of suggested riser height and tread run combinations adapted from the OSHA regulations is shown in Table 14-1. These are not the only combinations that can be used. It should be noted, however, that it is not permissible to "mix and match" height and run combinations within a single stairway. Using different height and run combinations greatly increases the chances that an individual will trip or miss a step leading to a potentially serious injury.

Table 14-1
Rise and Run Combinations to Give an Appropriate Stair Angle

Appropriate Angle to Horizontal	Riser Height Inches	Tread Run Inches
30.5	6.50	11.00
32.0	6.75	10.75
35.0	7.25	10.25
37.0	7.50	10.00
40.0	8.00	9.50
45.0	8.75	8.75
48.0	9.25	8.25
50.0	9.50	8.00

Adapted from Table D-1 29 CFR 1910.24

While there is no maximum width for stairways, there is a requirement that they be at least 22 inches wide. There is also a requirement that all stairways be equipped with handrails. The requirements differ slightly from those for guardrails in that they must be between 30 and 34 inches in height from the top of the treads. A constant height should be maintained for the entire length of the stairway. The placement of the railing(s) is dependent on the width of the stairway. The pertinent requirements are shown in Table 14-2.

Table 14-2
Requirements for the Placement of Handrails on Fixed Industrial Stairs

Width and Configuration		Location of Handrail
<44 in.	Both sides enclosed	One side preferably the right side as one is descending
<44 in.	One side open	Open side
<44 in.	Both sidesopen	Each side
>44 & <88 in.		Both sides
>88 in.		Both sides and one intermediate railing approximately in the middle.

Adapted from 29 CFR 1910.23(d)

RUNWAYS AND RAMPS

Ramps are frequently used to move personnel and materials from one level to another and should have the least practical slope to facilitate necessary movements. The recommended maximum angle of ascent for ramps is 15 degrees and should not exceed 20 degrees. If space and/or configuration considerations would lead to a slope of greater than 20 degrees, the use of a stairway or mechanical lift should be considered. In some locations, the maximum allowable angle for ramps is 10 degrees. For this reason, local building codes should always be consulted before a ramp is constructed.

Ramps should always be covered with a nonslip coating to help assure safe footing. This can be an abrasive coating such as sand mixed with paint or pressure-sensitive adhesive strips. A reason for omitting a nonslip surface on a ramp would be if dislodged abrasive or adhesive strips could fall off the ramp and harm process equipment or materials.

Runways more than four feet above the level below must be provided with railings and toe boards. In certain special situations the railing and toe boards can be omitted from one side of the runway. Runways used for filling tank cars and oiling machinery are examples of where the railing may be omitted on one side. In these special cases the runway must be at least 18 inches wide.

AISLEWAYS AND WALKWAYS

Aisleways are designed to provide for the safe and unimpeded movement of people and material inside a building and should be specifically marked so that proper traffic patterns can be maintained. It is imperative that aisleways be kept clear and free of obstructions at all times. Often employees will inadvertently store excessive amounts of equipment and material near their work station. These items may overflow into aisleways and block access to machinery and vital safety equipment such as fire extinguishers. Blocked or cluttered aisleways present a serious safety hazard especially in case of an emergency. Because of the potential hazards, periodic inspections should be conducted of all working areas and immediate corrective action taken to alleviate observed clutter or blockages.

If aisleways are to be used for vehicular traffic, the passage must be a minimum of three feet wider than the widest load that will be moved through the area. For example, if two way traffic were allowed in an aisleway, the minimum width would be the maximum combined width of the two vehicles, with loads, plus three feet. This could potentially lead to aisleways that are unnecessarily wide. Because of this, consideration should be given to designating aisleways as one way only for traffic to minimize space requirements.

Walkway is the term generally used to denote a passage that is used to connect two buildings. The preferred construction medium for walkways that have a high traffic level is concrete because it provides superior footing. For areas that have lower traffic levels, asphalt or crushed stone covered with gravel are generally acceptable. These also have lower installation and maintenance costs. If possible, walkways should be designed to follow the straightest path between the two buildings that they connect to eliminate or minimize shortcutting. Where walkways must pass in close proximity to busy roads or rail lines they should be separated from these potential hazards by a fence.

During winter months, prompt and efficient snow and ice removal should be a prime consideration. A great number of injuries occur each year from employees slipping and falling on snow or ice covered walkways. Where walkways must pass along the side of a building, care should be taken to ensure that they do not pass directly underneath the roof line of the building. Such a situation places employees who must use the walkway at increased risk from falling snow and/or icicles.

EXIT/EGRESS REQUIREMENTS

The minimum number and location of exits from a building is governed by the "Life Safety Code" which is issued as National Fire Protection Association (NFPA) Publication 101. As with most regulations, some jurisdictions may establish exit and egress requirements that are more stringent than those in the NFPA Life Safety Code. Applicable local regulations should be checked carefully before a building is designed or modified to ensure that the proper number of exits of the correct size and location have been included. Many plant managers have learned the hard way that it is very expensive to add additional exits after the completion of construction.

In all cases, the basic premise is that there must be sufficient exits in number, size, and location to ensure that a building can be evacuated rapidly and without loss of life in case of an emergency. The term "sufficient" is open to discussion. In general, a minimum of two separate means of egress must be provided from each floor. In high hazard occupancies, no part of a building or floor should be more that 75 feet from an exit, while for medium and low hazard occupancies distances of 100 to 150 feet are allowable. Where aisles or passages terminate in exits, the width of the exit must be at least that of the aisle or passage. This is to prevent employees from trying to push their way through the exit because of its narrow width. In such situations, employees could easily lose their footing and fall, resulting in injuries. All exit doors must open in the direction of intended travel. The exits must have appropriate signs and be illuminated. In no case must a door that is designed for emergency egress be blocked or locked from the inside.

FLOOR AND WALL OPENINGS/HOLES

The difference between an opening and a hole is one of definition and is determined by size. A hole is greater than one but less than twelve inches in its smallest dimension while an opening measures more than twelve inches in its smallest dimension. The reason for the differentiation between openings and holes is that openings present a much greater hazard for people falling through, while only tools and equipment can fall through holes.

Although the OSHA standards differentiate between openings and holes, both potential hazards should be handled in the same manner. All floor openings and holes must be guarded to prevent injuries to personnel and to prevent equipment and material from falling through. This guarding is usually accomplished through the use of railings and toe boards or through the use of appropriate screens or cover plates.

For permanent floor openings and holes that are generally accessible, it is standard practice to provide guard railings and toe boards. A standard railing is nominally 42 inches high with both top and intermediate rails. It can be constructed of either wood, metal, or other material so long as it can withstand a pressure of a least 200 pounds on the top rail. Toe boards must be at least four inches high and attached flush with the floor. They can be constructed of any material with openings no greater than one inch in any dimension. Openings and holes that are generally not accessible because of machinery or equipment may be guarded by an appropriate cover that leaves open spaces no greater than one inch in their greatest dimension.

Floor openings and holes created by construction, maintenance, or other activities present a potentially serious hazard. Since these openings are only temporary, personnel are not as aware of them as they are of permanent openings. For this reason, they must be either constantly guarded or appropriate railings and toe boards constructed and installed.

If there are wall openings where there is a drop of more than four feet to the next level, appropriate barriers for personnel and falling material must be provided. The requirements for guarding wall openings allow more latitude than for floor openings. Standard railings, half doors, or movable barriers may be used. Movable barriers are quite useful when it is necessary to move equipment or material through the wall opening on a regular basis. Any wall opening that could possibly be mistaken as a means of egress must be labeled "Not an Exit."

If the bottom of the wall opening is less than four inches high, appropriate toe boards must be provided. These may be removable if the opening is used for the movement of equipment and material.

It is also necessary to provide railings along all open-sided floors and platforms if the drop to the next level is greater than four feet. If personnel can pass beneath the area, or if there is moving machinery or equipment for which falling objects could present a hazard, then toe boards must also be provided.

LADDERS AND SCAFFOLDS

Portable Ladders

There are detailed specifications for the construction and safe use of portable ladders published by the American National Standards Institute (ANSI). The construction specifications generally are not of concern to an industrial user, since equipment is almost always purchased and not manufactured in-house. If portable ladders must be constructed in-house, ANSI should be consulted to assure that all applicable requirements are met. When purchasing portable ladders, be sure they are appropriate for the intended job. Manufacturers generally specify the type of service for which a specific ladder is intended.

After purchase and before use, all portable wooden ladders must be thoroughly inspected to assure that they are free from structural defects such as knots, cross grain, compression wood, and pitch pockets. After inspection, it is advisable to coat wooden ladders with a sealant, such as varnish or paint. This will help to deter decay and deterioration, but it will not normally mask any structural damage that may occur later. Metal and plastic ladders must also be inspected upon initial purchase to assure their structural integrity.

Equally important to the initial inspection are periodic re-inspections to assure that ladders continue to maintain their structural integrity. Such inspections should be conducted at least quarterly. A sample inspection checklist for portable ladders is shown in Figure 14-1. This checklist is designed to be used for a wide variety of ladder types. If any defects are found during inspections, the ladder should immediately be removed from service, tagged with a **"DANGEROUS—DO NOT USE"** sign, and steps taken to assure that it is not used until proper repairs have been made. In some cases it will be impossible to restore a ladder to proper working condition. In these instances, the ladder should be destroyed.

Under no circumstances should employees be allowed to take damaged equipment home for personal use. If an accident or injury were to occur, the employer could be faced with significant civil liability and/or potential litigation costs.

Figure 14-1
Ladder Inspection Checklist

	OK	REPAIR
GENERAL		
Cracked, split, or broken uprights, steps, rungs, or braces	____	____
Loose steps or rungs	____	____
Loose metal parts (nails, screws, bolts)	____	____
Visible wood or metal slivers or splinters on rungs, steps or uprights	____	____
STEPLADDERS		
Hinge spreader stop broken or missing	____	____
Hinge spreader loose or bent	____	____
Wobbly	____	____
Loose hinges	____	____
Broken or excessively worn steps	____	____
EXTENSION AND SECTIONAL LADDERS		
Wobbly	____	____
Worn or loose metal parts	____	____
Deteriorated ropes	____	____
Broken, loose, or missing extension locks including locks that do not seat properly	____	____
TROLLEY LADDERS		
Floor wheel brackets broken, missing, or loose	____	____
Wheels that bind or are out of adjustment	____	____
Worn or missing tires	____	____
Ladder and/or rail stops missing or broken	____	____
Rail supports broken or section of rail missing	____	____
FIXED LADDERS		
Damaged or corroded cage parts	____	____
Damaged, loose, or worn side rails or rungs	____	____
Damaged or corroded handrails or brackets on platforms	____	____
Weakened or damaged rungs set in brick or concrete	____	____
Corroded bolts or rivit heads	____	____
Base of ladder obstructed	____	____
TRESTLE LADDERS		
Wobbly	____	____
Loose or bent hinges or hinge spreaders	____	____
Hinge spreader stop broken	____	____
Misalignment of center section guide for extension	____	____
Broken or defective extension locks	____	____
FIRE LADDERS		
Obstructed storage	____	____
Improperly stored	____	____
Illegible markings	____	____

A large number of injuries occur each year from the improper use of portable ladders. Because of this, the following work practices regarding the placement and use of portable ladders should be routinely followed:

1. The maximum distance that the base of a ladder should be placed from the vertical is approximately one-fourth of the distance between the support points. Thus, for a ladder which has 12 feet between the ground and its upper support point, the base should be no more than 3 feet horizontally from the vertical. Single piece ladders should also extend a minimum of three feet above the top support point. These points are illustrated in Figure 14-2.

2. Never use ladders in a horizontal position as a substitute for a scaffold or a runway between two elevated locations.

3. Never place a ladder directly against a window pane or sash.

Figure 14-2
Placement of Ladders

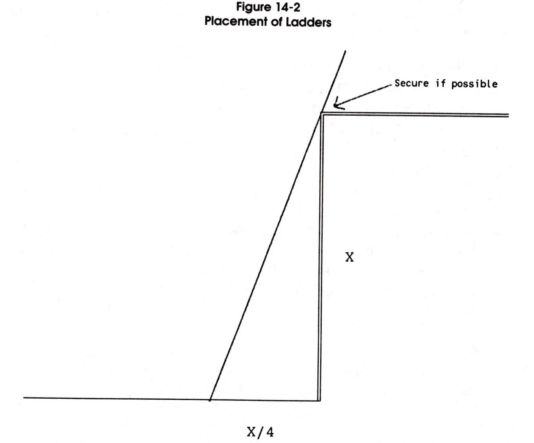

Fixed Ladders

As opposed to portable ladders, fixed ladders cannot be moved and are designed to provide access to a specific, elevated location. With the exception of those that provide access to towers, water tanks, and chimneys, all fixed ladders over 20 feet in length must be provided with cages or wells. The specific requirements for the construction of ladder cages are quite complex. For details of the appropriate dimensions and clearances that must be observed, consult Figures D-1 through D-11 of Paragraph .27(b) of Subpart D of the OSHA regulations. Because of these complexities, any construction, remodeling, or renovation project that requires the installation of fixed ladders should be under the direction of a professional engineer who is familiar with the applicable OSHA standards.

Fixed ladders on towers, water tanks, and chimneys are not required to have cages on runs that are over 20 feet in length. For these installations there is a requirement that there be a platform for individuals to rest every 250 feet. When fixed ladders are used on towers, tanks, or chimneys, appropriate ladder safety devices must be used. A ladder safety device is an appliance that will arrest the fall of an individual should he or she lose their footing. Many safety professionals advocate the use of ladder safety devices over cages in those instances where they are permitted since they provide a positive mechanism for arresting a fall.

Regardless of whether fixed ladders are equipped with cages, all fixed ladders must meet the following general requirements:

❖ Have a preferred pitch of 75 to 90 degrees.

❖ Be designed to bear a single concentrated load of 200 pounds.

❖ A ¾-inch rung diameter for metal or 1⅛ inches for wood.

❖ Have rungs at least 16 inches wide.

❖ Have rungs evenly spaced no more than 12 inches apart.

❖ Have hand or side rails that extend at least 3½ feet above the landing.

❖ Have a minimum clearance of 2½ feet on the climbing side of ladders with a 90 degree pitch and 3 feet for a 75 degree pitch.

❖ Have a clear width of at least 15 inches on each side of the centerline of the ladder unless enclosed within a cage or well.

❖ Have at least a 7-inch clearance in back of the ladder to assure adequate footing.

❖ Be painted, if metal, or appropriately treated to prevent deterioration if conditions indicate.

Scaffolds

A scaffold is defined as a temporary, elevated working platform used to support workers and material during construction and/or maintenance activities. Scaffolding is defined as the wood or metal framework that supports the scaffold. Metal scaffolding is recommended because of its greater ease of assembly and stability. Metal scaffolding is also easily disassembled and amenable to storage after use.

The following general regulations apply to all scaffolds and scaffolding:

❖ Construction materials must be able to support a minimum of four times the expected weight of the workers and the equipment that will be placed on them.

❖ Damaged or weakened structures must be repaired or replaced immediately.

❖ Wet and slippery conditions on scaffolds must be corrected as soon as they occur.

❖ When work must be performed above the individuals working on scaffolds, overhead protection must be provided for the individuals on the scaffold. This should not be more than nine feet above the working deck of the scaffold.

❖ When individuals may work below the scaffold, an 18 gauge × ½ inch wire mesh, or equivalent, screen must be installed between the toeboard and the guardrail.

❖ A safe means of access to the working deck must be provided. Most often this is a portable or fixed ladder.

❖ Components of scaffolding from different manufacturers must never be interchanged. This is a dangerous practice and can lead to failure of the structure.

❖ Scaffolding must never be erected or altered unless under the direct supervision of a competent individual.

❖ Footing must be on solid and level ground to prevent shifting and/or settling of the scaffolding during use. In some cases, the use of outriggers may be necessary to stabilize the structure.

❖ Scaffold deck planking must be inspected on a periodic basis to assure that it is free of structural defects.

❖ Guardrails, midrails, and toeboards must be installed on all open sides of structures over ten feet in height.

❖ Structures between four and ten feet in height and with a minimum platform dimension of 45 inches must have guardrails on all open sides.

❖ Guardrails must be between 36 and 42 inches high with rail support intervals not exceeding eight feet.

❖ Toeboards must be at least four inches high. This requirement may vary from state to state.

❖ Scaffold planking must extend a minimum of six inches and a maximum of 12 inches over the ends of the platform.

❖ Wood scaffolds greater than 60 feet in height must be designed by a qualified engineer. Tubular metal scaffolds greater than 125 feet must be designed by a registered professional engineer.

Wooden and metal scaffolds can be classified as light, medium, or heavy duty depending on their intended use and the amount of weight they will bear. They are classified as follows for a uniform weight distribution:

❖ Light Duty: 25 pounds/ft^2
❖ Medium Duty: 50 pounds/ft^2
❖ Heavy Duty: 75 pounds/ft^2.

The precise requirements concerning construction of wooden scaffolds are quite complex and beyond the scope of this chapter. Details of the construction requirements are given in Subpart D of the OSHA regulations. If it is necessary to use wooden scaffolds, they should be designed and constructed under the direct supervision of a competent engineer who is thoroughly familiar with the requirements of the OSHA standards.

The OSHA construction requirements for metal scaffolds are much less complex than for their wooden counterparts and merely specify the horizontal and longitudinal post spacing, number of working levels, and the maximum allowed height. These requirements can also be found in Subpart D of the OSHA regulations.

Metal scaffolds are normally purchased or rented from outside manufacturers or suppliers. These organizations usually guarantee that their products will meet or exceed regulatory requirements. In addition, most manufacturers and suppliers will be able to provide setup and teardown assistance, as well as engineering assistance for special situations that may arise. For these reasons, the use of metal scaffolds is recommended wherever possible.

Manually Propelled Scaffolds (Towers)

These working platforms are very similar to conventional metal scaffolds except that they are equipped with wheels and can be easily moved to where they are needed. Many such devices can also be easily broken down for storage. The requirements for these devices are essentially the same as for nonmobile metal scaffolds. As with nonmobile metal scaffolds discussed above, the manufacturer or supplier will generally warrant that these devices meet applicable regulations.

Because these devices are mobile, certain common sense safety practices need to be observed:

❖ Firmly secure or remove all material from the working deck before the scaffold is moved.

❖ Never allow individuals to ride on the scaffold while it is being moved.

❖ Keep wheel locks applied at all times except when the scaffold is actually being moved to a new location.

❖ If the working platform height exceeds four times the smallest base dimension, the scaffold must be appropriately stabilized or guyed.

CHAPTER
15

MATERIALS HANDLING
AND STORAGE

This chapter describes some common materials handling and storage problems, outlines the means of identifying and fixing these problems, describes equipment for materials handling, and concludes with some general guidance on safe storage practices.

COMMON MATERIALS HANDLING AND STORAGE PROBLEMS

Materials must be moved in all businesses. Often this is the least efficient and most troublesome aspect of manufacturing. Industrial engineers spend the majority of their careers attempting to reduce these movements, or at the least, to increase the efficiency of them. Materials handling and storage are of concern from a safety and health perspective for several reasons. First, because of the potential for strains, sprains, cuts, and other worker injuries associated with the movement of materials. Second, the movement and storage of materials many times results in spillage which may represent a hazard. Third, improper storage of materials can represent a physical threat to employees and property. Finally, incompatible storage could result in the potential for both fire and health hazards.

It is well known among safety managers that for most businesses manual materials handling is the most frequent cause of injuries to employees. Of these injuries, most involve the back. It has been estimated that 40 percent of all recorded absences from work can be attributed to back problems. Further, it is estimated that half of all back injuries are due to lifting. Even for those who lift properly or use mechanical assistance, crushing injuries and cuts are still commonplace events.

A second common problem in materials handling is spillage. This occurs primarily with bagged or boxed materials; however, speared drums and breakage of other containers is also possible. Spillage may also be evident with the manual filling of hoppers, blenders, or other bulk containers. This becomes a problem because the spillage could represent a slip or trip, fire, or health hazard. Further, improperly emptying containers can lead to significant exposures to the worker of potentially harmful materials. It is also quite common for spilled materials to be conveyed outside of the storage areas by the contaminated boots or shoes of workers or by the wheels of hand trucks, dollies, or fork-lift trucks. For materials that are dusts or powders and hazardous, vacuuming may then be needed rather than dry sweeping. Liquid spills may need to be cleaned using special sorbents while large spills may require an emergency response under the OSHA regulations (see Emergency and Disaster Planning, Chapter 13).

The next category of problems common to materials handling is inadequate or improper storage. This includes unstable stacking of loads on pallets (due to broken pallets or poor work practices), stacking that exceeds safe heights, storage on shelving that exceeds the weight limit of the storage racks, and damage to structural beams, supports or shelving from improper loading practices.

The final category listed in the introduction was incompatible storage. Some common examples of incompatibles are:

- ❖ Acids and bases
- ❖ Acid and chlorine bleach
- ❖ Acids and cyanide mixtures
- ❖ Corrosives and untreated metals (e.g., Aluminum)
- ❖ Fuels or solvents and oxidizers
- ❖ Fuels or solvents and peroxides
- ❖ Ammonia and chlorine bleach

Chemicals from the metal plating industry can be used as a "worst case" example. In metal plating, common components of the various plating baths are cyanide containing solutions and acids. Many times the chemical used to make these solutions are stored in the same area. However, if they were to be accidentally mixed (acid and cyanide salts), they could generate cyanide gas. This is the same material that is used in the gas chamber for executions. Obviously, they must be stored separately. This is an extreme but actual example of common incompatible storage. Therefore, it is important to determine the compatibility of materials before storing them in the same area.

IDENTIFYING AND CORRECTING MATERIAL HANDLING PROBLEMS

The identification of most of the problems associated with poor materials handling and storage is relatively straightforward for many of the areas previously identified. Simply put, visual inspections of warehouses and other storage areas should provide evidence of acceptable or unacceptable work practices. This includes improper lifting techniques. Accident and workers compensation statistics will also provide hard evidence of poor ergonomic practices for pushing, pulling, carrying and lifting of materials. Finally, incompatible storage can be determined by reviewing the incompatibilities section of the material safety data sheets for chemicals prior to their storage.

Correcting a problem is always more difficult than discovering that one exists. For worker injuries due to the movement of materials, retraining on proper lifting procedures and other materials handling training is usually the best solution. This solution assumes that the task is both reasonable and feasible. A detailed discussion on the evaluation of manual lifting tasks is given in the Chapter 10 on Ergonomics. For tasks that are very risky from an injury standpoint, mechanical assistance should be provided. Some general guidance for employees on manual handling follows.

Lifting Objects.   Evaluate the objects to be lifted. If they have edges that cut or damage your hands, wear gloves. If there is a choice, grip the material where it is least hazardous. If the object is greasy, wet, or slippery, clean it before lifting. Keep hands and gloves free of oil and grease. Do not attempt to lift more than can be carried without straining. If necessary, get help.

Follow the six basic steps for proper lifting. Step one: face the object and get as close as you can to it. Step two: place your feet far enough apart for good balance. Step three: bend your knees from the hips and squat, keeping your back as straight as possible (not arched). Step four: grip the sides of the object using your whole hands as a balance point. Step five: lift by straightening your legs using your thigh muscles to raise your body. Step six: bring your back and legs to a vertical position.

Carrying Objects.   Do not carry objects that block your vision ahead or to the sides. If you have to change your grip, set the object down and regrip. Do not change your grip while you are carrying the object. Do not hurry if you feel you cannot hold the object much longer. Put it down and rest and get assistance. Never walk backwards, always look ahead.

Setting Down Objects.   When you set down a heavy object to floor level, reverse the procedure described previously for lifting such objects. Do not drop, bounce, or lower it with an arched back. Do not set a heavy object into a position *below* floor level directly from a carry. It should first be lowered to floor level. Avoid awkward positions or full extension of your arms when setting down the package. If possible, try to slide the object onto surfaces so that your fingers do not become pinched.

Pushing and Pulling Objects.   When using equipment such as a handtruck or cart to transport materials, push instead of pulling the load, unless the equipment is specifically designed to be pulled. Pushing provides better control than pulling. Also, if you slip or trip while pulling, you run the risk of being struck by the load.

USE OF EQUIPMENT FOR MATERIALS HANDLING

Mechanical material handling is advantagous when compared to manual lifting and carrying. This type of equipment, however, is not without its own characteristic hazards. Some general guidance on the use of these devices follows:

Two-Wheel Handtrucks.   There are many varieties of two-wheel handtrucks. Some, for example, are especially designed for kegs, drums, and barrels. If several types are available, it is important to select the correct type for the materials to be hauled. Handles should have handguards. Trucks with foot brakes are the safest. For loading, the the load center of gravity should be kept as low as possible by placing the hearviest objects on the bottom. Two-wheel handtrucks should not be overloaded. Heavy or bulky loads should be secured. These trucks should be pushed instead of pulled, except when going up an incline.

Four-Wheel Handtrucks.   Handtrucks should be blocked while loading if they do not have a hand brake. The loads should be balanced to avoid tipping. With a push-type handtruck, the loads should not obscure vision of the worker using the handtruck unless a guide person is used. The truck should be pushed unless it is equipped with a pull-type handle. If the truck has a handle, it should be equipped with a spring to keep it in an up position when not in use.

Dollies.   A dolly is a small platform on low coaster wheels. It is usually best used for carrying single heavy objects short distances when it is impractical to lift such objects to a high truck bed. It is moved and guided by pushing on the load. Pulling on an attached rope is not advised unless a second person has a rope attached in the rear where braking action can be applied.

GENERAL GUIDANCE ON SAFE STORAGE PRACTICES

The following general guidance is provided for the storage of general classes of materials.

Boxes and Cartons.   Boxes and cartons should be stacked by cross-tying when piles are above head height. The safe height will depend upon the size and wieght of the containers. Cartons should be stored on pallets or other platforms to protect against moisture since wet carton will collapse. Wire or strap banded cartons and boxes should be stored so sharp ends don't protrude into walkways. Piles should be perpendicular to the floor, except for step back stacking. No boxes or cartons should protrude from the perpendicular line of the stack.

Drums.   When stored in large numbers, drums can be stored upright, upon each other on pallets (no more than 2 high), stacked in a pyramid row or stored in racks. When piled in a pyramid row, the bottom row must be securely blocked to prevent the row from spreading. Special care must be taken for drums containing flammables, corrosives, or other hazardous materials.

Bags and Sacks.   Bagged and sacked material should be placed on platforms or pallets to avoid moisture absorption. When bags or sacks are set against a wall in a single row, they should be pyramided to a safe height. The most economical storage from the standpoint of space is to cross-tie them to a safe height. Piles should be inspected for stability and also for ripped lower bags or sacks. If there is nearby truck traffic, it may be desirable to shield the lower bags against accidental ripping. Rodent damage may also cause piles to tumble. Bags should be removed from the top of the pile, however, employees should be warned against climbing the pile.

Lumber Stock.   Lumber should be stored in racks or piled on a firm foundation of heave cross pieces. Protection should be provided against the ground becoming soft from surface water. Soft ground may cause the pile to sink unevenly and topple over. As the pile increases in height, cross-tie pieces should be placed at regular intervals to provide stability and ventilation. The cross pieces should not protrude from the pile. High lumber piles should not be climbed, a ladder should be used instread.

Steel Piping and Bar Stock.   Pipes and bar stock should be stored in racks or in piles with the layers separated by wood strips with end blocks. Maximum floor loads should be determined before storing these materials in large quantities because of their heavy weight.

Sheet Metal. Sheet metal should be stored in flat piles or in horizontal or vertical slot-type racks that permit easy access. Sheet metal stock should never be stored by leaning bundles or loose stock against walls, pillars, or equipment. When bundles of sheet metal are piled, the bundles should be separated by wood strips or pallets for mechanical handling.

Heavy Machinery Spare Parts. Large machinery and equipment spare parts should be stored in rows in designated areas. The location should be as near to the source of use as practical. Walkway space should be left between rows to provide access and easy identification. Individual items should not be stored on top of each other. Shelving should have load limits clearly posted.

Wire and Cable. Wire and cable reels should be stored upended and large reels should be blocked to prevent rolling. Large bundles of wire should be stored in suitable racks or in piles in which the bundles are cross-tied. Such bundles should not be stored leaning upright against walls. Wire ends should not protrude into the surrounding area.

Gas Cylinders. Gas cylinders should be stored upright in cylinder racks. Individual cylinders may also be chained or clamped to a substantial structure such as a wall or pillar. Cylinders should not be allowed to stand free without being secured. They should be stored with the valve end up, never lying on their sides. The storage location should not be exposed to mobile equipment traffic, direct sunlight, or heat sources. Indoor storage areas should be well ventilated and posted for no smoking. Different gases should be stored separately. Oxygen cylinders should not be stored in a confined space with other gas cylinders, nor near oil, grease, or liquid flammables. Cylinders should be capped when not in use. Fuel and oxidizing gas cylinders must be separated by a fire wall or a distance of 20 feet.

Hazardous Liquid Chemicals. Hazardous liquid chemicals should be bulk stored in special locations designed for such storage. These storage locations should include diking. They should be well ventilated, free from temperature extremes, capable of being locked to prevent unauthorized access, and provided with suitable storage and handling facilities. Drums containing acids or other hazardous liquid chemicals should be stored in racks, never stacked. They should be provided with selfclosing spigots. To avoid errors in withdrawing materials, the area should be well illuminated and the drums should be clearly identified as to contents. Chemical-containing carboys should be stacked no higher than two tiers.

Flammable Liquids. Drums of flammable liquids should be racked for easy identification and access. They should be identified with a stenciled title or others means, and equipped with selfclosing spigots. To prevent static electricity from accumulating, the drums should be bonded and grounded.

CHAPTER
16
MACHINE GUARDING

The basic purpose of machine guarding is to prevent contact of the human body with dangerous parts of machines. When arms, fingers, hair, or any other body part enter into or make contact with moving machinery, the results can be disastrous and sometimes fatal. Methods of machine guarding vary greatly, depending on the machine and the types of hazards it presents. The intent of this chapter is to familiarize you with the hazards of unguarded machines, common safeguarding methods, and the safeguarding of machines.

This chapter includes illustrations and figures which have been adapted from *Concepts and Techniques of Machine Safeguarding*, published by the Occupational Safety and Health Administration (publication no. 3067). This is an excellent reference source which is available from the Superintendent of Documents, U.S. Government Printing Office, Washington, D.C. 20402.

HAZARDS OF UNGUARDED MACHINES

Unguarded machines can present two different types of hazards. The first type is mechanical, which includes hazards arising from the motion or operation of the machine. Examples of these machine motions and actions are presented in Diagram 16-1. The other type of hazard is one unrelated to the movement of the machine.

Mechanical Hazards

Machine Motions

Rotating Motions may present a hazard in two ways. The first of these is that protruding objects, such as burrs, screws and bolts, rough surfaces and the rotating motion itself, all tend to "grab" hair and clothing, causing the object to be pulled into the machine. Even smooth,

Diagram 16-1
Machine Motions and Actions

Rotating Motion:

Reciprocating Motion:

Transverse Motion:

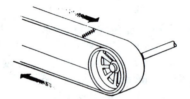

Cutting Action:

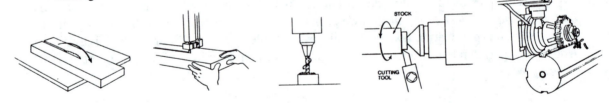

Punching, Shearing, and Bending Actions:

 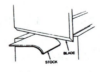

slowly rotating equipment can cause injury. The second type of hazard is called an in-running nip point. In-running nip points are created when machine parts rotate towards each other or when a rotating part rotates toward a stationary object. Nip points can present a severe hazard by pulling objects into the rotating parts or crushing the object against the stationary component. Common machinery which exhibits a rotating motion includes gears, pressure rolls, chains and sprockets, and spoked handwheels.

Reciprocating Motions are those that follow a back-and-forth or up-and-down pattern. These motions can be hazardous due to the possibility of a worker being struck by the reciprocating component or caught between the reciprocating component and a stationary object such as the floor, a wall, another machine component, or another piece of machinery. An example of a reciprocating motion would be an automated wood working machine equipped with a moving table. As the machine is put into operation the table moves back and forth processing the wood piece. The hazard is from being struck by the moving table.

Transverse Motions are those that follow a continuous, straight line, such as a conveyor belt. The hazards associated with transverse motions are similar to those of reciprocating motions. A worker may be struck or caught in a nip or shear point by the moving component. Workers can even be "carried away" or dragged by moving conveyor belts and other similar systems.

Machine Actions

The specific process performed by the machine can also present hazardous conditions. Examples include cutting, punching, shearing, and bending. The location of these actions is called the point-of-operation.

Cutting Action can be defined as one which cuts or removes material from an object. The hazard occurs as the cutting device approaches or contacts the wood, metal, or other material and as the pieces of the material being cut are projected toward the operator. Common machines which exhibit cutting action include abrasive grinding wheels, drills, circular saws, and planers.

Punching, Shearing, and Bending Actions occur as a powered ram or slide contacts the metal, wood, or other material for the purpose of blanking, drawing, stamping, or trimming. The most common machines which exhibit these actions are power presses, benders, and flying shears.

Nonmechanical Hazards

Nonmechanical hazards can also result from unguarded machinery. These include electrical power sources, high pressure systems, noise, chemical emissions, and contact with flying objects.

Electrical Power Sources for machinery can present a hazard if left unguarded. Contact with improperly grounded sources or damaged wiring can cause severe injury or death. These sources should be grounded and guarded, especially if located in high traffic areas. Standards for electrical power sources can be found in the National Electrical Code (NEC) and from the American National Standards Institute (ANSI).

High Pressure System malfunctions can result in the release of pressurized gases, gage blowout, and line rupture, producing explosions and flying objects. These systems should be guarded where possible to minimize the hazard.

Noise is an inherent hazard of virtually all machinery. Properly designed and installed guards can often assist in reducing noise levels. However, improperly installed guards can actually increase noise levels due to vibration or can channel the noise like a speaker. If used as part of a noise control strategy, be sure that guards are routinely inspected and maintained to ensure effectiveness.

Chemical Emissions can also be reduced by means of guarding. Lubricating oils, cutting fluids, coatings, cleaning fluids, and other process chemicals can easily be misted into the air by the motions of the machinery. For example, large cutting machines typically use nebulizers which atomize a fine spray of lubricant at the working face. In many cases, barrier guarding or shielding can assist in protecting the worker from exposure to mists.

Flying Objects, such as pieces of wood from cutting, metal trimmings from shearing, and objects which become entrained in rotating machines can present severe hazards to workers in many situations. Guarding can assist in reducing injuries by acting as a barrier between the machine and the worker.

COMMON SAFEGUARDING METHODS

There are several safeguarding methods to consider when you have determined that machine guarding is needed. These include guards, devices, location and distance, and feeding and ejection methods. The purpose of this section is to provide an overview of how to determine the most appropriate safeguarding method for some common machines. Diagram 16-2 summarizes this information along with the advantages and disadvantages of each safeguarding method.

Guards

Guards can be of several types. These include fixed, interlocked, adjustable, and self-adjusting.

Fixed Guards are guards which are permanent machine attachments. Unlike other types of guards, these do not move to accommodate the work being performed. These often have an access panel or transparent portions to view the operation of the machine. Special tools are normally required to remove the guard or to open the access panel. Because they provide only limited means of access and removal, these guards are preferred if their application is feasible.

Interlocked Guards operate on the principle that when the guard is opened or removed, the machine is immediately stopped and made inoperable. These guards are used in situations which require operators to have frequent access to the machine for setup, adjustment, or maintenance purposes.

Adjustable Guards are those that can be adjusted by the operator when the point-of-operation must be accessible because of varying sizes of work.

Self-adjusting Guards also accommodate different sizes of work at the point-of-operation, but due to the mechanics of the guard, they do not require the operator to adjust them. As the piece of work is inserted, the guard is pushed away, leaving an opening which is only large enough to accommodate the work piece.

Devices

Devices can also be used to safeguard machinery. This category includes presence sensing devices, pullback mechanisms, restraints, safety controls, and gates.

Diagram 16-2
Common Guarding Methods

Method	Graphic Representation	Advantages	Limitations
Fixed Guard		- Provides maximum protection - Prevents access to dangerous parts of the machine except during maintenance operations - Minimal maintenance required - Suitable to many operations	- Can limit visibility - Additional safeguards needed when opened or removed for maintenance of the machine
Interlocked Guard		- High degree of protection - Provides quick and easy access to machine	- Requires frequent maintenance and adjustment - Can be made inoperable
Adjustable Guard		- Makes varying sizes of work accessible to the point-of-operation - Wide variety of applications	- Degree of protection dependent on correct adjustment by operator - No protection during adjustment - Requires frequent adjustment and maintenance - Can limit visibility
Self-adjusting Guard		- Easier to purchase and fit to machine	- Lower degree of protection - Can limit visibility - Requires frequent adjustment and maintenance

Diagram 16-2 (cont'd)

Method	Graphic Representation	Advantages	Limitations
Presence-sensing Devices		- Allows access to point-of-operation and greater range of movement	- Usually expensive - Requires very frequent adjustment and maintenance to ensure effectiveness - Invisible if photoelectric or radiofrequency - May not provide protection in the case of mechanical failure - Only for use on slow speed machines which can be stopped easily
Pullback Devices		- Easy installation - Relatively inexpensive - Allows access to point-of-operation - Can be used on high speed machines - Provides protection in the event of mechanical failure - High level of protection for operator	- Limits range of motion for operator - Does not protect other employees in the work area - Requires adjustment for each operator - May obstruct operators workspace - Requires frequent inspection and maintenance - Operators hands not available to assist in operation - Enforcement of use necessary
Restraints		- Easy installation - Relatively inexpensive - Provides protection in the event of mechanical failure - High level of protection for operator - Can be used on high speed machines	- Limits range of motion for operator - Does not protect other employees in the work area - Requires adjustment for each operator - May obstruct operators workspace - Must use tools for placement of work - Enforcement of use necessary - Requires frequent inspection and maintenance

Diagram 16-2 (cont'd)

Method	Graphic Representation	Advantages	Limitations
Gates		- Easy to install - Relatively inexpensive - Can offer adequate protection for other employees in the work area	- Machine not easily accessible - Other means of protection must be provided when inside the gate - Can limit visibility
Location and Distance		- May be easy to arrange and/or install - Can be relatively inexpensive	- Other means of protection may be required during inspection and maintenance - May not offer adequate protection for other employees in the work area
Automatic and Semiautomatic Feeding		- Operator's hands not required in the danger zone - May increase production rate	- Often expensive to install - Usually requires other types of guarding around point-of-operation - Requires frequent maintenance - May not accomodate varying sizes of work
Automatic and Semiautomatic Ejection		- Operator's hands not required in the danger zone - May increase production rate	- Often expensive to install - Usually requires other types of guarding around point-of-operation - Requires frequent maintenance - May not accomodate varying sizes of work - May present additional hazards such as flying objects and high noise levels

Diagram 16-2 (cont'd)

Method	Graphic Representation	Advantages	Limitations
Safety Trip Controls		- Allows access to point-of-operation - Fast deactivation of machine	- Must be manually activated by the operator - Possible injury before deactivation - Position of tripping mechanism is critial for accessibility and effectiveness
Two-hand Control		- High level of protection with proper use - Allows operator to use hands after dangerous part of cycle is completed	- Can be modified by operator into a one hand operation - Does not protect other employees in the work area
Two-hand Trip		- Can be easy to install - Adaptable to multiple operators - Hands out of danger zone during start of machine cycle - No obstruction to hand feeding - Adjustment not required - May permit machine inching during setup and maintenance	- May be possible for operator to reach danger zone after tripping - Can be modified by operator into a one hand operation - Does not protect other employees in the work area - May not provide protection in the case of mechanical failure

Presence Sensing Devices commonly operate on photoelectric, radiofrequency, or electromagnetic principles to disengage the machine when something is detected in the zone of concern. The photoelectric device is comprised of a light source and a sensing mechanism. When the light field is interrupted, the machine will not operate. The radiofrequency device uses a radio beam which is a part of the control circuit of the machine. When an object is present in the danger zone, the capacitance field is interrupted, and the machine will not operate. Electromechanical devices utilize probes which enter the danger zone when the machine is actuated by the operator. If an object prevents the probe from fully entering the zone, the machine control circuit prevents the machine from performing the task. These sensing devices are commonly found on metals working presses.

Pullback Devices utilize cables which are attached to both the machine and the wrists of the operator to prevent hands from entering the point-of-operation during the power stroke. These devices are commonly used on power presses that require hand feeding. When the slide or ram is up, the operator has complete access to the point-of-operation. When the slide descends, however, the cables attached to the slide pull the workers hands away from the danger zone.

Restraints also utilize cables attached to the operator's wrist to prevent hands from entering the point-of-operation. The cables, however, are attached to a fixed object instead of the machine. By limiting the length of the cable, the operator's hands are prohibited from moving into the danger zone. Because the operator cannot reach into the point-of-operation, even when the machine is not operating, tools are often required for maneuvering the piece of work.

Safety Controls utilize involvement of the operator as a safeguarding method. These include safety trip controls, two-hand controls, and two-hand trips. Safety trip controls allow the operator to quickly disengage the machine in emergency situations by applying pressure to a body bar, pressing a safety tripod, or pulling on a safety tripwire. The positioning of these trip controls is of the utmost importance, as the operator must have easy access to the trip control mechanism. This method of safeguarding is more conducive to protecting the employee from further injury should an incident or accident occur. Two-hand controls require the operator to maintain hand pressure on two linked controls until the machine completes the dangerous part of the cycle. With the operator's hands on the controls, they are safely away from the point-of-operation (care must be taken to assure workers have not circumvented these controls to permit operation with only one button or control). Two-hand trips require the operator to press two buttons concurrently to start the machine. Unlike two-hand controls, however, the operator's hands are available for other tasks after the machine is engaged. In order to provide adequate protection using this method, the buttons must be located so that the time required for the operator's hands to reach the danger zone is greater than the time required for the machine to complete the dangerous part of the cycle.

Gates can also provide a high degree of protection to both the operator and other workers in the area, especially when interlocked to the machine cycle. When the gate is opened, the machine will not operate. However, gates that are not interlocked and serve only as a barrier, necessitate the need for other precautions to protect employees when working inside the gate.

Automatic and Semiautomatic Feeding and Ejecting Methods

Automatic and semiautomatic feeding and ejection of parts are other ways of safeguarding machine processes. These methods eliminate the need for the operator to work at the point-of-operation.

Automatic Feeding and Ejection operations require little involvement by the operator. On many systems, after the initial set up of the operation, the only task for the worker is that of monitoring. These may use roll stock systems, robots, forced air systems, gravimetric force, and other mechanical means to move the parts from location to location.

Semiautomatic Feeding and Ejection operations require some assistance by the operator during the process. Semiautomatic feeding may require the operator to place parts in a chute, plunger, magazine, or other device for subsequent feeding into the point-of-operation. Automatic and semiautomatic feeding and ejecting methods do not eliminate the need for machine guards or devices around the point-of-operation. In addition, other methods of safeguarding may be required during maintenance operations. Some systems may actually create other hazards such as excessive noise from air ejection blow offs, hazardous motions or contact with robots, and flying objects.

Location and Distance

Location and Distance can also be used to safeguard machinery. By analyzing the hazard and strategically locating machine parts and/or the operator, effective safeguarding can be accomplished. Examples include locating the operator control station several feet away from the point-of-operation, locating the dangerous machine operations several feet above the reach of workers, locating the feeding bin or mechanism far enough from the danger zone so that the worker's hands cannot enter the zone, and locating the machinery behind a structure or barrier which makes the machine inaccessible except for maintenance.

Miscellaneous Safeguarding Accessories

There are a variety of tools and methods that may help to lower the hazard potential created by certain machines, even though they do not provide full or complete machine safeguarding.

Special Devices or Tools for placing objects in power presses will allow the operator's hands and arms to remain away from the point of operation.

Spreaders and Nonkickback Devices help prevent work from being thrown back at the operator particularly with certain woodworking machines such as circular and radial saws.

Push Blocks and Jigs allow employees to keep their hands at a safe location when guiding wood or other materials during jointer and shaper operations.

SAFEGUARDING MACHINES

Determining When Machine Guarding Is Needed

In general, any machine which may cause injury to the operator or other employees working in the area must be safeguarded, and, according to OSHA (29 CFR 1910.212), "guarding shall be provided to protect the operator and other employees in the machine area from hazards such as those created by point of operation, ingoing nip points, rotating parts, flying chips and sparks." There are certain machines or processes which, by their nature, must always be guarded. These will be discussed in the following sections. In addition, custom machinery and processes should also be evaluated during design to identify hazards and to determine appropriate guarding techniques. In determining whether guarding is needed, it is necessary first to evaluate the hazards of the process. This will include the operator's relationship to the machine and the machine's location and relationship to other workers in the area. For example, if a machine is located in a pit covered with a metal grid and if the machine is controlled and monitored from an operator's station, location and distance should serve as an adequate safeguarding method for normal operations. However, if this machine were located at floor level, some form of guarding would be required. Knowledge of the types of mechanical hazards already discussed and common sense can usually determine if a machine poses a risk of injury. Nevertheless, it may be prudent to assume that <u>any</u> point of operation hazard on a machine that could be reached by a worker under any reasonable circumstances should be guarded. For this reason, most manufacturers will guard all points of access on machines.

GUARDING MACHINES

General Requirements for Guard Design

In order to provide the maximum protection for the operator and other workers, there are several general requirements that should be considered during the design stage of the guard or machinery.

Affixed to the Machine   Guards must be securely attached to the machine or, if not feasible, should be secured elsewhere.

Prevents Contact   The safeguarding method must prevent the workers arms, hands, and other body parts from contacting or entering the machine during the dangerous part of the machine cycle.

Tamper Resistant Design   Select guards which prevent removal or disengaging by the employee.

Durable Construction   Choose guards made of material which will be suitable for long-term operation and which require little maintenance. For example, plastic guards are often used instead of steel when corrosion is a problem.

Protects Workers from Flying Objects   Choose guards which prevent objects from falling into the machine and which prevent flying objects, such as wood chips and metal trimmings, from escaping the process.

Does Not Create New Hazards   Be certain that in the process of eliminating one hazard, you are not creating another. For example, robots used to automate and make an operation safer can strike employees or present a hazard at the point-of-operation if not effectively safeguarded themselves.

Does Not Create Interference   Be certain the guard allows the operator to perform the job without interference. Safeguarding methods are often disengaged or removed if they require additional effort on the part of the worker or make the jobs more difficult.

Allows for Lubrication   The guarding method should allow for lubrication of parts without removal of the guard.

Allows for Ventilation The guarding method should allow for normal cooling of the machine so as not to cause or present a fire hazard.

Point-of-Operation Guarding

It is beyond the scope of this chapter to discuss each safeguarding method for every machine. Only by evaluating the operation and the surrounding environment can the most appropriate method, or combination of methods, be chosen. The following OSHA machine guarding standards should also be consulted in detail to ensure compliance:

- ❖ 29 CFR 1910.212   General Requirements for all Machines
- ❖ 29 CFR 1910.213   Woodworking Machinery Requirements
- ❖ 29 CFR 1910.215   Abrasive Wheel Machinery
- ❖ 29 CFR 1910.216   Mills and Calenders in the Rubber and Plastics Industries
- ❖ 29 CFR 1910.217   Mechanical Power Presses
- ❖ 29 CFR 1910.218   Forging Machines
- ❖ 29 CFR 1910.243   Guarding of Portable Powered Tools

There are common safeguarding methods for many types of machinery and some of these are listed in Table 16-1. Nevertheless, it is important to evaluate the machine, the process, and employee work practices before installation of safeguarding equipment. It is also important that you not alter or modify the equipment manufacturer's own safeguarding schemes and devices without first consulting with them. Unauthorized alterations could result in additional legal liability for the user (products liability) by voiding warranties.

Power Transmission Guarding

Power transmission components are those components responsible for transmitting energy and transferring motion from the power source to the point-of-operation. The purpose of power transmission guarding is to protect employees from electrical power sources and mechanical components of the machine, including flywheels, cranks and connecting rods, tail rods, extension piston rods, shafting, pulleys, belt, rope and chain drives, belts, gears, sprockets, chains, friction drives, and clutches. Power transmission safeguarding is normally achieved with fixed guards because they provide the highest level of protection. Also, frequent access to the power transmission components is not normally required.

Requirements for guarding can be found in the OSHA Electrical Standards (29 CFR 1910 Subpart S) and the OSHA Mechanical Power-Transmission Apparatus Standard (29 CFR 1910.219). In general, any equipment located less than seven feet above the floor or on a platform must be guarded, and, in most cases, guards must be constructed of metal. Expanded metal, perforated or solid sheet metal, wire mesh with an iron frame, and iron pipe are examples of materials used for guard construction.

Periodic inspection of the guards should be performed to ensure that they remain securely fastened to the floor and/or machine and are free of damage. Damaged guards not only expose hazardous machine parts, but can also present a hazard due to sharp edges.

Control of Hazardous Energy Sources (Lockout/Tagout)

Maintenance and servicing of machinery often require that existing safeguardings be removed or disengaged to provide access to machine parts. During these operations, isolation

Table 16-1
Typical Machines and Common Guarding Methods

Machine	Common Guarding Methods
Abrasive Wheels	Fixed Guard
Bandsaws	Fixed Guard
	Adjustable Guard
Calenders	Safety Tripwire Cable
	Interlocked Guard
Circular Saws	Fixed Guard - Top Portion; Self-Adjusting Guard - Lower Portion
Jointers	Self-Adjusting Guard
	Semiautomatic Feed (Push Block)
Paper Cutters	Fixed Guard
Planers	Fixed Guard with Automatic or Semiautomatic Feed
Power Presses	Fixed Guard with Automatic Feed and Ejection
	Fixed Guard with Semiautomatic Feed (Chute or Plunger) and Ejection
	Fixed Guard with Sliding Die or Bolster
	Adjustable Guard
	Pullback Device
	Restraints
	Interlocking Guard
	Presence-Sensing Device
	Gate (Interlocked)
	Two-hand Control
	Two-hand Trip
Press Brake	Fixed Guard
	Presence-Sensing Device
	Pullback Device
Radial Saw	Self-Adjusting Guard
Router	Adjustable Guard
	Self-Adjusting Guard
Rubber Mill and Other Roll Systems	Pressure-Sensitive Body Bar
	Safety Tripod
	Interlocked Guard
	Presence-Sensing Device
	Fixed Guard with Automatic or Semiautomatic Feed
Shaper	Adjustable Guard
	Self-Adjusting Guard
Table Saw	Adjustable Guard
	Self-Adjusting Guard
	Fixed Guard with Automatic or Semiautomatic Feed
Veneer Clippers	Fixed Guard (with or without Automatic Feed)
	Two-hand Control
	Pullback Device

and deenergizing of the equipment by the person or persons performing the work is required to protect the employee(s) from unexpected start-up of the machine. This is called lockout/tagout. Lockout, which involves placement of a lockout device on an energy isolating device, is the most effective means to protect maintenance personnel. Energy isolating devices are mechanical devices that physically prevent the transmission or release of energy. Examples include manually operated electric circuit breakers, disconnect switches, and line valves. Tagout, which involves placement of a warning tag on or near the energy isolating device, can be adequate to protect personnel if all tagout procedures are understood and followed by all employees. Checklist 16-1 summarizes the requirements of the OSHA lockout/tagout standard (29 CFR 1910.147) and will assist you in developing and maintaining a lockout/tagout program.

Checklist 16-1
Lockout/Tagout Requirements

Energy Control Program

☐   Has an Energy Control Program been established?

Does the Program include:

☐   Energy control procedures?

☐   Employee training?

☐   Periodic Inspections?

Lockout/Tagout

☐   Is lockout used when an energy isolating device is capable of being locked out?

If tagout is used when an energy isolating device is capable of being locked out, can the employer demonstrate:

☐   That tagout devices are attached at the same location as a lockout device would be?

☐   That a level of safety equivalent to using lockout is provided to employees?

☐   That compliance with all tagout provisions of the standard has been achieved?

☐   Implementation of additional safety measures to reduce the likelihood of inadvertent energization?

☐   Is new equipment, or equipment which is replaced or undergoes major repair, renovation, or modification equipped with energy isolating devices designed to accept a lockout device?

Energy Control Procedures

☐   Have lockout/tagout procedures been developed and documented?

☐   Are the procedures actually used?

Do procedures outline the following:

☐   Scope?

☐   Purpose?

☐   Authorization?

☐   Rules?

☐   Techniques to be utilized?

☐   Means of enforcing compliance?

Do procedures include:

☐   A specific statement of the intended use of the procedure?

☐   Specific procedural steps for shutting down, isolating, blocking, and securing machines?

☐   Specific procedural steps for the placement, removal, and transfer of lockout/tagout devices?

☐   Specific requirements for testing to ensure the effectiveness of energy control measures?

Protective Materials and Hardware

☐   Are all necessary protective materials and hardware provided by the employer?

Are lockout/tagout devices:

☐   Singularly identified?

☐   The only devices used for energy control?

☐   Considered durable for the intended environment?

☐   Standardized within the facility in color, shape, size, print, and format?

☐   Does management ensure that lockout/tagout devices are not used for any other purpose than the control of energy?

☐   Are lockout devices substantial enough to prevent removal without excessive force or unusual techniques?

☐   Are tagout devices substantial enough to prevent inadvertent or accidental removal?

Are tagout devices:

☐   Nonreusable?

☐   Attached by hand?

☐   Self-locking?

☐   Nonreleasable with a minimum unlocking strength of 50 pounds?

☐   Do lockout/tagout devices identify the specific employee applying the device?

Do tagout devices:

☐   Warn against hazardous condition?

☐   Include a legend specific to the hazard (such as "do not start" or "do not open")?

Periodic Inspections

☐   Are inspections of the energy control procedures conducted at least annually?

☐   Are inspections performed by an authorized employee other than the person utilizing the procedures being inspected? (An authorized employee is defined as a person who locks out or tags out machines or equipment in order to perform servicing or maintenance.)

☐   Does the inspection include a review, between the inspector and each authorized employee, of the employee's responsibilities?

☐   If tagout devices are used, does the review also include training on the limitations of tags?

☐   Are all inspections documented including:

  ☐   The identity of the machine or equipment?

  ☐   The date of the inspection?

  ☐   The employees included in the inspection?

  ☐   The person performing the inspection?

Training and Communication

☐   Have all authorized employees received training?

Does the training include:

  ☐   Recognition of applicable hazardous energy sources?

  ☐   The type and magnitude of the energy available in the workplace?

  ☐   The methods and means necessary for control and isolation of energy?

☐   Have affected employees been instructed in the purpose and use of the energy control procedures? (Affected employees are those whose jobs require them to operate or use a machine or equipment on which servicing or maintenance is being performed under lockout or tagout, or whose job requires them to work in an area in which such servicing or maintenance is being performed.)

☐   Have all other employees whose work operations are or may be in an area where energy control procedures may be utilized received training?

Does this training include:

  ☐   Instruction about the procedure?

  ☐   Instruction about the prohibition relating to attempts to restart or reenergize machines or equipment which are locked out or tagged out?

☐   If tagout systems are used, do employees receive training on the limitations of tags?

Does this training include:

  ☐   The fact that tags are merely warning devices and do not provide physical restraint?

  ☐   The fact that tags are not to be removed unless authorized by the employee responsible for the tag?

  ☐   The fact that tags should never be bypassed, ignored, or otherwise defeated?

  ☐   The fact that tags must be legible and understandable by all employees to be effective?

☐   The fact that tags must be made of materials suitable for the environment in which they are used?

☐   The fact that tags may evoke a false sense of security?

☐   The fact that tags must be securely attached?

Is refresher training provided under the following circumstances:

☐   When a change in job assignments occurs?

☐   When a change in machines, equipment, or processes occurs?

☐   When a change in the energy control procedures occurs?

☐   When a periodic inspection reveals deviations from or inadequacies in the employee's knowledge or use of the energy control procedures?

☐   When the employer has reason to believe that there are deviations from or inadequacies in the employee's knowledge or use of the energy control procedures?

☐   Is training documented?

Does the documentation include:

☐   Each employee's name?

☐   Dates of training?

Application of Controls

☐   Does the authorized employee have knowledge of the type and magnitude of the energy, the hazards of the energy to be controlled, and the method or means of control before turning off the machine or equipment?

☐   Is the machine or equipment turned off or shut down using the specific procedures established for that machine or equipment?

☐   Are all lockout/tagout devices located and operated in a manner which isolates the machine or equipment from the energy source?

☐   Are lockout/tagout devices affixed by authorized employees?

☐   Are lockout devices affixed in a way which will hold the energy isolating device in a safe or off position?

☐   Are tagout devices affixed in a way which will indicate that the operation or movement of energy isolating devices from the safe or off position is prohibited?

☐   Is all stored energy relieved or otherwise rendered safe following the application of the lockout/tagout device?

☐   Prior to starting work on locked out or tagged out equipment, does the authorized employee verify that isolation and deenergization has been accomplished.

☐   If there is a possibility of reaccumulation of stored energy, does verification of isolation continue until servicing or maintenance is complete?

Release from Lockout or Tagout

☐ Are the following procedures completed before the removal of lockout or tagout devices:

    ☐ Inspection of the work area to ensure that nonessential items have been removed and that the machine is operationally intact?

    ☐ Inspection of the work area to ensure that all employees have been safely positioned or removed?

☐ Are affected employees notified that the lockout or tagout devices have been removed before starting of the machine or equipment?

☐ Are lockout and tagout devices removed by the employee who applied the device?

Additional Requirements

Is the following sequence of actions completed if lockout or tagout devices must be temporarily removed and the machine energized for testing or positioning:

    ☐ Clearing tools and materials from the machine?

    ☐ Removing employees to a safe location?

    ☐ Removing the lockout or tagout devices?

    ☐ Energizing the machine?

    ☐ Deenergizing the machine and reapply lockout or tagout devices?

☐ Are outside personnel informed of lockout/tagout procedures before engaging in maintenance and servicing operations?

☐ Are employees informed of the restrictions and prohibitions of the energy control program of any outside personnel performing maintenance and servicing operations?

☐ Is primary responsibility given to one authorized employee during group lockout/tagout efforts?

☐ Do provisions exist for this authorized employee to ascertain the lockout/tagout status of individual group members?

☐ Is primary responsibility given to one authorized employee during lockout/tagout efforts involving two or more groups of employees?

☐ Does each employee of a group affix a personal lockout or tagout device to a group lockout device before beginning maintenance or servicing?

☐ Does each employee of a group remove their personal lockout or tagout device from the group lockout device when their work on the machine or equipment is completed?

☐ Do procedures exist to ensure the continuity of lockout or tagout protection during shift or personnel changes?

OTHER CONSIDERATIONS

Once safeguards are installed and operational, there remain several items which require attention on an on-going basis. These include training, employee clothing, and inspection and maintenance of the safeguards.

Training

Employees should receive training on the machine and on the purpose of the safeguard before assignment to operate the machine. Refresher training should be provided on a periodic basis. At a minimum, training should include the purpose of the guard and the hazards created by removing or disengaging the guard. Some safeguarding methods, such as adjustable guards, restraints, pullback devices, and safety controls, may require more in-depth training. In these situations, the operator is an integral part of the safeguard procedure. Additionally, some safeguarding methods will require daily testing of the safeguard by the operator. For these systems, proper testing procedures must be conveyed to the employee.

Employee Clothing

It is management's responsibility to assure employees wear appropriate clothing when operating or working around hazardous machines. Loose, oversized clothing can easily catch on machine parts. Employees should wear chemical resistant clothing and shoes if the machine creates an exposure potential to lubricating oils or other substances. Employees with long hair may need to wear hats or hairnets if the long hair represents a hazard due to the proximity of moving machinery.

Inspection and Maintenance

Guards and safeguarding devices must be inspected and maintained to be continually effective. Broken or inoperable parts should be replaced. The general condition of the guard should be routinely evaluated. For example, rusting metal guards may be inoperative and will offer a false sense of security. Similarly, inoperable presence-sensing devices and interlocking guards can also present a false sense of security and can lead to serious injury. Some machines may require daily inspections with maintenance of formal logs (e.g., power presses), while others may need less frequent inspections (e.g., table saw) or simple visual inspections (i.e., guard present?).

CHAPTER
17

HAND AND PORTABLE POWER TOOLS

Hand and portable power tools are used extensively in almost every type of industrial operation, both in production and maintenance activities. They are also widely used by just about anyone who has a home workshop or tool box. Unfortunately this familiarity with hand and portable power tools can frequently lead to accidents, almost all of which could be prevented.

Each year over half a million injuries involving various hand and portable power tools occur in and around the home. No comparable data for industrial accidents are available because of the way industrial accident statistics are kept, but the number is substantial. Each year approximately 6 percent of all disabling workplace injuries involving compensation are directly attributable to hand tool accidents.

Injuries involving hand and portable power tools range from minor cuts and bruises to amputations of fingers and loss of sight. Occasionally these injuries are severe enough to result in fatalities. Thus, it is important for all of your employees to understand how such injuries can occur and how they can be prevented by using their tools in a safe manner.

Hand and portable power tools are designed to make jobs easier and faster, but these advantages are quickly lost when the tools are used improperly. Your workers should be thoroughly familiar with the proper use and care of the tools they work with in order to increase productivity and to minimize the chance of injury to themselves and others.

TYPES OF HAND AND PORTABLE POWER TOOLS

While there are hundreds of different hand tools, they can be grouped into the following general categories:

❖ striking tools
❖ turning tools

❖ metal-cutting tools

❖ wood-cutting tools

❖ material handling tools

❖ gardening tools

❖ screwdrivers

❖ pliers

❖ knives and miscellaneous cutting tools.

Striking tools include several types of hammers, mallets, and sledges. Turning tools include numerous types of wrenches. Metal-cutting tools include shears, snips, bolt cutters, wire cutters, hacksaws, metal chisels, and files. Wood-cutting tools include hand saws, drills, planes, axes, hatchets, mauls, wedges, and wood chisels. Material handling tools include crowbars and hooks. Gardening tools include shovels, rakes, hoes, and post-hole diggers. Miscellaneous cutting tools include scissors, scrapers, bits, and awls. This is by no means a complete list of the different types of hand tools, however, it is representative.

You should never assume that your employees know how to properly use these common hand tools. The large number of injuries that occur every year illustrate the fallacy of this assumption. Thus an important part of your in-plant job training program for new and transferred employees should be instructed in the proper use of hand tools.

Portable power tools are usually grouped according to their power source as follows:

❖ electric

❖ pneumatic

❖ gasoline

❖ hydraulic

❖ powder-actuated.

Many of the more common portable power tools, such as saws, drills, and grinders and lawn and garden tools, such as mowers, trimmers, and edgers may be electric or gasoline-powered. Chipping tools, impact wrenches, spray painting units, and some hammers, saws, drills, and grinders are examples of pneumatic tools. Hydraulic tools are typically used for compression work. Powder-actuated tools, such as nail drivers and other types of fasteners, are used for speed in penetration, cutting, and compression work.

HAZARDS OF HAND AND PORTABLE POWER TOOLS

The primary hazards of using various hand tools include striking or contacting part of the body with the hand tool or the work piece and projectiles flying off the tool or work piece into the eyes. The most common type of injury from the use of hand tools is a laceration, or cut, from a knife blade, saw, or other tool with a sharp surface or jagged edge. Contusions, or bruises, are also very common from striking the fingers with the tool, such as hitting your thumb with a hammer while you are holding a nail or when a tool slips as you are applying force to it. Such bruises may be severe enough to involve broken bones. Other more serious injuries include infections from puncture wounds, severed fingers, tendons, and arteries from knife or saw blade cuts, broken bones from various types of accidents, and loss of vision from projectile accidents. Almost all of these injuries can be prevented by following a few simple safety rules discussed later in this chapter.

In general, the hazards from portable power tools are similar to those from the corresponding stationary power tools. In addition, there are risks from handling the portable tools when they are energized. For power tools, the saw blade or other work surface is usually in motion, presenting the potential for much more serious injury than from hand tools. Also, accidents resulting from electrical shock, fires and explosions involving flammable gases and vapors, and falling tools are much more likely.

SAFETY TRAINING

One of the key elements of a safety program for hand and portable power tool use is training for all workers and supervisors who use these types of tools. This training should include at least the following topics:

- ❖ how to select the proper tool for the job
- ❖ how to use these tools properly
- ❖ procedures for inspection of tools
- ❖ procedures for storage of tools
- ❖ procedures for repair of faulty tools
- ❖ the importance of planning jobs ahead so that the correct tools are available

Initial training of workers prior to assignment at a job where they will be using hand or portable power tools is essential. In addition, supervisors should continually be spot checking workers to ensure that these tools are being used properly. Emphasizing the important points of safe hand and power tool usage during routine and special safety meetings may be necessary if improper tool usage is observed or if an injury or near miss has occurred.

SAFETY PRACTICES

There are several general worker safety practices which apply to the use of almost all hand and portable power tools. These include:

- ❖ Wearing safety glasses with side shields or other equivalent eye protection when using hand or portable power tools.

  Approximately 90,000 work-related disabling eye injuries occur annually in this country. The majority of these injuries could have been prevented by using appropriate eye protection. Whenever striking tools are being used or the cutting action of a tool may propel particles into the air at high velocities, appropriate eye protection must be worn.

  ANSI Standard Z87.1 spells out performance requirements for eye protection. The OSHA Eye and Face Protection Standard, 29 CFR 1910.133, requires the use of protective eye and face equipment in all cases where injuries can reasonably be prevented by their use. Specifically, eye protection is required whenever there is a hazard from flying objects.

❖ Using the right tool for the job.

Frequently your employees will be tempted to grab the nearest available tool to get a job done rather than to spend a few extra minutes to find the proper tool. Examples of using the wrong tools for a job include using a wrench instead of a hammer for pounding a nail, using a screwdriver instead of a pry bar for opening the lid of a container, and using pliers instead of the proper wrench for gripping a large pipe. These practices can lead to unnecessary injuries and property damage and should be strongly discouraged.

❖ Inspecting tools before use and not using damaged tools.

Inspecting tools before using them can help to prevent injuries caused by damaged, worn, cracked, or otherwise faulty parts. This would include items such as loose or split handles on hammers, broken plugs or frayed wires on electrical tools, and sprung jaws on open end wrenches. Inspection checklists are included later in this chapter.

❖ Knowing how to use the tool properly.

Incorrect use of tools is another frequent cause of accidents. Lists of things to do and not to do when using hand and portable power tools are also included later in this chapter.

❖ Carrying tools properly and storing them in a safe manner.

Many accidents have been caused by improper transportation or storage of tools. This includes events like tools falling from overhead shelves and striking people, cuts from knives and other sharp tools carried in pockets or left in tool boxes unprotected, and trip hazards from tools left lying on the floor.

TOOL INSPECTIONS

The importance of inspecting tools before using them has already been discussed in relation to preventing accidents. Thorough inspections should be done on all tools at least once a month by a qualified person, such as the maintenance manager or tool room attendant. In addition, your employees should be trained to always inspect their tools themselves before use. The following inspection checklists should be useful in this task:

INSPECTION CHECKLIST FOR HAMMERS AND OTHER STRIKING TOOLS		
	OK	Action Required
1. Is the tool free of oil, grease, and accumulated foreign material? (Pay particular attention to the striking surface and handle.)	☐	☐
2. Does the tool have any damaged edges or irregularities on the striking surface?	☐	☐
3. Is the head securely fastened to the handle?	☐	☐
4. Are there any visible cracks in the head or handle?	☐	☐
5. Do you have the proper size tool for the job to be done?	☐	☐

INSPECTION CHECKLIST FOR WRENCHES AND OTHER TURNING TOOLS

	OK	Action Required
1. Is the tool free of oil, grease, and accumulated foreign material? (Pay particular attention to the jaws and handle.)	☐	☐
2. Are the surfaces of the jaws worn, damaged, or otherwise out of proper alignment?	☐	☐
3. For tools with ratcheting mechanisms, does the ratchet operate properly and freely?	☐	☐
4. Does the adjusting nut of an adjustable wrench move freely?	☐	☐
5. Are there any visible cracks in the jaws or handle?	☐	☐
6. Do you have the proper size tool for the piece of work to be gripped or turned?	☐	☐

INSPECTION CHECKLIST FOR METAL-CUTTING TOOLS

	OK	Action Required
1. Is the tool free of oil, grease, and accumulated foreign material? (Pay particular attention to the cutting surface and handle.)	☐	☐
2. Does the hinge of shears, snips, and other similar tools move freely?	☐	☐
3. Are there any nicks, burrs, or imperfections in the cutting surface?	☐	☐
4. Are the blades of hacksaws installed properly and securely with the teeth pointing away from the handle?	☐	☐
5. Are the teeth of hacksaws sharp and well set (angled) to prevent binding?	☐	☐
6. Are there any visible cracks in the cutting surface or handle?	☐	☐
7. Do you have the proper size tool for the metal piece to be cut?	☐	☐

INSPECTION CHECKLIST FOR WOOD-CUTTING TOOLS

	OK	Action Required
1. Is the tool free of oil, grease, and accumulated foreign material? (Pay particular attention to the cutting surface and handle.)	☐	☐
2. Are the blades of saws sharp and the teeth well set (angled) to prevent binding?	☐	☐
3. For axes, hatchets, mauls, and similar tools, is the head securely fastened to the handle?	☐	☐
4. For planes, does the blade adjusting mechanism move freely and hold the blade securely?	☐	☐
5. Are there any visible cracks in the cutting surface or handle?	☐	☐
6. Do you have the proper size tool for the wood to be cut?	☐	☐

INSPECTION CHECKLIST FOR MATERIAL HANDLING TOOLS

	OK	Action Required
1. Is the tool free of oil, grease, and accumulated foreign material? (Pay particular attention to the handle.) Note: Shovel blades may be lightly oiled to help prevent material from sticking.	☐	☐
2. Are the tips of hooks sufficiently sharp to grip the work piece securely?	☐	☐
3. Are there any visible cracks in the tool?	☐	☐
4. Do you have the proper size tool for the work piece to be handled or moved?	☐	☐

INSPECTION CHECKLIST FOR GARDENING TOOLS

	OK	Action Required
1. Is the tool free of oil, grease, and accumulated foreign material? (Pay particular attention to the cutting or digging surface and handle.) Note: Garden tool blades and prongs may be lightly oiled to help prevent material from sticking and retard rust.	☐	☐
2. Are there any visible cracks in the cutting or digging surface or handle?	☐	☐
3. Is the head of the tool securely fastened to the handle?	☐	☐
4. Do you have the proper size tool for the task?	☐	☐

INSPECTION CHECKLIST FOR SCREWDRIVERS

	OK	Action Required
1. Is the tool free of oil, grease, and accumulated foreign material? (Pay particular attention to the tip and handle.)	☐	☐
2. Is the tip broken or worn?	☐	☐
3. Does the tip fit the screw snugly?	☐	☐
4. Are there any visible cracks in the tip, shaft, or handle?	☐	☐
5. Do you have the proper size and type screwdriver for the job?	☐	☐

INSPECTION CHECKLIST FOR PLIERS

	OK	Action Required
1. Are the pliers free of oil, grease, and accumulated foreign material? (Pay particular attention to the jaws and handle.)	☐	☐
2. Are the gripping surfaces of the jaws worn, damaged, or otherwise out of proper alignment?	☐	☐
3. Does the hinge of the pliers move freely?	☐	☐
4. For pliers with expandable hinges, does the hinge lock securely in place when the jaws are expanded?	☐	☐
5. Are there any visible cracks in the jaws or handle?	☐	☐
6. Do you have the proper size and type pliers for the job?	☐	☐

INSPECTION CHECKLIST FOR KNIVES AND MISCELLANEOUS CUTTING TOOLS

	OK	Action Required
1. Is the knife or other tool free of oil, grease, and accumulated foreign material? (Pay particular attention to the blade and handle.)	☐	☐
2. Does the hinge of scissors move freely?	☐	☐
3. Are there any nicks, burrs, or imperfections in the blade or cutting surface?	☐	☐
4. Is the blade or cutting surface sharp enough to cut through the work piece?	☐	☐
5. Does the knife have a handle guard to prevent your hand from accidentally slipping onto the blade?	☐	☐
6. Are there any visible cracks in the blade or handle?	☐	☐
7. Do you have the proper size knife or cutting tool for the job?	☐	☐

When using the following checklists for portable power tools, the checklist(s) for the corresponding hand tools should also be consulted for applicable items.

INSPECTION CHECKLIST FOR ELECTRIC POWER TOOLS

	OK	Action Required
1. Is the plug and insulation on the cord intact so that live wires are not exposed?	☐	☐
2. Is the tool approved for use in hazardous atmospheres? (as required)	☐	☐
3. For tools to be used in tanks or wet areas, is the tool low voltage or battery powered? Is there ground fault circuit interrupter protection for the circuit to be used?	☐	☐
4. Is the tool motor in good condition?	☐	☐
5. Is the ground prong in good condition (for three-wire grounded tools)?	☐	☐
6. If the tool isn't three-wire ground protected, is it double insulated?	☐	☐
7. Are there any visible cracks or defects in the tool housing?	☐	☐
8. Is there a trigger lock or guard to prevent accidental activation of the tool?	☐	☐
9. Are there effective guards whenever possible for all moving parts of saws, grinders, and similar tools?	☐	☐
10. Is there a blade brake on lawn mowers, hedge trimmers, and similar tools?	☐	☐
11. Do moveable guards operate freely?	☐	☐

INSPECTION CHECKLIST FOR PNEUMATIC POWER TOOLS

	OK	Action Required
1. Is the compressed air hose in good condition with no visible cracks, bubbles, or kinks?	☐	☐
2. Do the hose connections to the tool and to the compressor fit snugly so there are no noticeable air leaks?	☐	☐
3. If the air hose is recoilable, does it pull out and retract freely?	☐	☐
4. Are there pressure reduction devices (to less than 30 psig) on all fittings designed to blow compressed air for cleaning?	☐	☐
5. Is there a safety chain at all hose fittings to prevent whipping of the hose if a connection comes loose?	☐	☐
6. Is there a safety check valve in the air hose at or near the compressor connection that will shut off or bypass the air flow if a break occurs in the air hose?	☐	☐
7. Are there any visible cracks or defects in the tool housing?	☐	☐
8. Is there a trigger lock or guard to prevent accidental activation of the tool?	☐	☐
9. Are there effective guards whenever possible for all moving parts of saws, grinders, and similar tools?	☐	☐
10. Do moveable guards operate freely?	☐	☐

INSPECTION CHECKLIST FOR GASOLINE POWER TOOLS

	OK	Action Required
1. Are there signs of fuel leakage around the gasoline tank or fuel line?	☐	☐
2. Is the tool motor in good conditions?	☐	☐
3. Are there any visible cracks or defects in the tool housing?	☐	☐
4. Is there a handle or trigger lock or guard to prevent accidental activation of the tool?	☐	☐
5. Are there effective guards whenever possible for all moving parts of saws, trimmers, edgers, and similar tools?	☐	☐
6. Is there a blade brake on lawn mowers, hedge trimmers, and similar tools?	☐	☐
7. Is there a tip guard on chain saws?	☐	☐
8. Do moveable guards operate freely?	☐	☐
9. Are there fire extinguishers or other fire suppression equipment nearby?	☐	☐
10. Are mufflers in good condition?	☐	☐
11. Are spark plugs and wire connections in good condition?	☐	☐

INSPECTION CHECKLIST FOR HYDRAULIC POWER TOOLS

	OK	Action Required
1. Are there signs of fluid leakage around hydraulic lines, cylinders, reservoirs, pumps, or other system components?	☐	☐
2. Are hydraulic lines in good condition with no visible cracks, bubbles, or kinks?	☐	☐
3. Are all hydraulic line connections secure?	☐	☐
4. Are there any visible cracks or defects in the tool housing?	☐	☐

INSPECTION CHECKLIST FOR POWDER-ACTUATED TOOLS

	OK	Action Required
1. Are you properly trained and qualified to operate powder-actuated tools in accordance with the manufacturer's instructions?	☐	☐
2. Are there any unprotected people in the immediate area?	☐	☐
3. Have you informed all nearby people what you will be doing?	☐	☐
4. Are there any visible cracks or defects in the tool housing?	☐	☐

These checklists have been designed for general use. You should always consult the manufacturer's instruction manuals which come with some hand tools and most portable power tools for additional items to check or verify prior to using the tool.

GENERAL RULES FOR USING HAND AND PORTABLE POWER TOOLS SAFELY

The importance of using tools safely has also been discussed previously in relation to preventing accidents. The following lists of things to do and not to do when using these tools may also be useful when training your employees. Once again, instruction manuals from tool manufacturers, especially for portable power tools, should be your primary source of information.

When using all hand and portable power tools, DO:

❖ use tools only for the purpose for which they are designed and intended.

❖ always wear safety glasses with side shields or other equivalent eye protection.

❖ use special gloves or other protective gear whenever they are recommended by the tool manufacturer.

❖ select tools that fit the work piece securely (e.g., screwdrivers that fit snugly in the screw slot, wrenches that fit snugly around the nut, etc.).

❖ use non-sparking tools whenever a fire or explosion hazard exists.

When using all hand and portable power tools, DO NOT:

❖ use any tool unless you are familiar with its safe operation.

❖ use any tool unless your footing and balance are stable

❖ use a damaged or faulty tool.

❖ alter the basic configuration of the tool or use a tool which has been altered.

❖ use any handle extensions or adapters unless they are specifically designed for the tool.

❖ expose tools to excessive heat which can ruin the tool.

❖ Use hammers, wrenches, screwdrivers, or other tools in place of a pry bar or chisel.

❖ engage in any kind of horseplay.

When using hammers or other striking tools, DO:

❖ use a hammer with a striking face approximately ⅜" larger than the striking face of another tool to be struck such as a chisel, punch, or wedge.

❖ strike the work piece or other tool squarely with each blow from the hammer.

❖ use a ball peen hammer to strike other metal tools.

When using hammers or other striking tools, DO NOT:

❖ strike hardened steel surfaces with a hammer.

❖ use hammers to strike a surface at an angle or with a glancing blow.

❖ use a hammer to strike another hammer.

❖ use a hammer with a loose or cracked handle.

❖ use cold chisels to cut or split harder materials such as stone or concrete.

❖ use chisels or other cutting, stamping, or marking tools that are too short to keep your fingers well away from the surface of the tool to be struck.

❖ use a bricklayer's hammer to strike anything metal.

❖ use a wooden mallet to strike anything metal.

❖ use any tool with a mushroomed striking face.

When using wrenches or other turning tools, DO:

❖ use wrenches only when you are in a braced position in case the wrench or fastener should break or slip.

❖ use box or socket wrenches wherever possible in place of open-end wrenches.

❖ use only single or double square design wrenches on square-headed fasteners.

❖ use only special heavy duty, sledge-type box wrenches for stubborn nuts and bolts.

❖ use a properly calibrated torque wrench where the torque on a fastener has been specified or where it is important that all fasteners be uniformly tight.

❖ pull, not push, on wrenches whenever possible to get them to turn.

❖ use adjustable wrenches with the open jaw facing you whenever possible.

❖ use only specially designed pipe wrenches for grasping round pipes.

❖ tighten the jaws of adjustable wrenches as far as possible around the nut or bolt.

When using wrenches or other turning tools, DO NOT:

❖ use an English wrench for a metric fastener or a metric wrench for an English fastener.

❖ strike the handle or head of a wrench with a hammer or other tool to get the wrench to turn.

❖ use an open end or adjustable wrench to free stubborn nuts and bolts.

❖ use a pipe wrench on nuts or bolts.

❖ use a pipe wrench on valves or soft metal fittings.

❖ use hand sockets on power or impact wrenches.

❖ overtighten nuts and bolts.

When using metal-cutting tools, DO:

❖ secure the free end of the metal sheet or wire being cut so that it will not fly and injure someone.

❖ apply the cutting force at right angles to the work piece.

❖ use protective shields around punches and chisels whenever possible to avoid striking your hand with the hammer.

❖ tighten hacksaw blades securely in the frame.

❖ use hacksaw blades with the proper number of teeth per inch for cutting metal of various hardnesses.

❖ use pressure only on the forward stroke (i.e., away from your body) when using hacksaws.

❖ make sure that nobody, including yourself, is standing where they might be hit by flying pieces when cutting metal strapping.

❖ cut metal straps squarely, not at an angle.

❖ clamp the piece to be filed securely in a vise at a comfortable working height and position.

❖ use your thumb and forefinger of one hand to guide the point of a file while gripping the handle securely with your other hand.

When using metal-cutting tools, DO NOT:

❖ twist the blade, apply too much pressure, or saw too fast with a hacksaw.

❖ continue an old, unfinished cut with a new hacksaw blade.

❖ use any kind of cutting tool near live electrical equipment or wires.

❖ bend wires back and forth to get them to break.

❖ use snips or bolt cutters to cut metal that is too heavy or thick.

❖ use shears or snips that are too tight or too loose.

❖ use a metal chisel with a mushroomed striking face or dull point.

❖ use a metal file that does not have a secure handle.

❖ use a file that has metal chips or other loose debris on it.

When using wood-cutting tools, DO:

❖ always apply force or pressure away from the body when using saws, drills, planes, etc.

❖ secure the free end of the wood if possible so that it will not injure someone when it falls off.

❖ use a cross cutting saw for cutting across the wood grain and a ripping saw for cutting with the grain.

❖ use axes or hatchets with sharp, narrow blades for chopping hard wood and sharp, wider blades for chopping soft wood.

When using wood-cutting tools, DO NOT:

❖ use any rusted saw or other tool.

❖ use a saw that is too dull to cut with moderate force.

❖ use an axe, hatchet, or maul with a loose or cracked handle.

❖ use a wood chisel with a mushroomed striking face or dull point.

❖ cut metal with saws designed for wood only.

❖ strike axes or hatchets against metal, stone, or concrete.

❖ use an axe or hatchet as a wedge or maul.

❖ swing an axe or hatchet through the air unless there is ample clearance around you to do so without hitting anyone or anything.

❖ hold small pieces of wood in your hands while trying to cut them.

When using material handling tools, DO:

❖ use only the proper size and type of crowbar for the job.

❖ use a block of wood under the heel of a crowbar or pry bar if necessary to prevent slipping.

❖ be sure that the fulcrum of a pry bar or crowbar is on solid footing before using them to lift heavy objects more than several inches off the ground.

❖ keep hand hooks sharp enough to securely grip boxes.

When using material handling tools, DO NOT:

❖ use any other tool not designed for prying or lifting in place of a crowbar or pry bar.

❖ heat a crowbar or pry bar in an attempt to bend or reshape it.

❖ use any long hook unless the handle and point are bent in the same place.

When using gardening tools, DO:

❖ use the ball of your foot, not the arch, to push shovel blades and post-hole diggers into clay or other thick material.

When using gardening tools, DO NOT:

❖ leave rakes, hoes, or shovels lying around with the prongs or blade facing upward.

❖ use a shovel or other gardening tool that has a splinter in the handle.

❖ use a shovel or post-hole digger with a dirty or dull blade.

❖ wear thin soled shoes when using shovels or post-hole diggers.

When using screwdrivers, DO:

❖ drill a pilot hole, especially in hard wood, before driving the screw.

❖ use a vise or hold the work piece securely on a flat surface when driving screws.

❖ use only an insulated screwdriver if you must work near live electrical equipment.

❖ use a screwdriver that has a handle that you can grip securely.

When using screwdrivers, DO NOT:

❖ use one that is much too large or too small for the screw.

❖ use them for prying, punching, chiseling, scoring, scraping, or wedging.

❖ use one with rounded or twisted edges or tips.

❖ use pliers or any other tool on the handle to get extra turning power unless the shank has been specially designed for that purpose.

❖ use one that has a bent or broken handle or shank.

❖ use an uninsulated screwdriver near live electrical wires.

❖ use a screwdriver for testing the continuity of circuits.

❖ hold work in the palm of your hand when tightening or loosening screws in it.

❖ use a flat head screwdriver on a Phillips head screw.

❖ use one that has anything slippery on the handle.

❖ put any part of your body in front of the tip when driving a screw.

When using pliers, DO:

❖ use them for gripping small articles only.

❖ use only insulated ones if you must work near live electrical equipment.

❖ wear insulating electrician's gloves when working around live electrical equipment.

❖ use a guard over the cutting edge of cutting pliers whenever short ends of wire are cut.

❖ cut only at right angles when using cutting pliers.

When using pliers, DO NOT:

❖ use them as a substitute for a wrench.

❖ use them for gripping the heads of nuts or bolts.

❖ use them for hammering or pounding on anything.

❖ use them for cutting hardened wire unless they are specially designed for that purpose.

❖ use them if the hinge does not operate properly.

When using knives and miscellaneous cutting tools, DO:

❖ make the cutting stroke away from your body whenever possible.

❖ use tools small enough to be handled easily.

❖ put them away carefully to protect the cutting edges.

❖ wear protective gloves and other protective clothing whenever there is a possibility of cutting yourself because of the nature of the task.

❖ use them only on firm, slip-resistant surfaces.

❖ clean the blade with a clean towel or cloth with the sharp edge pointed away from your hand.

❖ use specially designed hooked knives for cutting corrugated cardboard.

When using knives and miscellaneous cutting tools, DO NOT:

❖ use one with a dull or damaged blade.

❖ use one with a damaged handle.

❖ ever put any part of your body in front of the blade or pointed tip of a cutting tool when applying force to the tool.

❖ leave them lying around unprotected where someone may be cut.

❖ ever engage in horseplay when you or someone else are using knives.

❖ ever throw knives or scissors through the air.

❖ carry a sheathed knife on the front part of your belt ahead of your hips.

❖ use them in the vicinity of other people unless they are well aware of what you are doing.

❖ run with knives, scissors, or other sharp tools in your hand.

❖ wipe the blade of a dirty or oily knife on your clothing.

❖ use them to cut anything where excessive force is required.

❖ use them as a substitute for can openers, screwdrivers, or ice picks.

❖ use your thumb or any other body part in opposition to the direction of an intended cut (i.e., put your thumb on one side of the material to be cut while applying force from the opposite side with your forefinger).

When using electric power tools, DO:

❖ cut the power to the tool before changing accessories like blades, bits, etc.

❖ suspend cords well above head height when it is necessary to have them across aisles.

❖ use battery-powered tools whenever possible to minimize the electric shock hazard.

❖ use insulated platforms, rubber mats, rubber gloves, or similar materials when using tools in wet locations.

❖ use only low voltage tools in wet locations or inside metal tanks.

❖ use only three-wire ground protected, double insulated, or battery powered tools.

❖ use a ground fault circuit interrupter-protected power supply whenever possible.

❖ use tools equipped with "dead man" switches if possible.

❖ use the proper type of blade as recommended by the tool manufacturer for cutting special materials such as plywood, composite board, concrete, etc.

❖ use a cross cutting blade for cutting across the wood grain and a ripping blade for cutting with the grain.

❖ wear an approved dust respirator and/or hearing protection whenever excessive dust or noise may be generated.

When using electric power tools, DO NOT:

❖ hang cords over nails, bolts, or sharp edges.

❖ hang cords over doors or through doorways.

❖ lay cords across the floor unless they are protected by wooden strips or special raceways.

❖ use any tool that is sparking or appears to have an electrical short.

❖ energize the tool until just before you are ready to use it.

❖ use any tool unless you are standing or sitting in a secure position.

❖ use any tool when you are hanging over the edge of a ladder or scaffold or are in an otherwise precarious position.

❖ use any tool with the blade guard removed or rendered inoperable.

❖ use a saw with a blade over 2" in diameter unless it has blade guards above and below the base plate.

❖ use any tool with a damaged cord or exposed wiring.

❖ use temporary wiring or extension cords unless absolutely necessary.

❖ use any tool unless the blade or bit is securely tightened.

❖ use excessive force on saws or drills to cut through hard materials.

❖ use an electric grinding wheel, buffer, or wire brush that wobbles or vibrates excessively.

❖ use electric grinding or buffing wheels if the gap between the wheel and the work rest exceeds 1/8".

❖ put any part of your body near the moving parts of an electrical tool unless the power has been cut.

❖ use any electric tool in an area where flammable gases or vapors may be present unless the tool is intrinsically safe.

❖ generate large amounts of wood dust around potential ignition sources.

When using pneumatic power tools, DO:

❖ use tools with vibration dampers wherever possible.

❖ suspend air hoses over aisles and work areas when it is necessary to cross them.

❖ use armored hose where physical damage is likely.

❖ wear an approved dust respirator and/or hearing protection whenever excessive dust or noise may be generated.

When using pneumatic power tools, DO NOT:

❖ lay the air hose across aisles or walkways where someone could trip over it.
❖ use the air hose for cleaning unless it has a pressure reduction device and you have effective chip protection in place.
❖ disconnect the air hose or tool until the compressor has been shut off.
❖ kink the air hose or subject it to physical damage.
❖ use an air line if you can hear air escaping through a leak.
❖ engage in horseplay using the compressed air hose.
❖ squeeze the trigger on air hammers, impact wrenches, etc. until the tool is in contact with the work.

When using gasoline power tools, DO:

❖ store gasoline and other flammable fuels only in approved, self-closing containers.
❖ disengage the spark plug wire or otherwise disable the motor before reaching inside the blade housing of a lawn mower to clear a jam, change the blade, etc.

When using gasoline power tools, DO NOT:

❖ use them if the guards have been damaged or altered.
❖ handle gasoline or other flammable fuels around potential ignition sources.
❖ use them without a muffler in place.
❖ engage in horseplay with chain saws or other gasoline power tools.

When using hydraulic power tools, DO:

❖ become familiar with their operation before attempting to use any of these tools.
❖ suspend hydraulic lines above head height over aisles and work areas when it is necessary to cross them.
❖ use armored hose where physical damage is likely.
❖ use hose designed to withstand the pressure you will be using.

When using hydraulic power tools, DO NOT:

❖ lay hydraulic lines across aisles or walkways where someone could trip over them.
❖ kink hydraulic lines or subject them to physical damage.
❖ put your hand over a pinhole leak in a hydraulic line in an attempt to stop it.

When using powder-actuated tools, DO:

❖ become familiar with their operation before attempting to use any of these tools.
❖ keep the immediate area clear of unprotected people.
❖ activate the charge only when the tool is in contact with the work piece.
❖ place the tool flat on the work piece, not at an angle.

When using powder-actuated tools, DO NOT:

❖ engage in horseplay with any of these tools.
❖ point them at anyone including yourself.
❖ ever look directly into the tool when a charge is in place.

Of course, the best sources of information on the safe use of hand and portable power tools are the manufacturer's instruction manuals and factory-trained instructors if available. The above lists of safe and unsafe work practices are not intended to be all-inclusive. Safe use of any tool will always include good common sense and judgement on the part of the operator.

TRANSPORTING AND STORING HAND AND PORTABLE POWER TOOLS SAFELY

The best place to store tools in most small industrial plants is a centrally located tool storage room where your employees can check tools out and in as they need them. If you have a tool room attendant, this affords a good opportunity for routine inspection, maintenance, and repair by a trained employee as well as inventory control. Such an arrangement also permits the concurrent issuance of the proper protective gear such as safety glasses, gloves, respirators, and hearing protection.

When employees are allowed to use their own tools and tool boxes, they may tend to forget about routine tool inspections. Also individual tool boxes may be misused as something to stand on, as benches or saw horses, for storing lunches, and all sorts of other misuses.

OSHA regulations require that you as an employer be responsible for seeing that the tools your employees use are safe. If any of your employees insist on using their own tools or if you require them to furnish tools, you are still responsible for an inspection program.

Transporting tools from the storage area to where they will be used can sometimes be a problem, especially if they are heavy or bulky or if multiple tools are required. Tool carts and specially designed tool chests can be used in these instances.

The following are a few general rules about the proper storage of hand and portable power tools:

❖ Always protect tools from physical damage to susceptible parts.
❖ Store sharp tools in a specially designed cabinet or cupboard or with a blade guard in place.
❖ Smaller tools should be organized and arranged in drawers or trays so they can be found easily when needed.
❖ Larger tools should be stored in their original containers if possible or on sturdy shelves with raised lips to prevent them from accidentally falling or being bumped off the shelving.
❖ Cords should be neatly wound and secured so they are out of the way.
❖ Very large tools such as lawn mowers should be stored in a separate room or outside shed.
❖ Drain gasoline or other flammable fuels from tools if they are to be stored for extended periods of time.

❖ Metal lawn and garden tools may be wiped with a light coating of oil after they have been cleaned to inhibit rust formation.

❖ Do not leave tools lying on the floor in aisles or walkways, on ladders, scaffolds, or on elevated platforms or cranes. It is especially important to store tools with sharp edges or points upright with the edges pointed into the storage area or wall.

The following are a few general rules about the proper transportation of hand and portable power tools:

❖ Never carry power tools by their electric cord, air line, or hydraulic hose.

❖ Never carry tools in your hand when ascending or descending ladders or very steep stairs. Use rope hand lines to raise or lower tools in this case.

❖ Never carry sharp or pointed tools such as knives, scissors, screwdrivers, and chisels with the edge or point upward or toward your body. These types of tools should be carried in a tool box or in a special carrying belt whenever possible.

❖ Give sharp or pointed tools to another employee with the handle toward the receiver. Never throw any tools at or toward another person.

❖ Never carry a tool in such a way that it obstructs your vision ahead, to the sides, or to the floor.

In conclusion, hand and power tools can increase productivity and make our jobs easier, provided that we are properly trained in their safe use.

CHAPTER
18
HAZARDOUS MATERIALS

In order to assure a safe working environment for the use of hazardous materials, you should:

❖ Identify all the hazardous materials present.

❖ Evaluate the conditions under which they are being used.

❖ Assess the hazards represented by each of these materials.

❖ Take necessary steps to protect workers and control the hazards.

❖ Train your employees in the proper identification and use of hazardous materials.

This chapter describes these steps in five major sections: types of hazardous materials, principles of hazard control, control of chemical hazards, control of physical hazards, and hazard communication.

TYPES OF HAZARDOUS MATERIALS

A hazardous material is a substance which has the *capability* to cause harm or damage. This definition implies that even though a material has some hazardous property, the hazard can be controlled. Each organization and major government agency seems to have its own system for classifying hazardous materials, i.e., OSHA, EPA, DOT, NFPA, etc. The OSHA/EPA hazard classifications are used in this chapter.

A hazardous material may be present in any one of several forms, i.e., dust, mist, fume, gas, vapor, liquid, or solid. In addition, a material may present either a physical or chemical threat, or both (see Table 18-1). These two major hazard classifications are as follows:

❖ Chemical—A hazard that is present due to a material's chemical nature. The chemical characteristics and what happens when the substance comes into contact with the human body are important.

❖ Physical—These are threats presented by properties of a material such as flammability or physical state (i.e. gas under high pressure). These are usually divided into fire hazards and all others.

Table 18-1
Chemical Physical Hazard Categories

CHEMICAL HAZARD		PHYSICAL HAZARD	
ACUTE:	Toxic Highly Toxic Corrosive Irritant Sensitizer Skin Hazard Eye Hazard	FIRE HAZARD:	Flammable Combustible Pyrophoric Oxidizer
		SUDDEN RELEASE OF PRESSURE HAZARD:	Explosive Compressed Gas
CHRONIC:	Carcinogen Liver Toxin Kidney Toxin Nervous System Blood Toxin Lung Hazard Reproductive Hazard	REACTIVE HAZARD:	Organic Peroxide Unstable Reactive Water Reactive

It is important to remember that a material can fall into more than one category and present more than one type of hazard. For example, benzene is flammable, can cause defatting of the skin (a cutaneous hazard), is fatal in 5 to 10 minutes at concentrations of 20,000 ppm (2%) and is a carcinogen. It is therefore a flammable liquid, a skin hazard, toxic, and a carcinogen, presenting both physical and chemical hazards. Benzene may require a combination of hazard control measures, depending on how it is being used. Each hazardous aspect of each material must be examined. One control measure may be effective for several hazards; however, it is unsafe to assume that this is the case.

CHEMICAL HAZARDS

Dose

Most toxicologists and other scientists who study the effects of toxic materials agree that there is a threshold dose below which most materials are not harmful. This concept of a toxic threshold is critical to the control of harmful exposures. All materials can be harmful if taken in large enough doses or by the wrong route of entry into the body. For example, table salt (sodium chloride) is routinely ingested with food, however, a very small amount can be a fatal dose if injected into the bloodstream. The key concept is to minimize exposures to potentially harmful chemicals to ensure that the toxic threshold is never reached.

Acute Health Hazards

The following classes of hazards tend to have acute effects, that is, effects are seen after a single exposure.

Toxic and highly toxic. These classes of substances can bring about death at high exposure levels. Adequate human toxicity data or the animal toxicity study dose ranges shown in Table 18-2 are used to classify materials as either "toxic" or "highly toxic." As implied by the class designation, highly toxic materials present a greater hazard than those in the toxic classification. Toxics can come in any form (i.e., solid, liquid, gas). Acrylonitrile, ammonia, aniline, boron trifluoride, epichlorohydrin, and ethylene oxide, for example, are all considered toxic substances under one or more of the criteria. Aldrin (a pesticide), benzene, ethyleneimine, hydrogen cyanide, and organomercurials are considered highly toxic as another example.

Table 18-2 illustrates the cutoff dosages that will bring about death in 50 percent of the test population which is described by the acronym LD_{50}. Dosages are given in milligrams of chemical per kilogram of animal body weight. Chemicals are administered in oral dosages and by inhalation to albino rats and by continuous skin contact to albino rabbits. These data are then used to classify chemicals as "toxic" or "highly toxic."

Table 18-2
Highly Toxic/Toxic Classifications

	DOSAGE ROUTE AND RANGE		
	ORAL	SKIN	INHALATION
CLASSIFICATION	LD50	LD50-24 hr.	LC50-1 hr.
HIGHLY TOXIC	≤ 50 mg/Kg	≤ 200 mg/Kg	≤ 200 ppm ≤ 2 mg/liter
TOXIC	50-500 mg/Kg	200-1000 mg/Kg	200-2000 ppm 2-20 mg/liter

Corrosive. The corrosive class of chemicals causes visible destruction and irreversible alterations (chemical burns) to living tissue where exposure occurs. Corrosives, though usually liquids, can also be found as gases, vapors, and mists. Some examples are glacial acetic acid, hydrochloric acid, caustic soda (sodium hydroxide), hydrofluoric acid, phenol, and sulfuric acid.

Irritant. Chemicals classified as irritants cause reversible inflammation where contact occurs. Standardized skin tests and eye tests using albino rabbits are used to identify irritants.

Irritants can be found in any physical form (i.e., solid, liquid, gas), although irritant solids are least commonly seen. Potential targets include the skin, eyes, mucous membranes, and respiratory tract. Examples of irritants include ammonia, ethanol, nitric oxide, sodium hypochlorite (bleach), and stannic chloride.

Sensitizer. These chemicals cause most people to develop allergic reactions after single or repeated exposures. Even small quantities below levels that are safe for others and, occasionally, below detectable limits may exert an effect once sensitization has occurred. Sensitizers may exist in all physical forms. Examples include bromine, formaldehyde, isocyanates, and ozone.

Skin or cutaneous hazard. These materials affect the skin, causing "dermatitis" or drying, rashes, and irritation. Chemical agents are the primary cause of dermatitis within the manufacturing industry. Some examples of skin hazards are acetone, chlorinated compounds, cryogenic compounds, cutting oils, fiberglass, and methyl ethyl ketone (MEK).

Occupational Skin Diseases

Primary irritation dermatitis. Primary irritation dermatitis is caused by mechanical agents like friction, physical agents such as heat or cold, and from chemical agents such as acids and bases.

Sensitization dermatitis. This results from allergic reactions. Sensitivity is established during an induction period lasting from a few days to a few months, after which even minute quantities of the sensitizer will produce a response. Some substances such as organic solvents, chromic acid and epoxy resins can cause both irritation and sensitization dermatitis.

Eye hazard. This class of material affects the eye. Almost any substance introduced into the eye will cause some irritation. Some substances, such as alkalis, are very difficult to flush out without causing severe damage. Eye irritants include acids, alkalis, and organic solvents.

Chronic health hazards. Tend to require long periods of exposure before their effects are seen. The following classes illustrate some of the typical symptoms and effects of chronic hazards.

Carcinogen. Carcinogens cause cancer in humans or animals. OSHA has developed a list of specific carcinogens which is contained in its regulations and are discussed later in this chapter. An OSHA carcinogen may also be a chemical listed as a carcinogen or potential carcinogen by either the National Toxicology Program (NTP) in its *Annual Report on Carcinogens* or by the International Agency for Research on Cancer (IARC) in its *Monographs on the Evaluation of the Carcinogenic Risk of Chemicals to Humans.* You must address risks associated with chemicals on both the OSHA carcinogen list and those listed in the aforementioned references. Carcinogens include acrylonitrile, asbestos, benzene, beryllium, formaldehyde, and carbon tetrachloride.

Liver toxin (hepatotoxin). This is a material which is toxic to the liver and which will cause damage such as liver enlargement or jaundice. Examples of hepatotoxins include carbon tetrachloride, ethanol, chloroform, nitrosamines, trichloroethylene, and vinyl chloride.

Kidney toxin (nephrotoxin). These substances cause kidney damage. Ethanol, halogenated hydrocarbons, and trichloroethylene are examples.

Nervous system toxin (neurotoxin). The central nervous system is affected by these substances. Changes can include narcosis, behavioral alterations, decreased motor function, and death. These effects can be produced by substances such as carbon disulfide, ethanol, mercury, and tetraethyl lead.

Blood (hematopoietic toxin). These chemicals affect the blood or blood-forming tissues. Effects can vary from anemia to decreased ability to transport oxygen. Examples include carbon monoxide, cyanides, metal carbonyls, nitrobenzene, hydroquinone, aniline, and arsine.

Lung hazard. These materials affect lung tissue causing coughing, tightness of the chest, and shortness of breath, as well as more serious effects such as permanent reductions in pulmonary capacity through scarring and other effects. Examples include asbestos, beryllium, coal dust, cotton dust, and silica.

Reproductive toxin. These substances can cause birth defects or sterility. Examples include lead, PBB's, PCB's, selenium, and vinyl chloride.

PHYSICAL HAZARDS

Fire Hazard

The two major classifications of fire hazards are combustibles and flammables. The major difference between combustible and flammable is the temperature at which ignition occurs. Flammables ignite at temperatures less than 100° F while combustibles ignite at temperatures between 100° F and 200° F. This means that you must take greater care when handling flammables than combustibles.

A combustible is a fire hazard at a slightly elevated temperature, but not at room temperature. Common examples include #1 fuel oil, mineral spirits, and methyl cellosolve.

A flammable material can be an aerosol, gas, liquid, or a solid. The definition varies for each type; however, in general, a flammable presents a fire hazard at or below normal room temperatures.

Flammable aerosol. An aerosol that, in a Consumer Product Safety Test, will project a flame more than 18 inches or has a flame extending back to the valve. The flammability usually depends on the propellant. Isobutane, for example, is a common highly flammable propellant.

Flammable gas. These gases form a flammable mixture with air at a concentration of 13% or less or have a range of flammable concentrations with air wider than 12%, irregardless of the lower limit range. Examples may include acetylene, butane, and propane.

Flammable liquid. A liquid with a flash point of less than 100° F. Some examples include acetone, gasoline, ethanol, methanol, and many organic solvents.

Flammable solid. A flammable solid is "likely to cause a fire" and can be ignited readily and will then burn vigorously and persistently. This category excludes explosives and blasting agents. OSHA specifies a Consumer Product Safety Test for this determination. Magnesium metal and nitrocellulose film are examples.

Oxidizer. Oxidizing agents may possess corrosive properties, can initiate or promote combustion, and can release oxygen and other gases when they come in contact with materials such as organic substances. This gas production represents an explosion hazard. Examples include chlorine, fluorine, oxygen, nitric acid, and hydrogen peroxide.

Pyrophoric. This class includes substances which can ignite spontaneously in air at temperatures below 130° F (54° C). Examples include common linseed oil, white phosphorous, and some catalysts.

Sudden Release of Pressure Hazard

Explosive. A chemical of this type can cause a sudden release of pressure, gas, and heat (explodes or detonates), when subjected to shock, pressure, or temperature change. These include TNT, nitroglycerine, and some peroxides.

Compressed gas. Compressed gases present both potential physical and health hazards. Compressed gas in a cylinder represents a potential rocket or fragmentation bomb due to the tank pressure and small valve opening. This classification does not consider the flammability or toxicity of the gas itself. Common compressed gases are acetylene, air, carbon dioxide, nitrogen, and oxygen.

Reactive Hazards

Organic peroxides. Peroxides contain an unstable -O-O- (oxygen) group in their chemical structure. They are extremely sensitive to shock, sparks, and other types of ignition, and some are extremely flammable. Peroxides are uncommon in manufacturing. However, several types of compounds are known to form them inadvertently, such as aldehydes, ethers, vinyl and vinylidene compounds, most alkenes, and compounds containing benzylic hydrogen atoms; more specifically, cyclohexene, cyclooctene, decalin, *p*-dioxane, diethyl ether, diisopropyl ether, tetrahydrofuran, and tetralin. These chemicals should be monitored for peroxide formation as they present an unexpected hazard. Consult a chemist or some other knowledgeable person if you are using or plan to use these substances.

Unstable/reactive. These materials will vigorously polymerize, decompose, condense, or become self-reactive under conditions of shock, pressure, or high temperature. Acrylonitrile, benzoyl peroxide, and butadiene behave in this manner.

Water-reactive materials. These chemicals will react with water to produce a gas that is flammable or toxic. Some examples include acetic anhydride, calcium carbide, metal hydrides, and sodium and potassium metals.

PRINCIPLES OF HAZARD CONTROL

Traditionally, there are three approaches to the control of hazardous materials: engineering, administrative, and the use of personal protective equipment. These are discussed in the preferred order of implementation.

Engineering Controls

There are various approaches to engineering control, focusing on controlling (1) the source of the contamination, (2) the work environment, or (3) contact with the worker. You can implement controls "after the fact" or better still, before the problems arise, as in the first engineering control option discussed below.

Design stage engineering. As a manager, emphasize the need for integrating safety into an operation from the very start. Opportunities may arise when constructing a new plant, designing a new process, or adding a new piece of equipment. This is usually more economical than retrofitting control measures later on. Don't overlook potential hazards associated with maintenance activities.

Substitution. Substitution involves the replacement of a hazardous piece of equipment, material, or process with a less hazardous one. Substitution of equipment is the most common, easiest, and usually least expensive option. Application of this technique is based on common sense, ingenuity, knowledge of the technology available, and experience with the particular manufacturing process.

Material substitution is the second most common approach. Care must be exercised to avoid inadvertent substitution of one type of hazard for another type. A classic example is the substitution of carbon tetrachloride for petroleum naphtha as a cold-cleaning solvent. Many years ago, petroleum naphtha was the solvent of choice; however, it presented a serious fire hazard. Carbon tetrachloride gained acceptance as a substitute because of its low flammability, low cost, and effectiveness. Problems arose when it was found that carbon tetrachloride causes serious health effects. Today, various other solvents, low in both flammability and toxicity, are used. The major problem with material substitution is finding an appropriate material that meets all requirements. Suppliers are often able to recommend closely-related, satisfactory substitutes. A second option is compounds which contain reduced percentages of the hazardous substance. A third option is a change of physical form such as one that is dust-free, i.e., a liquid, pellet, or granule. Concentrated solutions are often safer and more economical. You should investigate and evaluate the performance of these substitutes.

Determining a substitute process may or may not be a difficult task, depending on the circumstances. As a rule, the closer a process approaches a closed system, the less hazardous it is likely to be.

Isolation. Simply stated, isolation means putting a barrier between the hazardous equipment or processes and those who may be affected by it. Isolation may involve time, distance and/or actual physical shielding.

Equipment can be isolated with a physical barrier; however, you must consider the increased hazard risk should the need for human intervention arise. An example of isolation is enclosing a process by a glove box. This setup may also include other controls such as local exhaust ventilation. Process isolation is probably the most expensive and therefore the least popular method of hazard control. Remote control operations, master-slave manipulators, surveillance systems, and computer control processing are all possible solutions. For example, distance barriers are used to reduce the severity of explosions by separating large quantities of explosive materials. Isolation of the employee from the hazard entails the placement of the workers in an enclosure. This may or may not be practical depending on the frequency with which the workers must enter the contaminated process area.

Ventilation is an excellent and effective method for control of airborne contaminants, however, it can be expensive. Depending on the type of ventilation, it can remove hazardous materials at the source before the worker has a chance to inhale them. It may also be useful in controlling dermal contact, i.e., dermatitis is caused by a mist. There are two types of ventilation: local exhaust ventilation and general ventilation.

The classic method of controlling airborne hazards is local exhaust ventilation, where hazardous material is captured at its source of generation through the use of hoods and adjunct equipment. Ventilation must be properly designed to meet the specific needs of the process or operation. You must periodically monitor and maintain the ventilation system once it is installed. Two advantages of local exhaust ventilation are that it requires less airflow than dilution systems, thereby saving plant heating and cooling costs, and it allows for reclamation of reusable materials.

General ventilation, or dilution ventilation, adds or removes air from the work area to control airborne levels of hazardous materials. The ventilation may be as simple as open doors and windows or may involve roof ventilators, chimneys, fans, blowers, and exhaust fans in walls, windows, and roofs. This form of ventilation is only useful in the following situations:

❖ Small amounts of material are introduced into the air at a constant rate.

❖ The distance between the workers and the source of contamination is great enough to allow sufficient dilution.

❖ The material is not very toxic.

❖ The air being exhausted outside does not have to be filtered.

❖ The material is noncorrosive.

Considerations include the amount of makeup or replacement air, the location of the intake and exhaust stacks or vents, and the incidence of recirculating exhaust air and subsequent build up of contaminant levels.

Administrative Controls

Housekeeping. These activities include removal of dust accumulations and spill cleanup. This is of particular importance for flammable materials. OSHA regulations concerning sanitation can be found in 29 CFR 1910.141. Cleaning methods should not disperse material into the air. You should periodically shut down equipment for maintenance as part of your housekeeping program. Decontaminate, disassemble, and repair or conduct maintenance. Use proper personal protective equipment (PPE) and decontaminate all equipment and tools before removal from the work area.

Materials handling or transfer procedures. Transfer operations should be closed-system or should have exhaust systems to prevent worker exposure or contamination of the workplace air. Use containers to collect overfill spills or leaking materials between transfer points.

Leak detection programs. Programs that involve visual inspection and automatic sensor devices will allow quick repair and minimal exposure. Most sensor devices are specific for one or more materials and will trigger an audio or visual alarm. Ideally, the alarm should automatically begin corrective action or shut down the process.

Training. Training your workers to understand the hazardous materials encountered in their workplace and how to protect themselves is required by OSHA and state right-to-know laws under hazard communication standards. It can also be a very effective method for reducing the risk of injury or illness.

Modifying the work schedule. Reduce exposures by limiting the duration of exposure. Accomplish this by limiting the amount of time a worker is exposed so that OSHA Permissible Exposure Limits (OSHA PEL's) are not exceeded. Other workers may be rotated in to continue the activity. This strategy is, however, unpopular with most industrial hygiene and safety professionals, as it merely spreads the exposure and does nothing to control the source. It is nevertheless, commonly used for exposures such as lead and noise.

Personal hygiene. Personal hygiene entails keeping exposed areas of skin and clothing clean of contamination, whether resulting from normal work or accidental circumstances. Make appropriate cleaning agents and facilities such as regular showers, emergency eyewashes, emergency showers, and changing rooms available and conveniently located. A worker who is splashed with a chemical probably won't bother to wash it off if the proper wash basin is in an inconvenient location. Instead he or she may use undesirable means of cleaning such as the use of solvents. Some cleaning materials can contribute to dermatitis by drying the skin so it may be necessary to provide moisturizing agents.

Medical surveillance. Medical surveillance is useful for employee evaluation purposes and verifying adequacy of control measures. Use a physician who is familiar with the workplace, job tasks, and hazards present. The physician will then be able to give you informed recommendations regarding employees. Medical surveillance may be required under certain circumstances such as when respirators or certain hazardous substances are used (i.e., lead, radioactive materials). Medical examinations usually fall into one of three categories:

1. Preplacement—

   Initial employee medical examinations which allow the doctor to identify and recommend restrictions for individuals and supply baseline medical data. This data, when combined with information obtained from periodic exams, will allow the tracking of any deleterious effects of employment such as back problems, etc.

2. Periodic—

   Periodic examinations can range from full physicals to simple biological monitoring techniques (i.e., blood or urine analysis). This type of examination is useful as a supplement to environmental and industrial hygiene monitoring. Dose-response relationships can be established by combining this data. This information is useful as a risk indicator for given jobs. It is here that ineffective control measures can be identified by groups of workers exhibiting clinical or biological evidence of exposure. Hypersusceptible individuals may also be identified.

3. Exit—

   Medical examinations performed when employees leave your employment will give an indication of their health status at that time and be useful in the tracking of chronic disease development once exposure stops. This type of information has proven useful in litigation for compensation cases for diseases with long latency periods such as asbestosis.

Personal Protective Equipment (PPE)—A detailed discussion on PPE can be found in Chapter 12. PPE should only be used when other methods fail to reduce or eliminate the hazard or while proper engineering controls are being installed. Briefly, the types commonly used to control materials-related hazards include:

❖ Respiratory: When engineering controls are not feasible or are being instituted, appropriate respirators should be used to control exposures to airborne hazardous materials. As an employer, you are required to provide such equipment whenever it is necessary. Use of such equipment automatically requires you to implement a Respiratory Protection Program.

❖ Protective Clothing (Chemical/Thermal/Electrical): PPE of this type generally includes gloves, aprons, coveralls, suits, etc. There are many materials and types available depending on need.

❖ Head, Eye, Hand, Foot Protection: This type of PPE is required in any situation where there is a reasonable probability of injury. These items are worn for protection from physical injury and include hard hats, safety glasses, goggles, leather gloves, and steel-toed shoes.

CONTROLLING CHEMICAL HAZARDS

For each potentially hazardous material in your plant it is necessary to answer the following general questions:

1. How can the material contact the worker? What are the routes of exposure? Does the work process provide an opportunity for exposure?

2. What effects will it have on the worker?

3. What are the best control methods for the situation?

OSHA or other regulatory agencies may require you to implement specific control measures for certain substances, such as particularly hazardous substances or carcinogens which are regulated.

Routes of Exposure

A chemical can enter the human body either by ingestion, injection, inhalation, or absorption. The first is controlled by prohibiting eating, drinking, smoking, application of cosmetics, etc., in the work area and by encouraging good personal hygiene habits. Exposure by injection is unlikely in most manufacturing settings. Protective gloves, for example, are the best defense for a material that may enter the body from a puncture. Inhalation and absorption are the two most common routes in the manufacturing setting and are discussed below.

Inhalation. Inhalation is the most common route of entry into the body in the industrial environment. It is particularly important because of the rapidity with which a toxic material can enter the bloodstream and cause an adverse effect. Use process/worker isolation, local exhaust ventilation, housekeeping, and as a last or interim resort, respirators to control inhalation hazards. The choice of the most appropriate method will depend upon the exact compound and how it is used.

Absorption. Absorption through the skin can occur through cuts, abrasions and in some cases, even intact skin. The risk of absorbtion can be increased by skin disease. Intact skin affords a reasonably good deterrent for the entry of most chemicals into the body; however, there are a few substances that can be absorbed readily through the skin. Systemic damage can be caused by several organic compounds such as TNT, cyanides, aromatic amines, amides, and phenols.

Occupational dermatitis is prevented through cleanliness of the work environment and the worker. Process "cleanliness" is accomplished with control measures such as isolation, substitution with materials of low toxicity and irritant potential, and ventilation. Enhancing the cleanliness of your workplace through an efficient housekeeping program will also reduce the chance for occupational dermatitis.

Controlling Particularly Toxic Compounds

The use of carcinogens, radioactive materials, and highly toxic substances, often requires special control measures. OSHA has specific standards for approximately twenty-five compounds in addition to its list of air contaminants. These standards can be found in 29 CFR 1910.1000-1101. These specifically regulated substances are:

2-Acetylaminofluorene	1,2-Dibromo-3-beta-Naphthylamine
Acrylonitrile	3,3'-Dichlorobenzidine and its salts
4-Aminodiphenyl	4-Dimethylaminoazobenzene
Arsenic, inorganic	Ethylene oxide
Asbestos, tremolite	Ethyleneimine
anthophylite, and actinolite	Formaldehyde
Benzidine	Lead
Benzine	Methyl chloromethyl ether

bis-Chloromethyl ether	alpha-Naphthylamine
Chloropropane	4-Nitrobiphenyl
Coal tar pitch volatiles	N-Nitrosodimethylamine
Coke oven emissions	beta-Propiolactone
Cotton dust	Vinyl chloride.

OSHA regulations for these substances include requirements for exposure monitoring, personal protective equipment, medical surveillance programs, recordkeeping, regulated areas, warning sign posting, and hygiene facilities and practices. In general, if you are using one or more of these substances you may be required to do the following:

❖ Comply with all Permissible Exposure Limits (PELs).

❖ Conduct initial exposure monitoring and additional monitoring as needed to confirm compliance with PELs.

❖ Establish regulated areas that limit access to authorized personnel wherever airborne contaminant concentrations may exceed the PEL. Personnel will be required to wear appropriate PPE. No eating, drinking, smoking, or cosmetic application is allowed in these areas.

❖ Use engineering controls and work practices to eliminate or reduce contaminant levels to as low a level as possible.

❖ Use specific work practices outlined in the standards.

❖ Provide protective work clothing and ensure that contaminated clothing is removed or stored in a manner that does not contaminate "clean" areas. This usually requires the establishment of a change room with separate lockers for street and work clothing. You are also responsible for cleaning and maintaining work clothing.

❖ Provide shower facilities and separate eating facilities.

❖ Communicate hazards to the employees using labelling, MSDS's, and training.

❖ Post signs in regulated area identifying the hazard.

❖ Conduct medical surveillance as discussed in the Principles of Hazard Control section earlier in this chapter.

❖ Maintain accurate records of exposure measurements, medical data, and training. These records must be available to affected employees, former employees, and the Assistant Secretary of OSHA and his designated representatives.

Specific requirements vary and this list is by no means complete. Consult the OSHA standards for further information.

CONTROLLING PHYSICAL HAZARDS

Flammable, Combustible, and Explosive Hazards

Explosions can be caused by several types of materials: flammables, combustibles, explosives, highly reactive and unstable chemicals such as organic peroxides, and incompatible mixtures of chemicals. Control of these hazards is discussed in detail in Chapter 19, Fire Safety.

The basic theory of control is simple. For a fire or explosion to occur three ingredients are required: fuel, air, and a heat source. Remove one of these ingredients and you reduce the risk of fire and explosion. Removal or control of fuel and air are the least practical strategies in the majority of circumstances, but, there are steps you can take. You can limit quantities of these materials, ensure that they are stored properly, use distance barriers to separate quantities and reduce explosion severity, substitute materials, and ensure that good housekeeping is practiced. You can prevent the build up of flammable atmospheres with ventilation. Control of heat sources usually offers the best and simplest chance of success. Fire or explosion will not occur by the reaction of air and fuel alone, except in rare cases of self-ignition. Potential electrical, chemical, and mechanical sources of ignition must be controlled. You should ensure that flammables are never heated with an open flame. Control more than one parameter whenever possible.

Compressed Gas Cylinders

Compressed gas cylinders present two types of hazards: those due to the stored potential energy from the compression and those due to the cylinder's contents. It is therefore necessary to use proper handling techniques whenever working with gas cylinders.

Handling Cylinders

❖ Only accept cylinders approved for use in interstate commerce for transportation of compressed gases.

❖ Do not remove or change numbers or markings.

❖ Roll cylinders on bottom edge. NEVER DRAG THEM. Transport cylinders weighing more than 40 pounds with a hand or motorized truck.

❖ Protect cylinders from cuts and abrasions.

❖ Do not lift cylinders with an electromagnet. When handling by crane or derrick, carry them in a cradle or suitable platform and take extreme care not to drop them. Do not use slings.

❖ Do not drop cylinders or allow them to strike each other violently.

❖ Do not use cylinders as supports, rollers, or for any purpose other than that for which they are intended.

❖ Do not tamper with safety devices; when in doubt contact the supplier.

❖ Label cylinders as EMPTY when empty. Close valves and replace valve protection caps.

❖ Load cylinders to be transported in a way that minimizes movement.

Storing Cylinders

❖ Store in a safe, dry, well-ventilated place.

❖ Do not store flammable materials in the same area as compressed gases.

❖ Do not store near elevators, gangways, stairwells or anywhere else they may be damaged or tipped.

❖ Do not store oxygen cylinders within 20 feet of cylinders containing flammable gases or highly combustible materials unless they are separated by a fire partition with a $\frac{1}{2}$ hour minimum rating. Welding rigs are an exception.

❖ Store acetylene and liquified fuel gas valve end up in well ventilated areas with no other occupancy. Limit total capacity to 2000 cu. ft. of gas exclusive of that in use.

Store larger quantities under NFPA Standard 51 guidelines, and in a separate building or outdoors.

❖ Store cylinders on a level fireproof floor.

❖ Protect cylinders stored outside from the elements.

❖ Cylinders are intended for use at temperatures less than 130° F. Do not store near heat sources.

❖ Store so that stock is used on a first-in, first-out basis.

❖ Never allow a direct flame or electric arc to contact any part of the cylinder.

❖ Storage areas for flammable gases should be ventilated with no ignition sources. Contain wire in conduit. Lights should be in a fixed position, encased in glass or plastic, and equipped with guards. Locate light switches outside the room.

Using Cylinders

❖ Use cylinders in an upright position and tie them down to prevent tipping.

❖ Leave valve protection caps in place unless the valve is protected by a recessed head.

❖ Make sure regulator and cylinder valve threads match. Never force a connection.

❖ Open valves slowly. Those valves without a handwheel should be opened with the PROPER tool.

❖ Never use a cylinder without a pressure regulator or manifold.

❖ Before making a connection to a valve, "crack" the valve to remove any debris.

❖ Do not repair hardware. Return to the manufacturer for this purpose.

❖ Take leaking cylinders out of service immediately. Close the valve, tag the cylinder, and notify the supplier. Remove fuel gas cylinders to the out-of-doors and allow the gas to leak out.

❖ Never use oil or grease on an oxygen cylinder.

❖ Never substitute oxygen for compressed air.

❖ Never bring cylinders into unventilated rooms or confined spaces.

❖ Do not fill cylinders without the consent of the manufacturer.

❖ Before removing a regulator, close the cylinder valve, and release the gas from the regulator.

Reactive Hazards

Organic peroxide hazards. Use the same precautions given for controlling fire and explosion hazards plus the following:

❖ Prevent contamination of supply, i.e., don't return unused peroxides to the container.

❖ Dilute with inert solvents to reduce sensitivity to shock and heat.

❖ Watch for situations where peroxides are mixed with a volatile solvent - the solvent can evaporate and increase the concentration.

❖ Avoid using metal tools.

❖ Avoid all operations involving friction, grinding, and impact.

❖ Store at lowest possible temperature consistent with their solubility or freezing point. Do not store at temperatures at or below which the peroxide freezes or precipitates because this increases the sensitivity to shock and heat.

❖ Disposal must be evaluated on a case by case basis. Generally, stabilizing chemicals must be added and no other wastes may be mixed with the peroxides.

Unstable reactives. Handling of these chemicals must examined on an individual basis. In general, anticipate and take all necessary precautions in storage, use, and disposal of these substances to control contact with incompatible substance.

Water-sensitive chemicals. Use and store water-sensitive materials in a manner that prevents any accidental contact with water, i.e., areas where large quantities of these materials are present should not have automatic sprinkler systems.

HAZARD COMMUNICATION TO EMPLOYEES

The Hazard Communication Standard (HazCom Standard) is the regulation most important to the communication of hazards to employees. The standard is written in a performance manner, leaving specifics on Material Safety Data Sheets (MSDSs), labels, training, and related programs to the employer to develop. There may also be state level programs, however, the essence of these requirements rarely differs significantly from the federal standard. With respect to communicating hazards to your personnel you must do the following:

1. Make MSDSs available to all employees, designated representatives, and OSHA.

2. Instruct employees to label all portable containers containing hazardous substances not intended for their immediate use.

3. Train your employees and contractors about the main features of the Standard: labels, MSDSs, the written program, list of hazardous materials, and required training. Also tell them how to recognize, understand, and protect themselves from hazards they will encounter in their workplace.

How to Use Material Safety Data Sheets (MSDSs)

You, as an employer, are responsible for obtaining or developing an MSDS for every hazardous chemical used in your plant. One method of accomplishing this is to put a statement on all purchase orders or requisitions that MSDSs meeting the requirements of OSHA's HazCom Standard 29 CFR 1910.1200 must be supplied with shipments of <u>any</u> hazardous material. It is your responsibility to confirm that the supplied MSDSs are adequate. If they are not, the supplier must be informed, and any insufficient or deficient information corrected. The following minimum information is required on an MSDS, which must be written in English.

1. Specific identity of each hazardous chemical or mixture ingredient and common names.

2. Physical and chemical characteristics of the hazardous material such as:

    a. Density or specific gravity of liquid or solid

    b. Density of gas or vapor relative to air

    c. Boiling point

    d. Melting point

    e. Flash point

    f. Flammability range

    g. Vapor pressure.

3. Physical hazard data such as stability, reactivity, flammability, corrosivity, explosivity.

4. Health hazard data including acute and chronic health effects and target organ effects.

5. Exposure limits such as OSHA Permissible Exposure Limits (PELs).

6. Carcinogenicity of material: Is it an OSHA , NTP, or IARC listed carcinogen.

7. Precautions to be taken including use of personal protective equipment.

8. Emergency and first aid procedures. This includes spill cleanup information and EPA spill reportability information such as CERCLA Reportable Quantity or SARA listing.

9. Supplier or manufacturer data including:

    a) Name

    b) Address

    c) Telephone Number

    d) Data prepared.

No blanks are allowed in any section of the MSDS. As was pointed out, the OSHA HazCom Standard is performance oriented, therefore, there is no standard "accepted" format for MSDSs. Unfortunately for plants without a health and safety professional on staff, this makes employee use and training for MSDSs more difficult. It is not an overstatement to say that for 20 different suppliers you will find 19 varied formats for MSDSs. The layout, orientation, type fonts, order of information, categories, comprehensiveness, length, and almost any other attribute are widely variable. Some MSDSs are encyclopedic in treatment and may run to 20 pages while others will be only or two pages with very brief information.

With these caveats in mind, consider the function of MSDSs and end result you wish to achieve for your employees. The information on the various hazards of a chemical or mixture must be conveyed from the MSDSs to the employee in such a way that they will learn at least the following:

1. How hazardous is it?

    (a) Flammable?

    (b) Toxic?

    (c) Poisonous?

   (d) Explosive?

2. Will it cause cancer?

3. What organs does it effect (e.g., lungs, skin, etc.)?

4. Will it evaporate readily?

5. Will the liquid float or sink in water?

6. Will the gas or vapor rise or fall in air?

7. What are the exposure limits?

8. What protective equipment must be used?

9. What first aid measures are required if exposed?

10. What to do if there is a spill.

11. How to handle a spill.

12. Whom to phone or write to for more information.

Simply "making available" a set of MSDSs for employees to read may not satisfy the need for this necessary training.

Availability of MSDSs

Many plants maintain "books" of MSDSs available in work areas for easy employee reference. Some discretion must be used since one "book" of all 492 hazardous materials used in a 50,000 square foot plant of several hundred employees may not suffice. Keep this file or book in an organized way, such as alphabetically by product name. File only actively used materials with only the most recently available MSDS for that material. Place in a permanent file, not directly accessible to all employees, the "old" or obsolete MSDSs.

The paper work involved in keeping the file or collection of MSDSs updated can be formidable. An index or cross reference for each set of MSDSs should be maintained. Only MSDSs for materials used in a specific area or department should be made available to those employees in the department. Employees should be encouraged to read the MSDSs for each of the materials they use at work.

CHAPTER
19
FIRE SAFETY

This chapter is intended to provide a basic overview of fire safety. The first section of the chapter reviews the hazards of fires, fire chemistry, and causes of fire. This is followed by a discussion on the use and storage of flammable materials, fire prevention construction and inspections, types of portable fire extinguishers and training for fire fighting. Then this chapter covers inspection and testing of fire fighting equipment and recordkeeping. It should also be noted in this introduction that there are a considerable number of specific fire safety requirements for manufacturing contained in the OSHA regulations and in state or municipal codes, depending on the operations conducted. The details of these are not covered in this chapter. Nevertheless, the chapter will provide the reader with a broad overview for fire safety.

The National Fire Protection Association (NFPA) reported that large fire losses in 1991 exceeded 2.6 Billion dollars. Of these large fire losses, approximately 38 percent were at manufacturing or industrial operations with losses of almost a half billion dollars. More importantly, many of the fires resulted in the loss of life or serious injury to workers and others. One example was a Louisiana chemical plant fire that occurred in 1991 and resulted in the death of eight persons and over 120 injuries. Perhaps the most tragic example from 1991 was the infamous fire that occurred at the Imperial Foods Processing Plant in Hamlet, North Carolina on September 3, 1991. This fire resulted in the deaths of 25 employees with another 54 injured. The plant had blocked fire exits, locked doors, no fire suppression system and many other deficiencies which most likely caused the high death toll. OSHA and a number of other state and federal agencies responded with an extensive investigation that resulted in the owner of Imperial Foods pleading guilty to involuntary manslaughter and being sentenced to 20 years in prison. Additionally, the victims and their families collected over 15 million dollars through civil proceedings against Imperial Foods. Clearly, fire safety is very important. For plants handling large quantities of flammable materials (e.g., paint manufacturing, batch chemical operations using flammables, fuels manufacturing, etc.), fire safety is the single most important safety program area. For others, fire still usually represents an extreme hazard given the potential for economic damage and human tragedy.

THE HAZARDS OF FIRES

The major hazards of fires and consequently the need to prevent them are potently obvious. The major hazards are the loss of life or the injury to employees and others and direct property damage due to the fire. Less obvious consequences of even a small fire can include:

❖ Loss of production and potentially customers
❖ Destruction of business records and costs for reconstruction
❖ Direct costs for losses not fully insured or depreciated
❖ Increased insurance premiums
❖ Water and/or smoke damage to equipment and property.

FIRE CHEMISTRY

The chemistry of fires is a very complex and not fully defined science; however, a few basic principles are well-recognized and of benefit to our understanding especially for fire control. Fire or combustion is an exothermic (gives off heat) oxidative reaction that can involve solid, liquid, or gaseous fuels. Normally, the oxidizing agent is atmospheric oxygen. Flaming results from the vaporization of fuels (fuels heated to the point of giving off gases) and their ignition and subsequent oxidation. Not all fires exhibit flaming. Some solids may exhibit smoldering which is glowing combustion on the surface of the material. Combustion that occurs in confinement with the subsequent generation of pressure will result in an explosion. For example, flammable gases that are premixed with oxygen upon ignition will expand and result in considerable pressure. If confined, they may explode the container.

For a fire to occur, an oxidizing agent (e.g., oxygen), combustible material (e.g., fuel), and an ignition source (e.g., open flame) are essential. The combustible material must be heated to its ignition temperature before it will support combustion or the spread of flames. Burning will continue until:

❖ the combustible material is consumed or,
❖ the oxidant is depleted or below the necessary amount for combustion or,
❖ heat is removed or prevented from reaching the combustible materials not allowing for fuel vaporization or,
❖ the flames are chemically inhibited or cooled to stop the oxidation reaction.

Fire control is based on achieving some or all of the above conditions.

CAUSES OF FIRES

In the section on fire chemistry, the three essential elements of fire were described as the oxidizing agent, the combustible material and the ignition source. Fuels and the oxidizing agent (e.g., oxygen from the air) are readily available in most industrial environments. Fuels include any item that will burn, such as most materials of construction, furnishings, products, packaging, solvents, and other materials. Sources of the oxidizing agent can include some strong acids and chlorine but in most cases it is simply the oxygen content of normal air. Air can be provided to a fire by doors, windows, mechanical ventilation systems and in other ways. It is the ignition source that most efforts at fire control are concerned with controlling. In an analysis of historical fire data by Factory Mutual Engineering Corporation, it found the seven most common ignition sources were (in order of occurrence) electrical malfunctions, arson, friction (e.g., hot bearings), overheated materials, and cutting and welding.

USE AND STORAGE OF FLAMMABLE MATERIALS

A flammable material is any liquid, solid, or gas that will ignite easily and burn rapidly. Materials that are flammable are of concern due to their ability to render damage to property and, more importantly, to injure or cause the death of workers. With these two concerns in mind, it is important to exercise caution when dealing with flammable materials.

Flammable liquids are denoted Class I liquids by the National Fire Protection Association (NFPA) and have a flash point below 100°F. This means that at temperatures below 100°F, these liquids can release vapors in quantities sufficient to ignite if exposed to extreme heat or an ignition source. The vapors of such materials will expand when heated increasing their ability to flash. Examples of flammable liquids include gasoline (flash point of approximately -45°F) and benzene (flash point of approximately 12°F). Further subdivisions of flammable liquids are as follows:

❖ Class I A liquids have a flash point below 73°F and a boiling point below 100°F.

❖ Class I B liquids have a flash point below 73°F and a boiling point at or above 100°F.

❖ Class I C liquids have a flash point at or above 73°F but less than 100°F.

Flammable solids can be found in a number of different forms each of which may exhibit different properties. Examples of flammable solids are: dusts and powders (flour, cellulose, fine metals); spontaneously ignitable materials (white phosphorous, sodium hydride); and endothermic materials (fish meal).

Flammable gases can ignite and burn in normal atmospheric concentrations of oxygen (i.e., room air). Some examples of flammable gases are butane, methane, hydrogen, and acetylene. It is essential that flammable gases receive proper care and handling during use as well as during storage so that ignition of the gases are prevented.

Caution must be exercised during the handling and use of flammable materials. Primary attention must be given to the fire properties of the material, including flash point, ambient temperature, work activities, and potential ignition sources in the area of use. Precautions must be taken to prevent the uncontrolled release or exposure of the flammable materials, or their vapors, to heat or ignition sources.

Two important considerations must be made in storing flammable materials: the chemical properties of the material and the proper container to control the potentially hazardous properties. The container of choice must be leak-proof and must control, if not prevent, the release of flammable vapors. For portable containers, this usually means self-closing or sealed spouts. Initial consideration and planning of the storage location should include the types of materials being stored, the amounts stored, the chemical properties of the materials, the area in which storage will occur, and the activities or potential activities in the area.

To eliminate the potential for a static charge to become an ignition source, it is important that the storage container is properly bonded or grounded. Bonding is installing a conductor between two or more conductive objects so that the objects may jointly share static charges, such as between 55 gallon drums or portable metal containers. Grounding is a form of bonding in which objects are connected to the ground, thereby reducing static charges. Grounding is especially important when the storage container is an above ground storage tank.

FIRE PREVENTION CONSTRUCTION AND INSPECTIONS

Storage tanks, buildings and structures that will contain flammables must be constructed with fire-resistant materials and be of such design that fire hazards are eliminated, or limited to protect employee and property loss. Considerations involve the type of storage tank to be used (above ground, underground), location of the tank (inside, outside), and the proximity to work areas or occupied buildings. Flammable materials, especially liquids, have been stored underground in tanks for the last few decades as a fire protection measure. However, leaking underground storage tanks created such a severe national threat to drinking water that most tankage has been moved above ground. Storage below ground is still possible and safer from a fire protection standpoint, but now much more difficult and expensive.

Storage areas must be designed to account for worst-case emergency situations. To assist in the control of fire hazards, here are some additional items that are essential to fire prevention and control:

- ❖ proper construction and choice of storage container
- ❖ elimination of heat and ignition sources
- ❖ separation of incompatible materials
- ❖ adequate means of fire fighting (sprinklers, extinguisher, hoses, etc.)
- ❖ unobstructed means of egress for employees in the event of an emergency
- ❖ adequate aisle clearance for firefighters and equipment
- ❖ proper ventilation system for venting and reducing vapor buildup

Indoor storage areas must be constructed with fire-resistant materials and must be located in such an area that employees will not be trapped in the event of a fire. The storage room must be leak-proof, have appropriate floor level ventilation with six air changes per hour, have adequate aisle clearance, be equipped with a fire suppression system, and contain approved fire doors and exits. All ignition sources must be eliminated and potential ignition sources, such as electrical equipment and wiring, must be approved for use around flammable materials.

Outside storage areas should be located away from occupied buildings and employee work areas so that, in the event of a release, the liquid will not flow toward the building, or else it should be contained by a dike. Activities around and near the storage area shall be kept to a minimum so that ignition sources are prevented. The area must be kept clear of debris, high grass and other materials that have the potential to burn.

Annual inspections of flammable storage areas must be conducted and records of each inspection need to be kept on file. An effective means of conducting the inspection is the development of a checklist noting the observation of the following items:

❖ types of materials being stored

❖ types of storage containers being used

❖ condition of the storage area

❖ types and condition of fire suppression systems

❖ observation of aisles, ventilation, and exits

❖ condition of nearby electrical and mechanical equipment

❖ notation of nearby work areas and activities

TYPES OF PORTABLE FIRE EXTINGUISHERS AND TRAINING

Portable fire extinguishers are primarily used (1) in areas where fixed fire suppression systems are not available and (2) as supplements to fixed systems. The use of portable extinguishers is often required or recommended to supplement fixed systems. The insurance industry can testify that the prudent use of a portable fire extinguisher can save money. Records show that the damage costs from fixed water suppression systems can often exceed the costs from fire damages.

Classification of Extinguishers

There are four classifications of fire extinguishers. They are required to be designated by the class of fire they are designed to suppress.

❖ Class A—Used for combustibles, such as wood, paper and textiles. Class A extinguishers can be identified by a letter A in a triangle. If the triangle can be identified by color, it will be green.

❖ Class B—Used for flammable liquids, such as greases, paints and gasoline. Class B extinguishers can be identified by a letter B in a square. The square will be color coded red.

❖ Class C—Used for electrical fires from wiring, electrical powered machinery or Class A or B fires in near proximity to electrical equipment. Class C fire extinguishers can be identified by a letter C in a circle. If the circle is colored coded it will be blue.

❖ Class D—Used for combustible metals such as aluminum, magnesium and titanium. These fire extinguishers can be identified by a letter D in a five pointed star. If the star is color coded it will be yellow.

Location of Extinguishers

Fire extinguishers must be placed in readily visible areas unencumbered in any way from easy access and at intervals predicated on the anticipated class of fire and distance of travel by employees to the extinguisher:

❖ Class A Extinguishers—75 feet maximum travel

❖ Class B Extinguishers—50 feet maximum travel

❖ Class C Extinguishers—As per use as Class A or B

❖   Class D Extinguishers—75 feet maximum travel

Exemptions

Portable fire extinguishers need not be installed if (1) uniformly spaced standpipe systems or hose stations are installed, and (2) the employer has implemented a fire safety policy that requires the immediate evacuation of all employees.

Training

Training requirements for fire extinguisher use are predicated on the extent employees are expected to fight a fire. If employees are only expected to use an extinguisher on incipient and smoldering stage fires, the employer need only provide fire extinguisher education for the employee on information such as:

❖   how to recognize the types of fires

❖   how to fight a fire

❖   when to call for outside assistance

❖   when to abandon the fire fight

❖   when to evacuate the building or area.

If employees are expected to fight a fire to its conclusion, they must have complete "hands-on" training under the supervision of an approved fire fighting school, and must be outfitted in complete fire-fighting clothing and equipment. Much of this training can usually be arranged through local fire departments.

Inspection, Maintenance, and Testing

All portable fire extinguishers must be visually inspected monthly. The monthly inspections are to ensure that the extinguishers are charged, in the proper location and in operable condition. Records of inspection dates and the name of the inspector need to be available. All portable fire extinguishers must receive an annual maintenance check. The employer is required to record the dates of the annual maintenance checks and retain the records for one year following the shelf life of the extinguisher. Hydrostatic testing of portable fire extinguishers is to be conducted on a prescribed schedule. (See Table 19-1.)

RECORDKEEPING FOR INSPECTIONS

Records of the inspections conducted on fire equipment, fire safety training, and the evaluations of flammable material storage areas must be maintained and readily available to interested employees and authorized agencies. The option of either hard copy (paper) forms or a computer data base is left solely to the discretion of the individual companies. In some instances, it may be beneficial to retain both a hard copy and a computer data base form of documentation. These forms of documentation should be kept on file for at least six years.

Table 19-1
Hydrostatic Testing of Fire Extinguisher

Extinguisher Type	Test Interval (Yrs)
Soda Acid (stainless steel shells)	5
Cartridge operated-water/antifreeze	5
Stored pressure-water/antifreeze	5
Wetting agent	5
Foam (stainless steel shells)	5
Aqueous Film Foaming Foam (AFFF)	5
Loaded steam	5
Carbon dioxide	5
Dry Chemical in stainless steel	5
Dry Chemical, stored pressure with mild steel, brazed brass or aluminum shells	12
Dry Chemical, cartridge or cylinder operated with mild steel shells	12
Halon 1211	12
Halon 1301	12
Dry Powder, cartridge or cylinder operated with mild steel shells	12

CHAPTER
20

POWERED INDUSTRIAL TRUCKS

This chapter focuses on developing a systems approach to powered industrial truck safety in manufacturing plants.

To ensure that your plant's powered industrial trucks are used safely, take the following steps:

❖ Establish a safe work environment where the truck will be used.

❖ Select a truck that will meet or exceed the safety requirements.

❖ Select and train drivers to safely meet the requirements of the work to be done.

❖ Set up a comprehensive inspection and preventative maintenance program for all trucks.

❖ Periodically evaluate the effectiveness of your truck safety program.

These steps are discussed in detail in this chapter.

ESTABLISHING A SAFE WORK ENVIRONMENT FOR POWERED INDUSTRIAL TRUCKS

Before selecting a powered industrial truck, define the work tasks and operating environments that drivers will encounter. Be mindful of hostile environments, such as inclement weather or exposure to flammable liquids, vapors, or explosive dust which require specific safety considerations. The use of industrial trucks in hostile environments could result in accidents, injury, or even death.

Industrial truck accidents may not be frequent, but they usually are more serious than other types of industrial accidents. For example, an industrial truck unable to stop because of ice or wet, slippery conditions on a loading dock slides off the end of the dock and crushes the driver and also causes extensive damage to the truck. Or a driver takes a truck not rated for use in areas where flammable liquids and vapors are likely to be present into a flammable liquid storage and dispensing operation. The truck's electrical controls arc and ignite flammable vapors in the area and a flash fire and explosion follow, killing the driver and a production worker as well as shutting down production for several weeks to repair damage to the building and equipment. Both of these accidents could have been prevented by controlling the work environment where the industrial trucks were expected to be used.

The National Fire Protection Association (NFPA) classifies environments according to their hazardous conditions of the area. NFPA hazard classifications are listed in Table 20-1. For a complete description of hazard classification, consult NFPA-70, the National Electric Code.

Table 20-1
Hazardous Location Classifications

Classification	Description
Unclassified	Locations not possessing atmospheres as described under other classifications.
Class I	Locations in which flammable gases or vapors are, or may be, present in the air in quantities sufficient to produce explosive or ignitible mixtures.
Class II	Locations which are hazardous because of the presence of combustible dust.
Class III	Locations where easily ignitible fibers or flyings are present but not likely to be in suspension in quantities sufficient to produce ignitible mixtures.

A survey for environmental hazards for industrial truck operations should be conducted to identify the different hazard classification areas in your plant. If your facility has multiple hazard classifications, you must post warning signs at each area entrance restricting entry to industrial trucks and other equipment approved for use in that area.

Other environmental conditions you should consider are whether the industrial truck will operate indoors, outdoors, or both; and the condition of the terrain, ramps, railroad tracks, pavement, and soil types that will be traversed.

Following the completion of the survey for environmental hazards, you should develop an action plan to control or eliminate existing hazards that could affect the safe operation of industrial trucks. The action plan should identify the hazards that are to be controlled or eliminated, the action to be taken, the person responsible for the action, and a schedule identifying the proposed starting date for mitigation, the actual starting date, and most importantly the completion date for each action item. Table 20-2 provides an example of an environmental hazard action plan form.

Be sure drivers know that industrial trucks in environments other than those for which they were designed can be hazardous to property and life. Misuse of industrial trucks will also subject your company to possible OSHA citations.

Table 20-2
Industrial Truck Environment Survey Form

INDUSTRIAL TRUCK ENVIRONMENTAL HAZARD ACTION PLAN

LOCATION	CONDITIONS	ACTION PLANS	RESPONSIBLE PERSON	SCH. START	START DATE	COMPLETED
Dept. 145	Wet loading dock	Extend dock canopy 6 feet	J. Brown	6/7/91	6/7/91	7/6/91
Dept. 456	Class I Environment	Post warning signs at doors	M. Wright	6/7/91	6/8/91	6/8/91
Dept. 100	Uneven RR tracks drive gate 4	1. Obtain contractor bids for improving roadway.	I. Corn	6/9/91	6/9/91	6/15/91
		2. Let contract	I. Corn/J. Brown	6/21/91		
		3. Start crossing repair	J. Brown	7/1/91	6/12/91	
Dept. 200	No lighting on back shifts for outside drum storage area	1. Install 4 light poles and sodium vapor lights.	V. Watts	6/10/91		
		2. Terminate after dark pick up and deliveries to area until lights are installed.	M. Wright	6/7/91	6/7/91	6/7/91

SELECTING SAFE POWERED INDUSTRIAL TRUCKS

Powered industrial trucks are available in many different varieties. The trucks most commonly used in manufacturing are the forklift and platform styles, which are controlled by operators either standing or sitting on the vehicle or walking along with it. Industrial trucks are available with a wide selection of front-end attachments to increase versatility. However, we caution when selecting a truck to be sure that the vehicle matches the job it is intended to perform.

Selection of a walkie lift truck for jobs that require loads to be moved long distances or over uneven work surfaces is not recommended. Too often operators will ride the truck, which compromises their ability to control the vehicle and increases the likelihood of an accident. Platform trucks usually have smaller diameter wheels when compared to fork lift trucks. The smaller diameter wheels make it difficult to move loads over uneven surfaces, such as train tracks, without spilling the load or getting stuck.

When loads must be lifted above the driver's head, it is required that a safety cage be installed that will protect the driver from falling loads.

When moving loads that require unique material handling methods, such as large rolls of material or paper that must be rotated for clearance during transportation, compressed gas cylinders, or loads requiring extra long or wide set forks, be sure that the truck you select can meet these unique requirements. The purchase of a load rotating attachment for your fork truck or extra long forks can reduce the possibility that the load will be dumped or strike machinery and other obstructions during transportation. Loads should never be carried on one fork. If the load does not fit the forks of your truck in their narrowest position, place the load on a pallet or frame that can be safely lifted without the risk of falling. Select an industrial truck that has the capacity to handle the heaviest load that must be moved. Loads that are over the capacity of the truck cannot be dragged or pushed or the load shared by another truck. The addition of counter weights to balance the load is not permitted by OSHA.

OSHA's Powered Industrial Truck Standard 29 CFR 1910.78 does not permit any modification subsequent to purchase of industrial trucks without approval of the vehicle manufacturer. OSHA general safety standards for powered industrial trucks require that all powered industrial trucks purchased and used after February 15, 1972, meet the design and construction specifications of the American National Standards Institute (ANSI) standard B56.1-1969. Beyond this general requirement there are several regulatory issues that you must address when selecting an industrial truck.

Selecting Trucks Used In Hazardous Locations

The information you gather during your analysis of the work environment will determine the type of vehicle needed for use in each classification. A list of the types of trucks is found in Table 20-3.

Management is responsible for selecting the appropriate truck for safe use in areas of the plant having hazard classifications. If a truck will be used in multiple areas of the facility, it must be approved for use in the most hazardous area. Table 20-4 is a good reference source from OSHA for selecting the correct type of industrial truck for safe operation in your plant. This table lists the truck types of Table 20-3 that are approved for use in specific groups and divisions of the hazard location classifications given in Table 20-1.

Table 20-3 Powered Industrial Truck Destinations	
Designation	*Type*
D	Diesel Powered
DS	Diesel powered with additional safeguards
DY	DS with no electrical equipment and equipped with temperature limitation features
E	Electrical powered
ES	Electrical with additional safeguards to prevent sparks and limited-surface temperatures
EE	Completely enclosed motor
EX	Electric units acceptable in areas containing flammable vapors and dust
G	Gasoline powered
GS	Gasoline powered with additional safeguards
LP	Liquid petroleum powered
LPS	Liquid petroleum powered with additional safeguards

OSHA does not approve any industrial trucks for use in atmospheres containing hazardous concentrations of materials such as acetylene, butadiene, ethylene oxide, hydrogen or equivalent gases or vapors such as manufactured gas. However, approved EX vehicles may be used in areas where flammable liquids are contained in open vessels or flammable liquids, gases, or vapors are likely to form explosive or ignitible mixtures.

Other truck selection criteria include the size and weight of loads to be handled, the maximum lift height, and any required attachments for specialized use such as a platform for lifting personnel. OSHA does not permit adding ballast to lift loads heavier than the designed capacity of the truck, carrying compressed gas cylinders on truck forks, or using the forks to open train doors without a special attachment. In short, a powered industrial truck used to perform tasks beyond its design and rated capacity is not only unsafe, it violates OSHA regulations.

SELECTING DRIVERS AND TRAINING THEM

Be sure that all candidates for drivers of powered industrial trucks are given medical examinations to verify they have the physical attributes to safely handle the requirements for the job. Truck operators require the same physical skills as an operator of a highway vehicle, including good vision, depth perception, and hearing.

Potential operators must be trained before they may operate a truck. Training should include classroom work as well as a driver's test to demonstrate that the operator can handle the truck safely under controlled conditions. Be sure training records are maintained for each operator to verify the type and amount of training received. Operator classroom training should include these safety practices:

Table 20-4
Summary Table on Use of Industrial Trucks in Various Locations

CLASSES	UNCLASSIFIED	CLASS I LOCATIONS				CLASS II LOCATIONS			CLASS III LOCATIONS	
Class Description	Locations not possessing atmospheres as described in other columns.	Locations where flammable gases or vapors are or may be present in the air in quantities sufficient to produce explosive or ignitible mixtures.				Locations which are hazardous because of the presence of combustible dust.			Locations where easily ignitible fibers or flyings are present but not likely to be in suspension in quantities sufficient to produce ignitible mixtures.	
Groups in classes	NONE	A	B	C	D	E	F	G	NONE	
Examples of locations or atmospheres in classes and groups	Piers and wharves inside and outside storage. General industrial or commercial properties.	Acetylene	Hydrogen	Ethyl ether	Gasoline Naphtha Alcohols Laquer solvent Benzene	Metal Dust	Carbon black Coal dust Coke dust	Grain dust Flour dust Starch dust Organic dust	Baled waste cocoa fiber, cotton, excelsior, hemp, jute, kapok, oakum, sisal, spanish moss, synthetic fibers, tow.	
Divisions (Nature of hazardous conditions)	NONE	1			2	1		2	1	2
		Above condition exists continuously, intermittently, or periodically under normal operating conditions			Above condition may occur accidently due to a puncture of a storage drum	Explosive mixture may be present under normal operating conditions or where failure of equipment may cause the condition to exist simultaneously with arcing or sparking of electrical equipment where dusts of an electrically condutive nature may exist		Explosive mixture not normally present, but where deposits of dust may cause heat rise in electrical equipment or where such deposits may be ignited by arcs or sparks from or electrical equipment	Locations in which easily ignitible fibers or materials producing combustible flyings are handled, manufactured, or used	Locations in which easily ignitible fibers are stored or handled (except in the process of manufacturing

Authorized uses of trucks by types in groups of classes and divisions

Groups in classes	NONE	A B C D	A B C D	E F G	E F G	NONE	NONE
Types of trucks authorized							
Diesel:							
Type D	D**						
Type DS	DS				DS	DS	DS
Type DY	DY			DY	DY	DY	DY
Electric:							
Type E	E**					E	E
Type ES	ES				ES	ES	ES
Type EE	EE			EE	EE	EE	EE
Type EX	EX	EX	EX	EX	EX	EX	EX
Gasoline:							
Type G	G**						
Type GS	GS				GS	GS	GS
Lp-Gas:							
Type LP	LP**						
Type LPS	LPS				LPS	LPS	LPS

** Trucks conforming to these types may be also used

❖ Be sure the load is secure before you move it. Loads that are not properly secured or balanced may shift during transport and fall off the vehicle causing property damage or injury to personnel that may be in or near the transportation route.

❖ Travel with the forks or pallet about 4 inches off the floor. Traveling with the load in an excessively high elevation tends to block the driver's vision and make the load unstable. Such conditions increase the risk of accidents.

❖ Never travel if your view is obstructed. Moving large pieces of equipment often obstructs the driver's vision. Operating the truck in reverse can eliminate this problem. If this can not be done, a guide person should be used to provide direction to driver by using hand or voice communications.

❖ Don't exceed the established speed limits. Excessive speeds have caused as many industrial truck accidents as all other causes combined. A truck driven at excessive speeds is difficult to stop or maneuver and presents a high risk of turning over or losing the load.

❖ Know the maximum load limit of a lift and never permit an overload. Overloading an industrial truck makes the vehicle difficult to control, increasing the risk of accidents. It also increases maintenance costs and down time and shortens the effective life of the vehicle.

❖ Never allow others to ride with you. Industrial trucks are not passenger vehicles. With few exceptions they are designed to accommodate only the driver. Transporting workers on the forks or platform of a truck or on the load is a dangerous practice. Sudden stops or a shifting of the load can cause passengers to fall from the vehicles and can cause serious injuries. A "no passengers allowed" policy should be included in your driver's training program.

❖ Keep the load in the front on an upgrade and in the back on a downgrade. This system increases the stability of the vehicle in moving up and down ramps and other steep grades.

❖ Follow established traffic patterns, such as four-way stops at intersections. Stay within the aisle markings. Driver safety is no different for industrial truck operators then for over the road vehicle drivers. Failure to stop at intersections or driving a truck through plant areas not designated for truck traffic is a prime cause of industrial truck accidents and endangers the driver, other drivers, and workers in your plant.

❖ Sound the horn only when necessary; don't try to frighten pedestrians. The horn on an industrial truck is to be used for issuing a warning that a truck is approaching and to exercise caution. It is not a toy to be used to scare people or as a device to be used in place of good driving habits. Horns are not to be used at intersections in place of stopping and obtaining good visual assurance that the right-of-way is clear. The misuse of the horn or relying on it instead of good driving procedures can cause accidents.

❖ Observe back-up rules and be sure the alarm is working. Back-up alarms are not required on all vehicles. If you have vehicles that require such devices (those with obstructed rear vision) you must be sure they are in good working order at all times. Drivers should be instructed to be sure that personnel are clear of the rear of their vehicles prior to driving in reverse. Make a visual check to assure clearance, sound the horn, and move the vehicle slowly. Never place the vehicle in reverse and make sudden and swift starts.

❖ Never leave the truck unattended when the engine is running. An industrial truck should never be left unattended with the motor running or the key in the ignition. Your industrial trucks could be considered an "attractive nuisance" or a temptation for unapproved drivers to "joy ride" or attempt to move a load rather than wait for the driver to return. Unapproved or unauthorized drivers are a danger to themselves as well as their co-workers.

❖ Don't operate your truck beside another truck in the same aisle. Operation of vehicles in close proximity increases the risk of accidents from falling loads and collisions. Make sure your drivers wait their turn before entering a pickup or lay down area.

❖ When refueling your truck, shut the engine off and don't smoke. Be sure the refueling area is well ventilated. The refueling of propane, gasoline, and diesel trucks subjects the driver and/or the attendant to the risk of a fire or explosion from spilled or released fuels. A running engine or a cigarette can provide the ignition source for a severe fire and/or explosion. Drivers should be cautioned to avoid filling gasoline powered vehicles that have hot engines or manifolds. Hot metal can provide an ignition source.

❖ Batteries should be serviced only in designated areas and with the proper personal protective equipment. Batteries can release hydrogen gas, an extremely flammable gas that is easily ignited. Areas without good ventilation can allow gas to accumulate in explosive concentrations. Batteries also contain sulfuric acid, a corrosive liquid that can cause severe burns to the skin and eyes. Servicing should be done in an area having a safety shower and eye wash, chemical safety goggles, fire extinguisher, and good ventilation.

❖ Never engage in horseplay or "trick" driving. Pinning a fellow employee to the wall with the forks of a truck or turning corners on two wheels and other forms of horseplay are dangerous practices. Too often injuries are caused by these antics and are never reported by the driver or the injured person for fear of retribution. Horseplay and trick driving have no place in any plant and must never be tolerated.

❖ Don't use forks to align or straighten stacked material. Using the forks of a truck to align or straighten a material stack puts the driver and anyone in the area at risk to an accident from the stack toppling on them. Drivers should be taught to reposition the stack by resetting the load even if the entire stack must be taken down and reset.

❖ Never raise or lower a load while traveling. The raising or lowering of the load "on the fly" must not be tolerated. The load may be lost by dropping the forks too far and hitting the floor or dumping the load when it becomes unstable from the movement of the vehicle and the forks.

❖ Don't stand under a load while it's being raised or lowered. Never let anyone stand under the forks of a truck that are loaded or empty. The failure of a hydraulic component or the stopping of the vehicle motor could cause the lifting mechanism to lose pressure and the forks to rapidly fall, striking anyone under them.

❖ When the vehicle is parked, lower the forks, keeping them as close to the ground as possible. Leaving forks in an elevated position is dangerous and has caused injuries to employees that have tripped over raised forks or have walked into the forks left at head and face level.

❖ Never attempt to accelerate quickly or stop suddenly. Rapid accelerations and panic stops may cause the load to shift and spill.

❖ Check your machine thoroughly before starting it and report any malfunctions immediately. Before and after each shift, check the brakes, steering, controls, forks, hoists, warning devices, and lights. Operating a truck that is not in good operating condition is dangerous to the driver as well as other personnel in the plant. See Table 20-5 for an operator checklist.

❖ If a powered industrial truck operates on liquified petroleum gas, the following specific precautions must be taken:

❖ Before starting the motor, always check for gas leaks.

❖ When exchanging fuel containers, do not disconnect the tank until you have shut off the fuel supply and run the engine to make sure all the fuel in the system has been used.

❖ If a truck is out of use for an extended period of time, close the hand valve and make certain the fuel system is dry.

❖ Never park the truck near intense heat or flammable materials.

❖ Don't try to connect the gas tank when gas is escaping from the connecting point. Connections should be performed outdoors if possible.

INSPECTION AND MAINTENANCE

A comprehensive inspection and preventative maintenance program is necessary for safe operation of industrial trucks. Require your operators to inspect their vehicles prior to and following daily use. One good method to ensure that inspections are routinely made is to use a prepared checklist of specific safety items to be signed by the operator. Table 20-5 is an example of a typical pre-operational checklist.

You should hold all operators responsible for inspection of their vehicle prior to starting work. Vehicles with safety equipment or mechanical equipment that is inoperable or in questionable repair should be tagged out of service by the operator and reported to the operator's supervisor who then issues a maintenance repair order to the maintenance department requesting repairs.

When damage to a vehicle occurs during the work shift, be sure the operator immediately takes the vehicle to the maintenance department for repairs (if the vehicle can be driven safely). If the vehicle cannot be driven safely, shut off the vehicle, remove the key, tag the vehicle out of service, and report the condition to the maintenance department.

Operators responsible for refueling their vehicles must be trained in the proper procedures for changing batteries including the use of face shields, gloves, and aprons to protect them from battery acid.

Gasoline and diesel powered vehicles should never be refueled while their motors are running because of the danger of a fire or explosion if spilled fuel comes in contact with a hot manifold.

Operators of liquid petroleum gas (LPG) powered vehicles need to be trained in the proper handling procedures of LPG fuels. The importance of the indexing pin for cylinder installation for safe operation needs to be stressed. Operators should know that LPG cylinders have a dip tube connected to the discharge valve that curves to the top of the cylinder to assure that gas and not liquid LPG is discharged from the cylinder. This is an important safety factor in the event an LPG hose should fail. If the cylinder is not mounted correctly, liquid LPG could be discharged from the leak or failure, causing a greater volume of LPG to be released in the same time frame.

Table 20-5
Pre-Operational Check

DRIVER_____ TRUCK_____ DATE_____

VISUAL CHECK	OPERATIONAL CHECK
____1. Crankcase oil level	____1. Horn
____2. Engine belts	____2. Back-up lights/alarm
____3. Plug wires	____3. Steering
____4. Brake fluid level	____4. Service brake
____5. Hydraulic fluid level	____5. Parking brake
____6. Fuel level	____6. Motorola pedal (forward/reverse)
____7. Tires/wheels/rims	____7. Transmission (forward/reverse)
____8. Head & tail lights	____8. Seat brake
____9. Turn signals	____9. Seat safety switch
____10. Warning lights	____10. Hydraulic controls
____11. Hour meter	____11. Mast lift (up/down)
____12. Other gauges	____12. Mast tilt
____13. Forks (damage)	____13. Side shift/squeeze
____14. Mast chains & lines	____14. Hydraulic leaks
____15. LPG clamps/index pin	____15. Battery charge
____16. Safety belts & lines	
____17. Fire extinguisher	
____18. Overhead cage/guard	
____19. Safe operating capacity of forks & attachments	
____20. Other conditions (Please list)	

Be sure that all truck operators follow a rigid preventative maintenance (PM) schedule, as outlined in the manufacturer's service guide. An example of a general manufacturer's maintenance schedule is provided in Table 20-6. It lists some of the key maintenance requirements. The benefits realized from following the PM schedule are safe operation and extended service life.

Recordkeeping for All Truck Inspections and Maintenance

Be sure that your operators maintain accurate records of truck inspections, such as the operator's daily checklist and all maintenance performed on the vehicle. Maintain these records throughout the service life of the truck. These records are required along with training and accident records as discussed in the following section to determine safety trends.

EVALUATING THE SAFETY PROGRAM FOR POWERED INDUSTRIAL TRUCKS

You must be notified of all intended equipment or process changes that could impact the safe operation of industrial trucks in your facility, such as installing open topped mix tanks for flammable liquids. All proposed equipment and/or operational changes should require approval by Engineering, Operations, and Safety prior to equipment purchases, equipment installation/movement, or process, chemical, or procedural changes. After all approvals are obtained the signed approval forms should be maintained. Immediate feedback on environmental changes is critical for safe truck operation.

Table 20-6
General Industrial Truck Manufacturer
Maintenance Schedule

MAINTENANCE SCHEDULE	HOURS OF OPERATION
Engine	
Oil and filter change	x,xxx*
Replace plugs	xx,xxx
Hydraulic System	
Fluid change	xx,xxx
Hose inspection	x,xxx
Cylider seal inspection	x,xxx
Drive Systems	
Brake linings-inspect	xx,xxx
Brake hoses-inspect	xx,xxx
Transmission-inspect and adjust	xx,xxx
Transmission fluid-change	xx,xxx
Steering system lubrication	xx,xxx

* Each truck manufacturer has their own maintenance schedules. Therefore actual hours of operation are not included in this table so as to not be misleading.

Review all accident records involving trucks when the accident occurs and on a periodic basis to determine if accident trends are developing. Standard accident investigation forms should be adequate for this procedure. Review all training, inspection, and PM records with the accident records to determine if they are interrelated. For example, you might find that a specific truck was involved in four accidents caused by brake failure and the same vehicle had three inspection reports for inadequate or improper brake operation. You should give this information to the maintenance department so the vehicle can be removed from service and new brakes installed. An effective industrial truck system can reduce accidents and property losses as well as improve overall plant performance.

CHAPTER
21
ELECTRICAL SAFETY

Electricity has helped make modern society possible. It provides light for employees to work during nondaylight hours and powers machinery that produces our products. Although electricity is indispensable to our modern way of life, there are also potentially severe hazards associated with it. This chapter will discuss:

❖ Basics of Electricity and its Hazards

❖ National Electrical Code

❖ Selection and Installation of Electrical Equipment

❖ Lockout-Tagout Procedures

❖ Electrical Safety Training

❖ Maintenance and Repair of Electrical Equipment

BASICS OF ELECTRICITY AND ITS HAZARDS

There are three basic concepts important to an understanding of electricity: Voltage, current, and resistance.

One of the easiest ways to visualize the flow of electrical energy through a wire or conductor is to think of water flowing through a pipe. Voltage, which is measured in volts, can be visualized as the pressure of the water in the pipe. Current, which is measured in amperes, is the total amount of water flowing past a point per unit of time. Resistance, which is measured in ohms, is the friction or any other impediment that tends to retard the flow of the water.

The primary factor in injury from electrical sources is the amperage or current density that flows through the body. In general, the greater the current and the longer it is able to flow through the body, the greater will be the injury. Experimental studies show that as little as 100 milliamps (mA) of 60 Hertz (Hz) alternating current can be fatal, while it has been estimated that the greatest hand held current from which the average individual can release himself is in the range of 16 mA. These current densities are well below those that can be found in average household electrical circuits, indicating that severe electrical injuries can arise from very ordinary electrical sources. When dealing with the much higher energy sources that are commonly found in the industrial setting, the potential hazards increase dramatically and due caution needs to be exercised.

Dry skin has a fairly high resistance to the passage of an electrical current, thus providing some measure of protection. The presence of moisture, however, dramatically lowers the skin's resistance allowing for much easier entry into the body where the flow of current can cause harm. Once an electrical current breaches the skin barrier, the body offers little resistance to the flow and several types of injuries can occur.

❖ Severe external burns that generally are extremely slow healing. These can be received by individuals that are at some distance from a source. Failure of switches, large fuses, and other equipment can lead to intense arcing that can injure individuals several feet away from the incident.

❖ Severe internal burns resulting in the destruction of tissue, muscle, and nerves along the path of the current flow.

❖ Severe muscle contractions and spasms that can break bones and/or interfere with breathing to such an extent that asphyxiation is possible.

❖ Disruption of the normal rhythm of the heart. This normally leads to fibrillation, which is the uncoordinated contraction of the fibers of the heart muscle. When this occurs, blood flow ceases and death ensues unless rapid steps are taken to resuscitate the victim.

❖ Temporary paralysis of the respiratory control center of the brain. This also leads to cessation of respiration.

Because of the extremely serious nature of electrical injuries, it is imperative that first aid and resuscitative measures be started immediately after the injury and continued until the victim either revives or is pronounced dead by medical personnel. If resuscitative measures are not started within the first five minutes after an electrical accident, the victim's chances of survival are very poor.

NATIONAL ELECTRICAL CODE

The National Electrical Code (NEC) was first published in 1895 by the National Board of Fire Underwriters, now the American Insurance Association. The National Fire Protection Association (NFPA) assumed sponsorship and control of the NEC in 1911 with the code being officially endorsed by the American National Standards Institute (ANSI) in 1920. The NEC is currently prepared by the NFPA National Electrical Code Committee whose stated duties are to prepare documents on "...minimizing the risk of electricity as a source of electric shock and as a potential ignition source of fires and explosions." It is currently issued by the NFPA as ANSI/NFPA 70.

The introduction to the NEC states that the "code is purely advisory as far as the NFPA and ANSI are concerned but is offered for use in law and for regulatory purposes in the interest of life and property protection." Because the NEC is probably the single most comprehensive document dealing with the installation and use of electrical equipment, it has, in fact, become law in most jurisdictions with all new electrical construction and installation being required to conform to the code. Because the NEC is considered the minimum standard, many states and municipalities have electrical codes that are more stringent than the NEC. Management, therefore, needs to make their personnel aware that additional requirements, above and beyond those specified in the NEC, may need to be taken into account when designing, constructing, and installing electrical equipment and facilities.

The NEC is very comprehensive, with the current annotated edition [with explanations] encompassing over 1,100 pages. The code is divided into nine main chapters. The main headings are given below.

1. General

2. Wiring Design and Protection

3. Wiring Methods and Materials

4. Equipment for General Use

5. Special Occupancies

6. Special Equipment

7. Special Conditions

8. Communication Systems

9. Tables and Examples

It must be stressed that the NEC is not to be used as a do-it-yourself manual for untrained individuals, but rather as a guide to be used by highly trained individuals such as engineers, electrical contractors, inspectors, and knowledgeable instructors.

SELECTION AND INSTALLATION OF ELECTRICAL EQUIPMENT

Selection

The main point to remember when selecting electrical equipment is that the proper equipment must be chosen for a given job. If the wrong or improper equipment is selected, it will probably not operate at peak efficiency and could present significant safety hazards for the employees who must use or operate it. In selecting and installing electrical equipment it is extremely important that all applicable standards and codes be followed. As stated above, the NEC is the primary code that must be followed, however, local jurisdictions may adopt more stringent requirements that need to be addressed if applicable. Remember that adherence to the NEC is mandated by the OSHA standards and also in the coverage of most insurance companies for property damage insurance.

There are a number of organizations that can provide information about electrical equipment and its suitability for a specific application. Some of these are listed in Table 21-1.

When requesting information, assistance, or literature from these organization it is necessary to provide them with as much background information as possible.

Table 21-1
Sources of Assistance in the Selection of Electrical Equipment

American National Standards Institute	Underwriters Laboratories, Inc.
1430 Broadway	333 Pfingsten Road
New York, NY 10018	Northbrook, IL 60062
(212) 354-3300	(708) 272-8800
National Fire Protection Association	Factory Mutual System
Batterymarch Park	1151 Boston-Providence Turnpike
Quincey, MA 02169	Norwood, MA 02062
(617) 770-3000	(617) 764-7900

Installation

All electrical equipment and fixtures must be installed in accordance with the NEC and any local electrical codes. It is, therefore, vital that all electrical installation work be performed by or under the supervision of individuals who are knowledgeable in applicable codes.

Because of the potential safety hazards associated with electrical equipment, major equipment such as generators and transformers should be installed in areas of the plant that have limited access. Ideally this type of equipment should be installed in separate rooms that can only be accessed by appropriately trained individuals. If it is necessary to locate major electrical equipment in an area where there is the potential for vehicular traffic, appropriate barriers must be installed.

All electrical equipment must be equipped with a means so that it can be positively isolated from its energy source and placed in a zero energy state for repair and maintenance work. For many pieces of equipment isolation is easily accomplished by merely unplugging the equipment from its electrical outlet. Larger equipment may be equipped with switches, fuses, or circuit breakers that serve as the primary isolating device. Fuses and circuit breakers also provide protection against external power surges that could damage the equipment and against internal shorting and malfunctions that could present a hazard to employees using the equipment.

All switches, particularly knife switches, should be installed so that there is no chance of their closing by themselves. This means that the handle should open from the top downward. If a switch were to open from the bottom upwards, there is a possibility that, while the equipment was isolated, the switch could swing down and activate the equipment.

Fuses and circuit breakers are designed to burn out or trip at a predetermined current. It is therefore important that equipment be connected to a circuit that has an adequate current rating, otherwise the fuse or circuit breaker will constantly burn out or trip. The extra load that is exerted when most electrical equipment is started should also be factored into the circuit current rating. If, during equipment operation, a fuse frequently blows or a circuit breaker trips, it is generally an indication that there is a short in the line or the equipment or that the circuit does not have a sufficient capacity. In these instances, the entire system, equipment and circuit, needs to be thoroughly checked for the source of the problem. Remember that fuses must never be replaced unless the equipment being served by the fuse is deactivated and the line has been isolated.

Control panels and operating boards should be designed and laid out in a logical and orderly manner. The input of the employees who will have to use the panel or board can often be of considerable value since they are generally knowledgeable about operating parameters and controls that are most critical and which need special consideration in their placement.

Ground fault circuit interrupters (GFCI) are devices that detect very small amounts of current leakage, often below the threshold of human perception, to ground potential. The installation of these devices on major pieces of equipment can often detect an incipient problem before it reaches a magnitude which can result in a severe electrical injury. For this reason, newer equipment often has these devices and consideration should be given to installing them on older equipment during a major outage or maintenance function. They are also required in certain areas under the NEC standards.

Electric motors are the heart of many pieces of equipment and care must be exercised selecting and installing them. A motor should be properly sized for a given application. If it is too small, it will always be under a heavy load and will tend to wear out sooner than a properly sized unit. Over-sized motors, on the other hand, are initially more costly and cost more to run while adding nothing to process or equipment efficiency.

Care should also be given to the location of motors. They should be located away from areas where there is pedestrian and vehicular traffic. They should also be located so that they will not be exposed to excessive dust or moisture. These latter two factors can lead to reduced time between scheduled preventive maintenance activities and potentially shorter life span because of the buildup of foreign material. In some instances, it may be necessary to enclose motors to protect them from excessive exposure to foreign materials. If this is necessary, care must be taken to assure that there is adequate air flow through the enclosure to provide for proper cooling. It is best to consult the manufacturer for their specific recommendations in cases where enclosures are being considered.

Some manufacturing processes may develop potentially hazardous [flammable or combustible] atmospheres. In these instances, it is necessary that all electrical equipment, wiring, switches, etc., be designed and installed so that it can be used safely. The details of the precautions that must be taken where hazardous locations may exist are beyond the scope of this chapter. The reader should consult Articles 500–503, (Hazardous Locations) of the NEC for the detailed requirements.

During the design phase of construction, all processes should be thoroughly evaluated to determine if they will have the potential to be classified as a hazardous location. After this determination has been made, consideration should be given to locating all such processes in one area of the plant. This will normally result in an overall cost savings.

Equally important to the precautions that are required for hazardous atmospheres are the requirements that all electrical equipment be grounded. In the event that there is a break in the insulation on electrical conductors inside a metal or other conducting enclosure, the enclosure will be at the same potential as the conductor. This situation presents the same hazard for individuals who might make contact with the enclosure as it would if they were to contact the live conductor. For this reason, it is necessary to assure that there is a path of adequate current carrying capacity from the enclosure to ground to dissipate this potential hazard. Grounding requirements vary depending on the magnitude of the voltage involved and the reader is referred to the NEC for details.

Grounding requirements also apply to small cord operated power hand tools. All such equipment manufactured today is equipped with a three prong cord unless exempt because of design. The third prong is for grounding of the equipment to the plant's electrical system. For this reason, the grounding prong should never be removed or a two prong to three prong adaptor used in cases where the current electrical system is not equipped with an internal grounding system. Updating of the plant electrical system is advisable in these instances.

LOCKOUT-TAGOUT PROCEDURES

The lockout-tagout standard was originally issued by OSHA in 1989 and amended in 1990. The purpose of the standard is to prevent accidents and injuries arising from the release of stored potential energy or from the unexpected energization of equipment during the servicing and maintenance of electrical and mechanical equipment. The major, and practically only, exemption to this standard is for plug and cord connected electrical equipment if operation can be completely controlled by unplugging the cord and the cord remains under the exclusive control of the individual performing the service or maintenance work.

Under the standard, employers must establish an energy control program that consists of control procedures, training, and periodic inspections to ensure that machines and equipment are isolated from their energy sources before any maintenance or servicing operations are performed if there is the possibility that the device could unexpectedly become energized, start up, or release stored potential energy and thereby cause an accident or injury. As discussed here, these lockout-tagout procedures apply only to electrical energy; however, it should be remembered that the lockout-tagout standard applies to all forms of energy and that the appropriate practices and procedures must also be developed to provide for the control of mechanical energy (see Chapter 17).

Energy control procedures must contain at least the four following elements:

1. A specific statement of the intended use of the procedure.

2. Specific procedural steps for shutting down, isolating, blocking, and securing machines/equipment to control hazardous energy.

3. Specific procedural steps for the placement, removal, and transfer of lockout or tagout devices and the responsibility for them.

4. Specific requirements for testing a machine or equipment to determine and verify the effectiveness of lockout-tagout and other energy control devices.

A simple procedure that has been developed by OSHA as an example to assist management in the development of an acceptable program is shown in Appendix 21-A at the end of the chapter.

ELECTRICAL SAFETY TRAINING

In addition to the training required under the provisions of the lockout-tagout standard, additional training is required under OSHA's standards for electrical safety. These can be found in Subpart S of 29 CFR 1910. The specific training requirements can be found in Paragraph 1910-332. Training is required for all employees who face the risk of electrical shock that is not reduced to a safe level through adherence to requirements for installation of electrical equipment. Typical job classifications for which training may be required are shown in Table 21-2.

Table 21-2
Job Classification That May Require Training Due to
Above Normal Risk for Electrical Accidents*

1. Electricians

2. Electrical and Electronic Engineers

3. Electrical and Electronic Assemblers

4. Electrical and Electronic Technicians

5. Industrial Machine Operators

6. Material Handling Equipment Operators

7. Mechanics and Repairmen

8. Painters

9. Riggers and Roustabouts

10. Stationary Engineers

11. Welders

NOTE: Workers in all job classifications except electricians may not need to undergo training if their work or the work of those they supervise does not bring them or their employees close enough to exposed parts of live electrical circuits operating at 50 volts or more for a hazard to exist.

Adapted from Table S-4 of 29 CFR Subpart S.

The requirements are broken down into those for "qualified" and "unqualified employees". A "qualified" employee is defined as one who is familiar with the construction and operation of the equipment and the hazards involved. All other employees are considered to be "unqualified"; thus, a particular employee might be qualified under one set of circumstances and unqualified under another. For both types of employees the training can be either classroom or on-the-job. It should be noted that the degree of training is determined by the extent of the potential hazards involved, so that, to some extent, the training is performance oriented.

Specific minimum training requirements for "qualified" employees are listed below.

❖ Skills and techniques necessary to distinguish live exposed parts from other parts of electrical equipment.

❖ The skill and techniques necessary to determine the nominal voltage of exposed live parts.

❖ The clearance distances specified for working with or in the vicinity of various voltage lines. These are shown in Table 21-3.

❖ The safety related work practices, as they apply to the employee's specific job that are set forth in Paragraphs 331 through 335 of Subpart S of 29 CFR 1910.

Table 21-3
Approach Distances for Qualified Employees
Alternating Current

Voltage Range (Phase to Phase)	Minimum Approach Distance
300 V and less	Avoid contact
300 V to 750 V	1 ft. (30.5 cm)
750 V to 2 kV	1.5 ft. (46 cm)
2 kV to 15 kV	2 ft. (61 cm)
15 kV to 37 kV	3 ft. (91 cm)
37 kV to 87.5 kV	3.5 ft. (107 cm)
87.5 kV to 121 kV	4 ft. (122 cm)
121 kV to 140 kV	4.5 ft. (137 cm)

Adapted from Table S-5 29 CFR 1910 Subpart S.

The provisions and requirements of Paragraphs 331 through 335 are quite extensive; therefore, anyone responsibe for developing an electrical safety training program should be familiar with this material and must determine which of them apply to their operations and must be included in the training.

The training provisions for unqualified employees who may be exposed to electric shock hazards are somewhat ambiguous and are not as stringent as for qualified employees. In addition to the job related safety provisions shown in the last training item for qualified employees, unqualified employees must be trained in "any electrically related safety work practices that are necessary for their safety."

Although there is no specific mention in the standard, an electrical safety program cannot be considered to be complete unless a sufficient number of employees are trained in appropriate resuscitation and life-saving techniques. It is, therefore, recommended that at least two individuals, who have completed an approved cardiopulmonary resuscitation (CPR) course, be assigned and be readily available on each shift if there is the potential for electrical accidents.

MAINTENANCE AND REPAIR OF ELECTRICAL AND ELECTRONIC EQUIPMENT

As with all other types of equipment, electrical equipment must be properly maintained if it is to provide its intended use in a safe manner. Also, using electrical equipment that is damaged and in need of repair is one of the major causes of electrical accidents and injuries. To assure that electrical equipment is kept in good working order, a preventive maintenance schedule should be developed and implemented.

Because of the potential for injuries to maintenance and repair individuals, it is imperative that all electrical equipment be isolated from its energy source before any work is undertaken. This is accomplished by having an effective lockout-tagout program in place. It is also imperative that individuals who perform maintenance and repair work on electrical equipment be thoroughly trained in appropriate work practices and procedures as well as the potential hazards associated with the work.

When performing repair or maintenance work, always be sure to follow the manufacturer's recommended procedures as well as observing their recommendations for the frequency of any periodic preventive maintenance. If there is any doubt regarding repair procedures or the frequency for preventive maintenance, the manufacturer should be consulted.

APPENDIX 21-A

SAMPLE MINIMAL LOCKOUT PROCEDURE

The following sample lockout procedure has been suggested by OSHA to assist employers in developing an acceptable energy control program. It should not be used verbatim, but should be modified to fit local plant circumstances. It will probably be necessary to develop multiple procedures because of the diversity of most company's operations.

It must be remembered that this procedure, as locally modified, will be acceptable only if the energy source is capable of being physically "locked out". In cases where lockout of the energy source is not possible, paragraphs must be added to assure that proper protective equipment is utilized, additional training is mandated, and that additional and more rigorous inspections are required. Depending on the circumstances, quite elaborate procedures may have to be developed.

LOCKOUT PROCEDURE FOR

_____

*(Indicate company name for single procedure or
specific equipment if multiple procedures)*

PURPOSE

This procedure establishes the minimum requirements for the lockout of energy [electrical] isolating devices whenever maintenance or servicing is performed on machines or equipment. It shall insure that the machine or equipment is stopped, isolated from all potentially hazardous energy [electrical] sources and locked out before any servicing or maintenance where the unexpected energization or start-up of the machine or equipment or releases of stored energy [electrical] could cause injury.

COMPLIANCE

All employees are required to comply with the limitations and restrictions imposed on them during the use of lockout. The authorized employees are required to perform the lockout in accordance with this procedure. All employees, upon observing a machine or piece of equipment which is locked out to perform servicing or maintenance shall not attempt to use that machine or equipment.

_____

_____

(Specifically spell out the company's enforcement policy for violation of the procedure)

SEQUENCE OF LOCKOUT

Prior to removing a machine/equipment from service for servicing or maintenance the following steps shall be taken.

1. Notify all affected employees that servicing or maintenance is required on a machine or equipment and that the machine or equipment must be shut down and locked out to perform the servicing or maintenance.

_____

_____

(The employees and/or job classifications that will be affected by the lockout and how to notify them should be listed here.)

2. The authorized employees shall refer to [appropriate company procedure] to identify the type and magnitude of the energy [electrical] that the machine or equipment utilizes, shall understand the hazards of the energy [electrical], and shall know how to control the energy.

_____

_____

(The magnitude of the energy source(s), the associated hazards, and appropriate control measures applicable to the specific procedure should be given here.)

3. If the machine or equipment is operating, shut it down through normal procedures.

_____

_____

(Detail how to shut off the device and the location of the controls that will perform this task.)

4. Deactivate the energy [electrical] isolating device(s) so that the machine or equipment is isolated from the energy source(s)

_____

_____

(List the types and locations of energy isolating devices that are applicable.)

5. Lockout the energy isolating device(s) with assigned individual lock(s).

6. Any stored or residual energy [state type] must be dissipated or restrained by methods such as [state type such as bonding, grounding, etc and how it will be accomplished.]

7. Ensure that the equipment is disconnected from its energy source. Make sure no personnel are or will be exposed; then verify the isolation of the equipment by attempting to operate through the normal controls or [specify other test methods if applicable]. **Caution:** The operating controls must be returned to the neutral or "off" position after verification of isolation.

8. The machine/equipment is now locked out and servicing or maintenance can proceed.

RESTORING EQUIPMENT TO SERVICE

After servicing or maintenance is completed and the machine/equipment is ready to be returned to normal operation, the following steps shall be taken.

1. Check the machine/equipment and the area immediately around the machine/equipment to ensure that non-essential items have been removed and the machine/equipment and/or components are operationally intact.

2. Check the area to ensure that all non-essential personnel are in a safe place or are well clear of the area.

3. Verify that all operating controls are in the "off" position or are in neutral.

4. Remove the lockout devices and re-energize the machine/equipment.

5. Notify affected employees that the servicing or maintenance has been completed and that the machine/equipment is ready for use.

CHAPTER
22
WELDING AND CUTTING

To ensure that your plant's welding and cutting operations are conducted in a safe manner, take the following steps:

❖ Identify the types of welding activities and their associated hazards.

❖ Select and implement control measures to address these hazards.

❖ Provide training in safe work practices.

The remainder of this chapter provides the information necessary to be able to accomplish these steps.

THE WELDING PROCESSES

There are many different types of welding, only some of which may be present at your plant. Most of the common types are described in the following paragraphs.

Arc Welding and Cutting

Arc welding uses electricity to melt the base metal and electrode which forms a bond upon cooling. Arc cutting, on the other hand, uses electricity to cut metal. It is the most common welding process and has many variations using different types of electrodes, fluxes, shielding gases, and other equipment. Arc welding and cutting are used for nonferrous metals, stainless steel, and steels with high chromium or tungsten content. Electric shock and those hazards common to all types of welding are of concern.

Oxyfuel Gas Welding and Cutting

In this process, heat from burning gases is used to melt the base metal. Welding rods are only used when filler material is required and are usually similar in composition to the base metal. Flux may or may not be used. Oxygen welding and cutting are used for plain carbon, manganese, and low-chromium content steels. The common welding hazards, such as metal fume generation, must be addressed as well safe handling of compressed gas cylinders and flammable gases.

Resistance Welding

In this type of welding, electric current is passed through pieces of metal bonding them at the contact point due to the heat generated by electrical resistance. Hazards associated with this welding method include lack of guards on the equipment at the point of operation, flying hot metal particles, improper handling of materials, and unauthorized adjustments and repairs which may result in eye injury, burns, and electric shock. Again, electricity and those hazards shared by all welding types are concerns.

Brazing

In this process metals are heated and joined together by a molten filler metal at a temperature greater than 450^0 C (840^0 F). This process is like soldering. The filler metal may come in the form of wire, foil, filings, paste, powder, slugs, or tape. Unless done in a vacuum, fluxes must be used to prevent oxidation and resultant weakening of the bond. Hazards associated with this process are dependent upon the type of heat source and will be similar to those listed for arc and oxygas welding.

Table 22-1 summarizes the major emission hazards from common welding and cutting processes.

Table 22-1 Welding Processes and Associated Hazardous Agents

Process	Hazardous Agent
Brazing: Cadmium Filler	Cadmium
Electron Beam Welding	X-Rays
Gas Metal Arc Welding (GMAW) w/ Aluminum or Aluminum/Magnesium	Ultraviolet (UV) Radiation, Ozone
GMAW w/ Stainless Steel	Chromium Fume, Nickel Fume, Ozone
GMAW w/ Carbon Dioxide	Carbon Monoxide
Gas Tungsten Arc Welding w/ Aluminum or Aluminum/Magnesium	UV Radiation
Oxygas Welding & Cutting	Carbon Monoxide, Nitric Oxide, Nitrogen Dioxide
Shielded Metal Arc Welding (SMAW), w/ low-Hydrogen Electrodes	Fluorides, UV Radiation
SMAW w/ Iron or Steel	Iron Oxide Fume, UV Radiation
SMAW w/ Stainless Steel	Chromium Fume, Nickel Fume, UV Radiation, Ozone
Plasma Arc Cutting/Welding	Noise, Ozone
All Welding Methods	Welding Fume, Gases: Oxides of Nitrogen, Ozone, Heat, Electromagnetic Radiation

HAZARDS OF WELDING AND CUTTING

There are many hazards associated with welding operations. The intense energy required to change a metal from a solid to a liquid produces a variety of toxic fumes, dusts, gases, and vapors. Welding also exposes workers to noise, vibration, heat, electricity, and radiation. One simple way of addressing these hazards is by dividing them into two categories: those hazards affecting a welder's health and those creating an unsafe plant environment.

Health Hazards and Contaminants

As a manager it is important to understand the contaminants produced by each welding process and the associated safety hazards. Focusing upon this knowledge will allow you to plan and implement necessary control strategies. Information related to specific health effects of welding is of greater use to your safety professional, industrial hygienist, or outside expert, who would use this information to conduct activities such as exposure monitoring, medical evaluations, and hazard assessments. For information purposes, Table 22-2 outlines the wide array of effects the welding processes can have upon the human body. These hazards can generally be controlled through the use of ventilation and respiratory protection.

Table 22-2
Health Hazards Associated with Welding

Pulmonary Effects

    Nonmalignatory Respiratory System Effects

    1) Metal Fume Fever

    2) Pneumonitis

    3) Pulmonary edema

    4) Pneumoconiosis & siderosis

    5) Bronchitis & decrease in pulmonary function

    Respiratory Cancer

Kidney and Urinary Tract Cancer
Reproductive Effects
Cardiovascular Effects
Gastrointestinal Effects
Ophthalmologic Effects
Dermatologic Effects
Auditory Effects
Musculoskeletal Effects

Welding Fume

A fume is formed when a solid, such as a metal, is heated, vaporized and cools into particles when released into the air. There may or may not be a chemical reaction with oxygen in the air to form an oxide of the metal. Welding fume is particularly hazardous because the particles are extremely small and, therefore, readily inhaled. The largest source of this contaminant is the filler metal. Fume may also originate from the base metal, base metal coatings, flux, and electrode coatings, as well as from fluxes and fillers. Some fume generation may also occur from the pulverizing of materials while cleaning welds and brazes by brushing or grinding. The chemical makeup of the fume generally depends on the materials present in the base metal, the flux, and the filler materials. The fume, however, may have a different chemical form so it is necessary to address each situation separately. Some components commmonly found in welding materials include:

Aluminum	Manganese
Beryllium	Nickel
Cadmium	Potassium
Calcium	Silica
Chromium	Sodium
Fluorides	Titanium
Iron	Vanadium Pentoxide
Lead	Zinc
Magnesium	

Gases

Several gases are associated with or can be produced during the welding process. These gases can produce a variety of health effects.

Carbon Monoxide.　Carbon monoxide can result from use of carbon dioxide as a shielding gas. This gas will act as a chemical asphyxiant and cause a person to suffocate by replacing oxygen in the blood.

Fuel Gases.　These are usually asphyxiants which will displace oxygen in the air and cause suffocation. They will be present due to leaks in gas cylinders and associated hardware. A dangerous situation will arise when working with leaking equipment in confined spaces or areas with poor ventilation.

Oxides of Nitrogen.　Arcs or high temperature flames cause atmospheric oxygen and nitrogen to combine and form oxides of nitrogen. These can include nitrogen dioxide and nitrous oxide which irritate the respiratory system.

Ozone.　Ultraviolet light from the welding process and atmospheric oxygen react to form ozone. The type of base metal can affect the rate of production. Stainless steel work is of particular concern. This gas acts as a respiratory system irritant.

Organic Decomposition Products.　Degreasers such as trichloroethylene and tetrachloroethylene can be present on the metal surface. Ultraviolet radiation reacts with the solvent vapors to produce a number of irritant and toxic gases. Some of these include:

dichloroacetyl chloride	trichloracetyl chloride
chlorine	phosgene
hydrogen chloride	

Other compounds can be produced when welding operations heat materials. For example, residual oil on steel may produce acrolein, a very strong irritant.

Physical Agents

Physical agents such as heat, noise, and radiation can present a number of hazards which primarily target the exterior regions of a worker's body. The majority of these are controlled through the use of personal protective equipment. These agents include:

Electromagnetic Radiation.   The levels of electromagnetic radiation, also known as nonionizing radiation, depend on the type of welding process, the tip size, flame type, and filler metal material. The electromagnetic radiation spectrum includes ultraviolet, visible, and infrared, radiations. Infrared radiation can be absorbed by the worker's clothing and skin, raising the skin temperature and contributing to the body's heat load. Ultraviolet radiation, primarily associated with arc welding, causes damage to the skin and eyes when large doses are received. Skin burns caused by UV radiation are similar to sunburn.

Electricity.   Electric shock can result from arc and resistance welding, especially when equipment is in poor repair. Contributory factors often include a reduction in resistance through the welder due to sweat or other wet conditions. A major shock can paralyze the respiratory system and cause ventricular fibrillation and death, while a smaller shock can lead to accidents such as falling.

Heat.   Welding materials such as hot metal, sparks, and "slag," can cause burns on contact. Heat stress due to excessive temperatures in the work area combined with heat from the welding operation can also be a potential problem.

Ionizing Radiation.   Ionizing radiation, such as X-Rays, are produced by electron beam welding equipment. This type of welding is not commonly used.

Noise.   Plasma arc welding and cutting present the greatest hazards of excessive noise exposure; however, other types can also produce high noise levels.

Asbestos

Exposure to asbestos can occur during the use of asbestos-containing materials, disturbing asbestos insulation, or working near operations which use asbestos.

SAFETY HAZARDS

As a manager, you must be aware of and address the potential hazards associated with welding operations which can result in injuries and fatalities. A brief description follows. Details on controlling these hazards are contained in the section on training later in this chapter. Specific chapters addressing each topic may also be found in this book.

Preventing Fires and Explosions

One of the most obvious hazards associated with welding is its potential for causing fire and explosion. The simplest methods of controlling this hazard are to ensure that work areas are free of flammable and explosive materials, to provide adequate ventilation, and to use welding shields to limit travel of sparks.

Some additional steps to reduce the risk of fire and explosion are:

- ❖ Assemble, install, and maintain all equipment using competent personnel. Establish a program for routine scheduled maintenance and leak testing of equipment.
- ❖ Establish permanant welding areas whenever possible. Otherwise, use ANSI Z49.1 and NFPA 51B to determine what fire protection is necessary.
- ❖ Require that any work performed outside permanent established welding areas be approved with a "Hot Work" Permit. Refer to the section concerning "Hot Work" permitting in Chapter 19 for details.
- ❖ Do not permit welding and cutting operations in or near rooms containing flammable or combustible vapors or on or inside closed containers until all fire and explosion hazards have been eliminated.
- ❖ Use welding shields to prevent sparks and other potential ignition sources from travelling outside the work area.
- ❖ Initiate fire watches when work begins and continue for at least 30 minutes after its completion.
- ❖ Ventilate work areas thoroughly and conduct frequent gas testing with an explosimeter or a combustible gas meter.
- ❖ Provide local exhaust ventilation where natural ventilation is inadequate to prevent buildup of gases or vapors. Exhaust equipment should be explosion proof.

Electric Shocks

This safety hazard is associated with operations that use electricity to generate heat, such as arc and resistance welding and cutting. Often, proper installation and maintenance of equipment and keeping the worker and the work area dry will be the only necessary precautions. Steps to prevent electric shocks include:

- ❖ Have all equipment assembled and installed by competent personnel.
- ❖ In confined spaces, protect cables from falling sparks.
- ❖ Never use a bare hand or wet glove to change electrodes.
- ❖ Never stand on a wet or grounded surface when changing electrodes.
- ❖ Ground the frames of welding units.
- ❖ Receptacles of power cables on portable welding units should make it impossible to remove the plug without opening the power supply switch. An alternative is to use receptacles which have been approved to break full load circuits of the unit.
- ❖ Repair frayed or damaged cables immediately. Keep all equipment in top condition.
- ❖ Keep cables clean and free of grease.
- ❖ If utilizing long lengths of cable, suspend them overhead whenever possible. If run along the floor, be sure they do not create a tripping hazard, become damaged, or tangle.
- ❖ Keep cables away from high tension wires and power supply cables.

Welding and Cutting Drums, Tanks, and Closed Containers

The greatest hazard associated with welding and cutting of vessels is the potential for fire and explosion due to the buildup of vapors inside the vessel. These vapors are typically present due to the vessel's contents. When the vessel is to be entered, there may also be a confined space hazard.

1. Determine proper protective equipment for all phases of the project and require use of that equipment.

2. Remove all ignition sources.

3. Remove the bung (as applicable).

4. Remove all solid and liquid materials. Allow to drain on a steam rack for at least 5 full minutes.

5. Steam clean drums for at least 10 minutes. If they contained shellac, turpentine, or similar substances, allow the vessel to steam longer.

6. Fill the drum part way with caustic soda ash or appropriate agent, rotate, and hammer lightly with a wooden mallet to remove scale.

7. Flush with boiling water for at least 5 minutes. Wash outside of container.

8. Air dry the drum and inspect it. If it is not clean, repeat entire procedure.

9. Test the container with a combustible gas meter. If the vessel is to be entered, also test for oxygen deficiency. If operations continue for an extended period of time repeat atmosphere tests periodically (may require conformance to the OSHA confined space standard).

Compressed Gas Cylinders and Hardware

Compressed gas cylinders present both mechanical and chemical hazards, depending on the particular gas. Welding gases are primarily simple asphyxiants, so from a health standpoint the hazard is minimal unless work is conducted in a confined space. Because these gases are flammable, however, there is danger of fire and explosion. These gases are at very high pressures within the cylinders and represent a physical hazard should the valving or cylinder become damaged. This means that special handling is required for the cylinders, regulators, and related hardware.

Cylinders

❖ Inspect all cylinders and hardware for proper connections and operation prior to beginning work.
❖ Test all connections for leaks.

Hose and Hose Connections

❖ Oxygen and acetylene hoses should be different colors. Red is generally used for fuel and green for oxygen. Black is generally used for inert gas and air hoses.

❖ Do not use unnecessarily long lengths of hose and protect them from traffic through the work area. Whenever possible suspend long lengths overhead.

❖ Check for leaks periodically by immersing hoses in water. Repair leaking hoses immediately as they represent fire hazards and are wasteful. Repair by splicing. DO NOT USE TAPE.

❖ Protect hoses from flying sparks and slag, hot objects, and grease and oil.

❖ Store in a cool place.

❖ Do not use single hoses with more than one gas passage where a wall failure could result in mixing.

❖ Do not tape more than 4" of each 12" of hose together when parallel lengths of hose are taped.

❖ Do not use metallic coverings on hose. Metal reinforced hose, where the metal is neither exposed to the gas nor the outside atmosphere, is acceptable.

❖ If flash back occurs and burns the hose, discard the burned section and purge the new hose before connecting.

Torches

❖ Select proper head or mixer tip or cutting nozzle and attach it properly.

❖ Shut off the gas at the valve before changing torches. DO NOT crimp the hose.

❖ Do not use matches; use a friction lighter or pilot flame.

❖ Do not put a torch down unless it is shut off.

Work Above Ground

❖ Use a platform with railings, or a safety belt and lifeline for work more than 5 feet above ground.

PERSONAL PROTECTIVE EQUIPMENT REQUIRED FOR SAFE WELDING AND CUTTING

The use of personal protective equipment (PPE) is critical to the control of welding hazards. Specifications for the PPE discussed below are available or referenced in the OSHA Standards 29 CFR 1910 Paragraphs 132-140, and 252-254. More information on PPE is also available in Chapter 12.

Required (OSHA) personal protective equipment to be worn by your welders is described in the following paragraphs.

Skin Protection: Clothing

Clothing is important to protect the welder from exposure to heat, ultraviolet radiation, and sparks. Items should be fire/flame resistant to protect the welder from burns and should include:

❖ Flame-resistant gauntlet gloves.

- ❖ Long sleeve shirts, preferably wool or leather.
- ❖ Fire-resistant aprons, coveralls, and leggings or high boots.
- ❖ Welders performing overhead work should wear fire resistant shoulder covers, i.e., capes, head covers, and ear covers.
- ❖ Workers welding on metal alloys containing toxic materials such as beryllium, cadmium, chromium, lead, mercury, or nickel, should wear clothing that is laundered each day. Street cothing should be stored separately. Fire resistance treatment must be provided after laundering. Laundry personnel should be made aware of the possible contamination.
- ❖ All clothing should be free of oil and grease, and should have no front pockets or upturned sleeves or cuffs. Cuffs and collars should be kept buttoned.

Eye and Face Protection

Eye and face protection is necessary for protection from ultraviolet radiation and flying sparks and debris. All eye and face protection must comply with 29 CFR 1910.252 (e) (ii). Arc welding and cutting requires:

- ❖ Welding helmets equipped with UV filter plates.
- ❖ Although allowed by OSHA regulation, hand-held screens should NOT be used since they may be held incorrectly.

Oxygas and brazing requires:

- ❖ Safety glasses with side shields or goggles with suitable filter lenses.

Resistance welding, mechanical cleaning, and chipping requires:

- ❖ Safety glasses with side shields or goggles with suitable transparent lenses.

Respiratory Protection

Respirators are useful in limiting the worker's exposure to certain air contaminants. The exact type to be worn depends on the welding contaminants and their concentration. Whenever a carcinogen or conditions immediately dangerous to life or health are present (i.e., work in a confined space with a possible oxygen deficiency), use the following:

- ❖ Self-Contained Breathing Apparatus (SCBA) with full facepiece operated in a pressure-demand mode or other positive-pressure mode, or
- ❖ An airline respirator that includes an escape bottle with a full facepiece operated in pressure-demand or positive-pressure mode.

Hearing Protection

The need for noise protection will depend on the nature of the work being performed. Generally, if a normal conversational voice can not be heard, hearing protection may be required. More information on noise control and hearing protection is available in Chapter 24, Noise and Vibration.

Precautions for Special Situations

Confined Spaces

- ❖ Wear respiratory protection appropriate for the situation when entering vessels. Protection requirements can range from none to a self-contained breathing apparatus with full facepiece operated in pressure-demand mode or positive-pressure mode. Also follow the OSHA confined space requirements (24CFR 1910.146).
- ❖ Safety harnesses should be attached in such a manner that the welder's body cannot become jammed in a small exit opening.
- ❖ If work is interrupted, disconnect the power and remove the electrode from the holder or turn off torch valves outside the confined area and remove the torch and hose from the area, as applicable.

Container Cleaning Operations

- ❖ Wear head and eye protection, rubber gloves, boots, and aprons when handling steam, hot water, and caustic solutions. When handling dry caustic soda, wear appropriate PPE.
- ❖ Wear proper thermal protective clothing when handling hot drums and equipment.

TRAINING PERSONNEL TO WELD SAFELY

At a minimum, train your workers in all relevant work practices outlined in the OSHA standards, i.e., 29 CFR 1910.251-254 "Welding, Cutting, and Brazing". This standard regulates:

- ❖ installation and operation of welding systems
- ❖ compressed gas cylinder and hardware use
- ❖ fire prevention and protection
- ❖ welding and cutting containers
- ❖ protection of personnel
- ❖ health protection and ventilation
- ❖ industrial applications

In addition, train your employees in the following:

- ❖ The work practices aimed at the control of specific welding hazards which will be encountered, such as those given earlier in this chapter.
- ❖ The safety and health standards established by the American Welding Society.
- ❖ Specific work procedures you develop, i.e., how to conduct operations in confined spaces.
- ❖ Hazards associated with welding, including the contaminants associated with it, the symptoms of exposure, and steps to take to control the hazards.
- ❖ The use of all applicable personal protective equipment.

❖ The safe handling of compressed gases.

❖ General safe work practices; e.g., when work is completed mark "hot metal" or post a warning sign.

❖ Good housekeeping practices; e.g., don't throw rod stubs and electrodes on the floor, discard properly. Return tools to their proper locations to prevent tripping hazards.

❖ Good personal hygiene practices.

❖ Emergency procedures.

CHAPTER
23
PRESSURE VESSELS

Pressure vessels are generally separated into two distinct types: boilers and unfired vessels. Each type has similar but unique hazards and requirements. Boilers are used to generate heat which is used to produce steam or heat other fluids that can do useful work, such as generating electricity or providing heat to an area. Unfired pressure vessels are used to contain a process fluid, liquid, or gas and do not have the problems associated with direct contact of burning fuel impinging on the vessel surfaces. Unfired pressure vessels can operate at pressures above or below normal atmospheric. Heating or cooling is accomplished through the heat generated by chemical reactions taking place in the vessel or through the application of a heating or cooling medium such as electricity, steam, cold water, or other fluid. Unfired vessels can include, but are not limited to, compressed air tanks, chemical reactors, steam jacketed kettles, etc.

In this chapter the following topics will be discussed:

❖ Hazards of Pressure Vessels
❖ Design Considerations
❖ Required Inspections
❖ Maintenance and Recordkeeping
❖ Operator Training

HAZARDS OF PRESSURE VESSELS

Unfired Pressure Vessels

There are two categories of hazards associated with unfired pressure vessels: pressure failures and leakage failures. Pressure failures are generally catastrophic. There is a blast effect from the very rapid expansion of pressurized process materials and a fragmentation effect from the pieces generated from an explosion or implosion of the vessel. Both of these effects can have severe consequences and result in considerable property damage and personnel injuries.

Leakage failures generally are not associated with the main body of a pressure vessel but rather with its fittings and attachments such as piping, valves, gaskets, and flanges. Failure of these fittings can range in seriousness from almost no consequence to extremely serious depending on the magnitude of the leak, its location, and the inherent toxicity of the material contained by the system. A small valve or flange leak of a low toxicity material into a relatively open area would present little immediate hazard to employees. On the other hand, the same leak of a highly toxic chemical into a confined space would represent a very serious hazard to employees. If flammable or combustible materials are present when either type of vessel failure occurs, there is also the potential ignition hazard of the process material.

Boilers

In addition to the two categories of failures and their associated hazards discussed above for unfired pressure vessels, boilers present the added hazard of always having a combustible fuel present. Most accidents, fires, and explosions occur when boilers are being brought "on" and "off" line.

In boilers the primary "process" fluid is water which can be present in several forms including high temperature water, low or high pressure steam, and supercritical steam as in the boilers of electrical generating plants. Exposure to live steam can cause very serious burns with supercritical steam being particularly dangerous in the event of a leak. A supercritical steam leak is invisible and, if an employee were to walk through such a leak, the force could easily sever a limb.

DESIGN CONSIDERATIONS

The Boiler and Pressure Vessel Code issued by the American Society of Mechanical Engineers (ASME) covers all facets of the design, manufacture, installation, and testing of most types of boilers and unfired pressure vessels. At the current time the ASME Code contains 11 sections and occupies several feet of shelf space. While the ASME Code was originally developed as a voluntary set of guidelines for steam boiler manufacturers, over the years it has been adopted as a mandated requirement in most jurisdictions. It should be noted that the following types of vessels are specifically *excluded* from the ASME Code:

❖ Fired process tubular heaters
❖ Pressure containers that are an integral part of rotating or reciprocating machinery or which serve as pneumatic or hydraulic cylinders
❖ Piping systems and piping components
❖ Small hot water storage tanks
❖ Any vessel with an internal or external operating pressure of less than 15 pounds per square inch-gauge (psig)

Because the ASME Code does not apply to vessels that operate at a pressure of less than 15 psig, the American Petroleum Institute (API) developed a consensus standard for this type of vessel. *Recommended Rules for Design and Construction of Large, Welded, Low-Pressure*

Storage Tanks is issued as API Standard 620. These two standards cover most of the applications that are found in the majority of plants.

Several procedural steps are spelled out in the ASME Code to assure that all applicable design and manufacturing provisions of the code are met and adhered to. These include:

❖ ASME certification of the manufacturer.

❖ Third party inspection and verification that all design and construction requirements have been.

❖ Marking of each manufactured vessel with the ASME stamp which includes the manufacturer, maximum and minimum working temperatures and pressures, serial number, and date of manufacture.

❖ Preparation of a data report for each vessel which includes the basic material specifications, allowable temperatures and pressures, and the required construction, assembly and installation certifications.

When contracting to have a vessel built and installed, it is extremely important that the design team be provided with as much information as possible. This would include but not be limited to:

❖ Detailed information on what will be contained in the vessel and the process chemistry.

❖ Maximum and minimum expected working temperatures and pressures.

❖ Whether temperature and/or pressure maxima and minima will be cyclical and the frequency of the cycling.

In an effort to save money, companies will sometimes attempt to purchase used or second hand vessels and to adapt them to their current needs. If this action is being considered, management should be aware that there can be many pitfalls and problems. A vessel that is to be located in an area where the ASME Code is applicable must be inspected by an authorized inspector prior to purchase and a written report must be obtained stating that the vessel meets the requirements of the intended service. There are numerous instances where a company has purchased, moved, and reinstalled a used vessel prior to inspection only to find that it did not meet ASME criteria for the intended application and, therefore, could not be certified or used.

Of critical importance to the safe operation of any pressurized system is the presence of emergency relief devices or safety valves. These devices are designed to protect the vessel from accidental overpressuring and greatly diminish the possibility of an explosion or other serious accident. While these devices are covered in the ASME Code, certain design aspects are common to all types of pressure operations such as:

❖ All pressure vessel must be equipped with appropriate relief devices.

❖ Piping on the discharge side of a relief device should have the minimum number of turns and should be as short as possible.

❖ Exhaust ports or discharge lines from relief devices should be configured to discharge into a "safe" area.

❖ The incorporation of an expansion tank or "knockout tank" is recommended as an intermediate device between the vessel and its ultimate point of exhaust if long pipe runs are required or to prevent the release of toxic or noxious chemicals to the environment.

❖ Relief devices must be set to activate at a point that is below the lowest maximum allowable working pressure of <u>any</u> vessel in the system.

❖ Self closing relief devices should be mounted in a vertical position to assure that they will close properly after opening.

❖ Relief devices that are equipped with shut off valves on either the inlet or outlet must have a means to lock the valve in the open position.

The locations where pressure vessels are installed is also an important consideration in the overall design process. Care must be taken to assure that all surfaces of the vessel and any auxiliary equipment that may also be under pressure are accessible for inspection. Cases are known in which pressure vessels have been installed such that the surrounding support members and ancillary equipment prevented inspection of portions of the vessel. It is in these uninspected areas where failures have occurred, causing considerable property damage.

REQUIRED INSPECTIONS

Most organizations do not have the appropriately trained personnel to undertake thorough and detailed inspections and evaluations of pressure vessels. Depending on the nature of the equipment that is being used, the requirements for periodic evaluations and inspections can be extremely complex and time consuming. For these reasons, it is generally recommended that pressure vessel inspections be contracted to a firm or individual with the appropriate training and/or certification. The general rule is that pressure vessels must undergo thorough inspections at intervals that are no greater than one-half of the estimated remaining corrosion life, or a maximum of ten years. Vessels in service where the probability of corrosion is high require more frequent inspections. These include de-aerators, amine and ammonia service vessels, and hydrogen sulfide service vessels.

In addition to mandated, periodic vessel inspections that must be performed by trained or certified inspectors, there are tasks that should be conducted in-house on a periodic basis. These deal primarily with pressure relief systems and include:

❖ All relief devices, piping, and valves should be inspected to assure that they are open and free of obstructions before the process is first pressurized and after every process shutdown or cycle if the process is not continuous.

❖ All relief devices should be removed from service for preventive maintenance and checking of relief set points as required by the ASME Code or local ordinances but no less frequently than the following schedule:

➢ One year for critical operations, to be determined by local management, and for all processes that are subject to plugging or that handle corrosive materials.

➢ Two years for all other processes.

➢ Three years for storage vessels.

MAINTENANCE AND RECORDKEEPING

Maintenance activities on pressure vessels often require personnel to enter the equipment. By the nature of their construction, entry into pressure vessels is generally through very small or restricted openings, thus they are considered to be confined spaces. Before any entry into a pressure vessel is made, it is imperative that appropriate precautions be taken. These include:

❖ Assure that oxygen content is at least 19.6 percent.

❖ Assure that a flammable atmosphere is not present.

❖ Assure that potentially toxic chemicals are not present at concentrations above mandated or recommended limits.

❖ Provide individuals with appropriate personal protective and safety equipment.

❖ Have an attendant stationed at the entryway and utilize a life line. Have a rescue team available.

❖ If appropriate, perform continuous atmospheric monitoring inside the vessel during repair or maintenance procedures.

Most vessels will be permit-required confined spaces.

All maintenance and repair work must be performed in accordance with the manufacturer's recommendations or requirements. This is particularly true of any welding or other "hot work" that might alter the metallurgy of the vessel and potentially compromise its design strength.

Good recordkeeping is an important aspect of pressure vessel safety. In addition to the initial manufacturer's data report and the mandated periodic inspection reports, records should be kept of the vessel's service history. These should include:

❖ The type and composition of the fluids that are handled in the vessel, including dates of any change.

❖ Actual operating temperatures and pressures, including dates of any changes.

❖ Type of service: continuous, intermittent, or irregular.

❖ Discontinuities in service and the reasons.

❖ Vessel history which includes alterations, re-ratings, and repairs, including the dates they were performed.

OPERATOR TRAINING

Adequate operator training in both routine and emergency procedures is one of the most important aspects for the safe operation of boilers and unfired pressure vessels. All employees who are being trained as either operators or helpers should be thoroughly aware of all facets of the process. Because familiarity with the equipment is of paramount importance in operating safety, all employees who will have responsibility for pressure vessels must undergo an extensive period of closely supervised on-the-job training.

It is advisable that a detailed process flow diagram be available at all times as well as written process procedures. These procedures should spell out, in detail, the exact routine that is to be followed for each process cycle and steps that are to be taken in the event of unforeseen incidents and emergencies. In order to properly detail incident and emergency procedures, it is generally necessary to perform a fault tree or "what if" analysis so that all potential events can be covered.

Many organizations utilize a checklist system wherein the operator must sign off after each step in a cyclical process. A separate checklist is generally used for each cycle. The checklists then become a part of the vessel's or processing unit's permanent operating history. These documents, when coupled with operating logs, can be valuable tools for detecting gradual changes in process operating parameters which can sometimes be an indication of potential problems. They are also valuable in the investigation of accidents and unusual occurrences.

CHAPTER
24

NOISE AND VIBRATION

Noise is commonly defined as unwanted sound and is produced by many industrial processes. Excessive amounts of noise, either inside or outside the work environment, can increase stress levels, interfere with communication, disrupt concentration, and most importantly, cause varying degrees of hearing loss. Exposure to high noise levels will also adversely affect job performance and increase accident rates.

Hearing loss due to occupational noise exposure has been well documented since the eighteenth century, and since the advent of the industrial revolution, the number of exposed workers has greatly increased. However, due to improvements in manufacturing equipment, worker protection, and government standards limiting noise exposure, the picture of hearing loss trends is not as bleak as it was in past decades. OSHA requires that noise exposures be controlled in all manufacturing environments. An effective Hearing Conservation Program can reduce the potential for employee hearing loss, reduce workers' compensation costs due to hearing loss claims, and lessen the financial burden of noncompliance with government standards. An effective program can be developed by taking the following steps:

1. Become familiar with the various types of hearing impairment and know their causes.

2. Determine the source and areas of high noise generation.

3. Comply with the OSHA Occupational Noise Exposure Standard (29 CFR 1910.95).

4. Develop a proactive approach.

5. Periodically evaluate the effectiveness of your Hearing Conservation Program.

The means and methods to complete these steps are provided in this chapter.

TYPES AND CAUSES OF HEARING IMPAIRMENT

The extent of hearing impairment is determined by audiometric testing, which measures a person's hearing threshold. However, because hearing impairment can have numerous causes, it is important to distinguish hearing impairment due to occupational noise exposure from other causes. The most common types of hearing impairment, along with common causes, are outlined below:

❖ Presbycusis: Hearing loss which is due to the natural degeneration of auditory nerve cells during the aging process.

❖ Tinnitus: Hearing impairment characterized by a ringing sound in the ears. Can be caused by middle ear conditions or infections and drugs such as aspirin, quinine, and alcohol.

❖ Sensorineural Hearing Loss: Hearing loss caused by damage to the structures of the inner ear. Can be caused by noise exposure.

❖ Conductive Hearing Loss: Hearing loss due to a middle or outer ear disorder. This type of hearing loss is not caused by exposure to noise but is a result of sound being prevented from reaching the inner ear. Examples include physical blockage of the auditory canals by a small object or ear wax and punctured ear drums.

Sensorineural hearing loss is the type of hearing impairment that is of most concern from an occupational viewpoint. It can be either temporary or permanent. Temporary loss may result from exposures to high level noise, such as operating a chain saw for several hours, or from short exposures to very high level noise, such as standing in the proximity of a jet during takeoff. Permanent hearing loss can result from prolonged exposure to very high level noise, from repeated exposure to high level noise, and even from short exposure to noise during explosions or blasts (acoustic trauma). Temporary hearing losses repeated over an extended amount of time tend to progress to permanent loss of hearing.

It is also important to note that sensorineural hearing loss can be caused by factors arising outside the work environment. Such causes can include: hobbies, such as hunting and listening to loud music; diseases, such as tumors and measles; hereditary or prenatal damage; and drugs which can damage the inner ear structures. It is important to obtain a thorough medical questionnaire to identify these risk factors. Not only can illegitimate workers' compensation claims be avoided, but employees can also be made aware of the methods available to reduce their risk of hearing loss when not at work.

THE OSHA OCCUPATIONAL NOISE EXPOSURE STANDARD

The OSHA Occupational Noise Exposure Standard was originally promulgated in May 1971. It was amended on several occasions, with the current standard becoming effective in June 1983. The standard is published in full in the Code of Federal Regulations (CFR) under 29CFR1910.95. This discussion of the standard is intended to provide general guidance for compliance.

Diagram 24-1 summarizes the general requirements of the standard and provides a simplified means of determining your obligations under the standard. To use Diagram 24-1, follow these steps:

1. Identify any areas or production processes which may overexpose an employee or employees to noise.

2. Obtain personal noise dosimetry measurements for employees assigned to these areas.

3. Review Diagram 24-1 to determine which requirements of the standard are applicable to the results obtained.

Sound is produced by pressure changes in air and can be caused by vibration or turbulence. Pressure changes produce wave fronts which are transmitted through the air. Sound pressure level, or noise level, is measured in decibels (dB). Noise measurements are usually obtained using different weighting scales, with the "A" scale being the most common because it most closely reflects the human ear's response to different frequencies. Measurements taken using the A-scale would be expressed as dBA.

Identification of high noise level areas is a continuing process. New equipment must be evaluated after installation, and existing processes should be periodically reviewed to identify changes in noise levels which may be due to a change in production rate and the aging or deterioration of equipment. When attempting to identify high noise areas, consider the following:

❖ Difficulty in oral communication may indicate high noise levels.

❖ Experiencing muffled sounds after leaving an area may indicate exposure to high noise levels.

❖ Increased trends in hearing loss in a particular work area may indicate high noise levels.

❖ If in doubt, obtain instantaneous noise measurements in areas of question during a representative work shift to ascertain approximate noise levels. Err on the conservative side and proceed to Step 2 above if the levels are 80 dBA or greater.

Any of these conditions necessitate the need to conduct personal noise dosimetry measurements. Noise dosimeters are instruments which integrate (measure and record) the sound levels over an entire work shift. Most have microphones attached to the worker's collar in the vicinity of the ear and are worn during a complete work shift. Exposures are typically measured as percent of the allowable dose. One hundred percent is equal to an 8-hour average sound level of 90 dBA, which is the OSHA Permissible Exposure Limit. A dose of 50 percent is equal to an 8-hour average sound level of 85 dBA, which is the OSHA Action Level. The reason that 5 dBA represents a 50 percent decrease in dose is that 5 dBA can be thought of as one-half of the the sound power since decibels are logarithmic functions.

The results of the personal noise dosimetry measurements will indicate your requirements under the standard. If the results are all below a 50 percent dose, all that you are required to do is to maintain the noise measurement records and remonitor the employees if a change in the production rate or process occurs which may expose the employees to higher noise levels.

If the results indicate that exposures are above a 50 percent dose, employees are required to be included in a Hearing Conservation Program, which is outlined below:

❖ Recordkeeping

—Retain exposure measurements (two years)

—Retain audiometric test results (duration of employment):

—Name and job classification of employee

—Date of audiogram

—Examiner's name

—Date of the last acoustic or exhaustive calibration of audiometer

Diagram 24-1 Flow Diagram of the Noise Standard

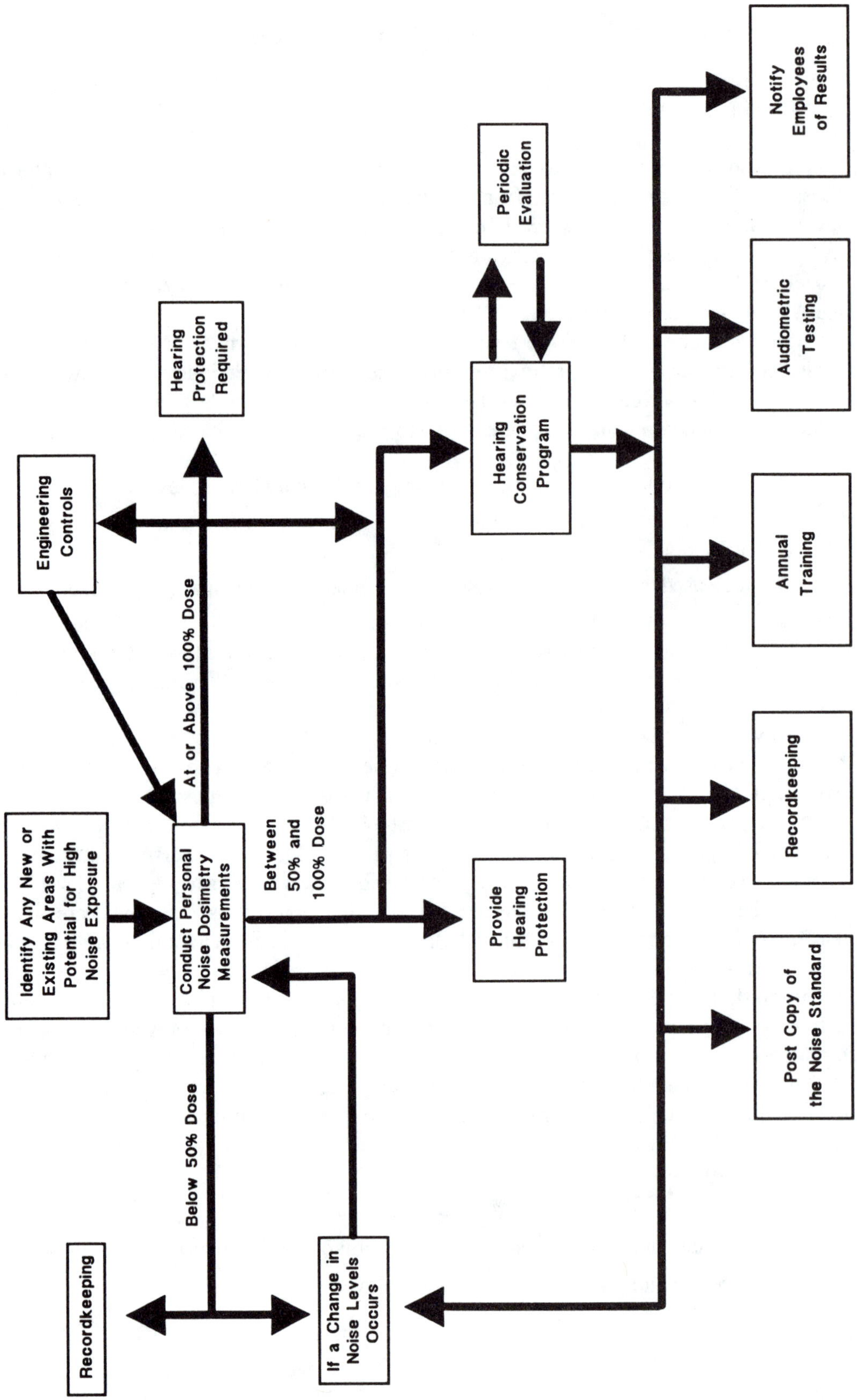

—Employee's most recent noise exposure assessment

—Measurements of the background sound pressure levels in audiometric test rooms

—Provide records upon request to employees

—Transfer all records to successor employer if company ceases to do business.

❖ Hearing Protection (Provided and/or Required)

—Provide at no cost to employee

—If required, ensure their use

—Allow employees to select from a variety of protectors

—Provide training in use and care of protectors

—Ensure proper fit and supervise correct use

—Evaluate hearing protector attenuation.

❖ Notify Employees of Personal Noise Dosimetry Results

—Notify in writing

—Notify employee in a reasonable time period.

❖ Audiometric Testing

—Obtain baseline and annual audiograms

—Precede audiogram by 14 hours of no occupational noise exposure

—Ensure equipment and persons performing tests are appropriate

—Ensure audiograms are evaluated properly

—See Diagram 24-2 for additional requirements and employer rights.

❖ Annual Training

—Ensure and document employee participation

—Ensure training materials are updated to be consistent with changes in protective equipment and work processes

—Inform employees of the following:

–Effects of noise on hearing.

–Purpose of hearing protection, advantages, disadvantages, and attenuation of various types, and instructions on selection, fitting, use, and care.

–Purpose of audiometric testing and explanation of the test procedure.

—Provide a copy of the OSHA standard.

—Ensure access to training materials.

❖ Noise Standard Posting

—Post in work areas affected by the standard

—Post in any area frequented by affected employees.

Diagram 24-2 Flow Diagram of Audiometric Testing

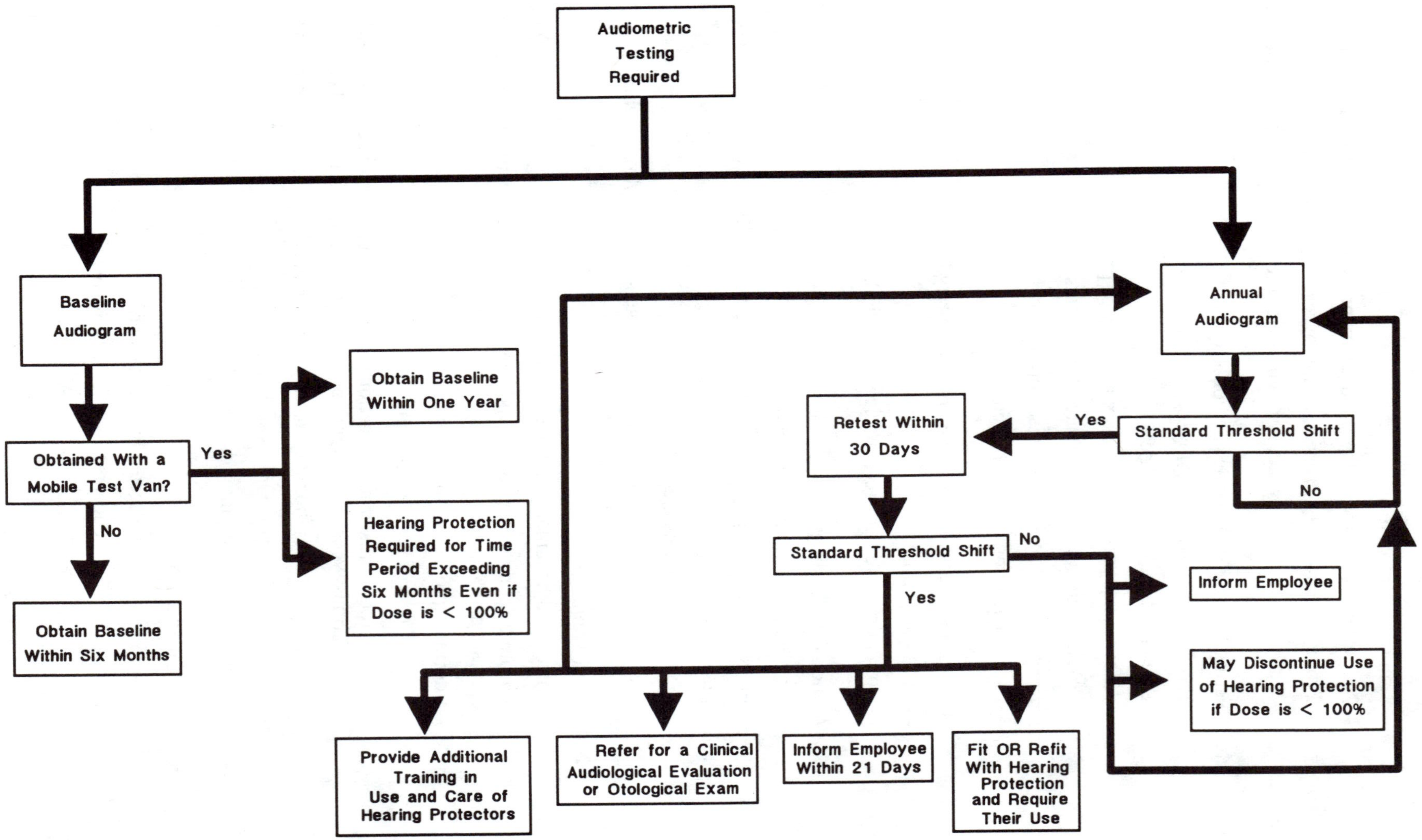

These requirements apply to all employees exposed to an average noise level of 85 dBA or greater. In reviewing the chart, you will note that between a 50 percent and 100 percent dose, it is only required that hearing protection be *provided* to affected employees. However, at a dose of 100 percent or greater, hearing protection must not only be provided, but the employer is also responsible for enforcing its proper use. The employer is required to provide a "suitable variety" of hearing protectors, usually interpreted to mean at least one type of ear plugs and one type of ear muff. Hearing protection must be evaluated to ensure that the attenuation is adequate to reduce the noise exposure below an average noise exposure level of 85 dBA. The most convenient method of evaluating the hearing protectors is by using the Noise Reduction Rating (NRR) supplied by the manufacturer. The procedure is as follows:

❖ Convert the dose percent to the 8-hour average noise level by using Table A-1 of the OSHA Standard. Certain conversion values are presented in Table 24-1. (The OSHA standard should be consulted for additional values.)

❖ Subtract 7 dB from the NRR.

❖ Subtract the remainder from the 8-hour average noise level to obtain the attenuated noise level with the use of the hearing protection.

Table 24-1
Conversion from Percent Dose to Average Sound Level*

Dose %	Average Sound Level	Dose %	Average Sound Level
10	73.4	200	95
20	78.4	250	96.6
30	81.3	300	97.9
40	83.4	350	99
50	85	400	100
60	86.3	450	100.8
70	87.4	500	101.6
80	88.4	600	102.9
90	89.2	700	104
100	90	800	105
150	92.9	900	105.8

* *Selected Values from 29 CFR 1910.95, Table A-1, "Conversion from Percent Noise Exposure or Dose to 8-Hour Time Weighted Average Sound Level (TWA)". For additional values, refer to Table A-1.*

The following examples will illustrate this calculation:

Example #1

Background Information: A press operator is monitored by noise dosimetry and at the end of the workshift it is determined that his exposure was 200 percent of the permitted dose. The package of hearing protection utilized by the operator states that the NRR of the hearing protection is 20.
Calculation:

1. In Table 24-1, a 200 percent dose is equal to an average sound level of 95.

2. Subtract 7 from the NRR (20).
   (20 - 7 = 13).

3. Subtract the remainder (13) from the average sound level (95).
   (95 - 13 = 82).

Conclusion: The hearing protection utilized by the press operator is adequate (82 dBA is less than the permissible exposure limit of 90 dBA).

Example 2:

Background Information: An airline mechanic is monitored by noise dosimetry and at the end of the shift it is determined that her exposure was 900 percent of the permitted dose. The NRR of her hearing protection is 21.
Calculation:

1. 900 percent dose = 105.8 dBA.

2. Subtract 7 from the NRR (21).
   (21 - 7 = 14).

3. Subtract the remainder (14) from the average sound level
   (105.8 - 14 = 91.8).

Conclusion: The hearing protection utilized by the airline mechanic is not adequate (91.8 dBA is greater than the permissible exposure limit of 90 dBA).

In addition to using the NRR to evaluate attenuation, there are more accurate, but involved, methods. These methods were developed by the National Institute for Occupational Safety and Health (NIOSH) and a copy of the procedure may be obtained from them.

Another requirement for employees exposed at or above a 100 percent dose is that engineering controls must be utilized when feasible. To the extent that engineering controls do not lower exposure to below a 50 percent dose, hearing protection is to be used to *supplement* engineering controls. Controls may include isolation of the worker or equipment, insulating materials, and noise damping of the equipment. A more detailed discussion of engineering controls is presented later in this chapter.

Audiometric testing is conducted to determine what level of hearing loss, if any, has occurred. A baseline audiogram is obtained to document the hearing ability of an employee prior to the employee's beginning work in a particular environment. Annual audiograms are obtained so that trends in hearing loss, *compared to the baseline,* can be determined. The standard requires that the baseline audiogram be obtained after 14 hours of no occupational noise exposure. This "quiet" period is necessary to ensure that any *temporary* hearing loss does not interfere with hearing ability. The *proper* use of *adequate* hearing protection may be used as a substitute for providing a "quiet" period. Audiograms must be obtained by qualified persons with appropriate equipment (see the standard for specific requirements.) Hearing thresholds are obtained at a minimum of 250, 500, 1000, 2000, 3000, 4000, and 6000 Hertz (Hz). Hertz is the term used to describe the number of cycles or pressure variations per second of time. We hear this as pitch. The 8000 Hz frequency is also commonly used in audiometric testing measurements. The frequencies most vital to communication are between 300 to 4000 Hz. Hearing loss at 6000 and 8000 Hz does not represent a communication handicap for the worker, but may indicate an early trend toward hearing loss at the lower speech frequencies. Diagram 24-2 summarizes the procedures you should take after obtaining the audiograms.

Annual audiograms are evaluated and compared with the baseline audiogram to determine if a Standard Threshold Shift (STS) exists. An STS is defined as a change in hearing threshold, relative to the baseline audiogram, of an average of 10 dB or more at 2000, 3000, and 4000 Hz, in either ear. Evaluation can be performed by a qualified technician. However, problem audiograms must receive further review by an audiologist, otolaryngologist, or qualified physician. The following should be provided to the person performing the evaluation:

❖ A copy of the standard

❖ Baseline and most recent audiogram

❖ Calibration records and background sound pressure levels of the audiometer

If a determination is made that an employee has experienced an STS, it may be an indication that noise control measures are inadequate. Therefore, the employer is required to take the following steps:

❖ Inform the employee of the STS within 21 days

❖ Fit or refit with hearing protection and *require that the employee use it.*

❖ Provide additional training in the care and use of hearing protectors.

❖ Refer the employee for a clinical audiological evaluation or otological examination. This may determine if medical pathology of the ear exists due to factors other than noise exposure.

❖ Any hearing level shifts averaging 25 dB or more must be recorded on the OSHA 200 Log for illness and injuries. An OSHA directive to field managers, dated June, 1991, states that citations will be issued if these shifts are not recorded.

Evaluation of audiograms should also take into account the effects of presbycusis, or hearing loss due to aging. Appendix F of the standard details the methods and values for age correction of audiograms.

Annual training for those employees who are included in the Hearing Conservation Program is also required under the OSHA Noise Standard. The required training components are listed in the Hearing Conservation Program outline.

The OSHA Occupational Noise Exposure Standard is, indeed, a very involved document. However, with proper planning, qualified personnel, and the guidelines previously set forth, the Hearing Conservation Program at your facility should accomplish its purpose—to protect your employees from occupational hearing loss.

THE PROACTIVE APPROACH

Using a proactive approach will not only increase effectiveness, but also will increase employee morale, improve employee/management relations, and improve the "bottom line" by lowering workers' compensation insurance costs, by improving employee productivity, and by helping to avoid OSHA citations stemming from gray areas of the standard. When the requirements of the standard are met, much information is compiled, and unfortunately, in some cases, filed until a workers' compensation claim is submitted. Knowing how to use this information is the key to an effective Hearing Conservation Program. The following procedures can assist you in developing a proactive program:

❖ Involve the employees, especially in the *concept* of the Hearing Conservation Program.

❖ Develop an annual audit program.

❖ Consider engineering controls as the first option to protecting employees.

❖ Evaluate audiograms thoroughly.

❖ Provide training that is interesting and informative.

❖ Review company policies regarding the use of hearing protection.

The first, and most vital, procedure in using the proactive approach is to involve all levels of personnel in contributing to the success of the Hearing Conservation Program. Unfortunately, this is often the most difficult step to accomplish, and the means to do this vary greatly depending on the management style of the company, existing employee attitudes, and the resources available. Although there is no magic formula, there are several guidelines you can follow:

❖ One person, who is knowledgeable about hearing conservation should have responsibility for the proper administration of the Hearing Conservation Program. Avoid situations in which individual program components are handled solely by different departments or persons. This type of arrangement obscures the concept of the program as a whole. It also requires a great deal of communication between the responsible parties, and is susceptible to communication failure. For example, in large plants proper communication between the engineering department, who designs controls, the medical department, who obtains and evaluates the audiograms, the safety department, who performs the monitoring, and the purchasing department, who orders the hearing protection, would be almost impossible without the direction and coordination of the single responsible person.

❖ Instruct supervisors to periodically discuss hearing loss and noise issues during production meetings.

❖ Identify high noise level areas with highly visible warnings. Posted warnings not only assist in reinforcing the hazard for the employees working in the area, but also inform visitors and employees who may enter the area infrequently of the hazard. See

Diagram 24-3
Examples of Detailed Noise Signs

CAUTION!
Noise Hazard

Noise Levels May Exceed 90 dBA
May Cause Hearing Loss
Hearing Protection Required

CAUTION!
Noise Hazard

Noise Levels May Exceed 85 dBA
May Cause Hearing Loss
Hearing Protection Recommended

Diagram 24-3 for possible content. For additional visibility, floor markings, denoting high noise level areas, should also be used.

❖ Involve union representatives, if applicable, in the Hearing Conservation Program and in gaining employee involvement.

❖ "Advertise" the fact that the company maintains audiometric testing records and also makes them, and the training materials, available for review.

These guidelines are only a few of the numerous ways in which you can involve employees in the program and can also increase its visibility. Use what you know about the management style of the company and employee attitudes to get everyone interested in making the program run effectively.

Instituting an annual audit program is one way to ensure that the Hearing Conservation Program is maintained. The audit should assist in identifying components of the program which require additional attention by both management and employees. The following components should be included in the audit process:

❖ An investigation of management attitudes, ensuring that all levels of management are included.

❖ An investigation of employee attitudes, including their perception of the noise levels, their hardships in complying with program requirements, their understanding of the reason for their inclusion in the program, and any ideas they have for improving the quality of the program and for making the requirements less burdensome.

❖ An evaluation of the company's compliance with the OSHA Occupational Noise Standard, including the adequate use of hearing protection.

❖ An evaluation of the effectiveness of the way in which the program components are linked together, looking at adequate communication, how responsibilities are met, and the timeliness of meeting the requirements.

❖ Identification and subsequent monitoring of potentially high noise level areas throughout the entire facility. Although identification should be a continuing process, an annual identification and monitoring mechanism will assist in identifying areas previously missed as well as providing documentation of employee exposure levels.

The audit process will point out problem areas and will also ensure that the program receives adequate attention. Many programs, while properly instituted, deteriorate over time due to personnel changes, low program visibility, and an obscuring of the benefits of an effective program and exactly what is required.

Another proactive procedure is providing engineering controls as the *first* method of reducing employee exposure levels. Not only should existing equipment be evaluated for possible engineering control applications, but new equipment should be evaluated *before purchasing*. This can be done by developing purchase specifications in regards to noise levels. It is much easier to modify equipment before installation than it is to make modifications once the equipment is in production. However, be sure that even if noise specifications are followed, the new equipment and surrounding environment are evaluated. Noise levels can vary greatly once the equipment has been installed, especially if there is other high noise level equipment in the area. A summary of engineering control methods is presented in Table 24-2.

Table 24-2
Common Engineering Controls

Source:

Preventative Maintenance
Speed Reduction
Substitution of Machines
Substitution of Processes
Vibration Isolation
Damping of Vibration
Enclosure
Sound Directivity
Mufflers

Air Path:

Absorptive Material
Sound Barriers
Increase Distance Between Source and Receiver

Receiver:

Enclosure

Thorough evaluation of all employee audiograms can provide useful insight into the effectiveness of the Hearing Conservation Program, can assist in early detection of hearing loss, and can identify those individuals who may be at an increased risk of hearing loss. Although the standard requires that audiograms be reviewed to determine if an STS has occurred, the proactive approach should also include the following:

❖ Trends in hearing loss should be reviewed. A hearing loss trend in a particular area may indicate that the hearing protection used by these employees is not adequate for their particular environment. It may also indicate that these particular employees have not received adequate training in the use of hearing protection, have not been properly fitted with hearing protection, or that the use of hearing protection is not enforced in that particular area.

❖ Thoroughly evaluate, or have evaluated by an expert, audiograms on a case by case basis. Early hearing loss can be detected by reviewing the hearing thresholds at the 4000, 6000 and 8000 compared with the baseline audiogram at these same frequencies. A shift in hearing ability at these frequencies may indicate that the individual is not wearing hearing protection, is not using protection correctly, is wearing protection which does not attenuate the noise stemming from his particular job, or that the employee has been improperly fitted with hearing protection. Early hearing loss in one ear only may indicate medical pathology or hearing loss due to a recreational activity, such as hunting.

❖ Establish proper follow-up procedures. This is required for Standard Threshold Shifts, but should also be done for early or slight hearing loss cases.

The thorough evaluation of audiograms is one of the most important ways to gain information regarding the effectiveness of the Hearing Conservation Program. The key is to use this information to investigate the reasons for hearing loss. Only by determining the problems and correcting them can hearing loss be reduced.

Effective training is one of the most important means to motivate employees and to involve them in the Hearing Conservation Program. In addition to covering the requirements of the standard, the following guidelines should make your training program more effective:

❖ Inform the employees that the training is provided to help them protect themselves from a social handicap, not just to comply with the standard.

❖ Limit the use of videotapes. A discussion format, which allows interaction with the trainer, will help the employees retain the information presented.

❖ Include as many illustrations and demonstrations as possible, such as: noise attenuation with a properly fitted ear plug versus one which is too small; the damage to the inner ear resulting from overexposure to noise; extent of hearing loss versus communication ability; and a graphic depicting how the incidence of hearing loss has decreased since the initiation of the program.

❖ Vary the content of the training sessions, so that additional information can be provided and the training sessions do not become routine experiences.

❖ Compile employee opinions regarding the training material and their suggestions for improvement.

The annual training program can be a very effective tool, not only to educate your employees, but to get them involved in the success of the Hearing Conservation Program.

The final step in developing a proactive approach is to review your hearing protection policies and procedures. With the exception of engineering controls, in many cases hearing protection is the only method available to prevent employee overexposure to noise. The following guidelines can assist you in developing policies and procedures which lend themselves to a more protective situation:

❖ Decide whether the company will either discipline workers who do not wear hearing protection when required, or reward those employees who wear it. In most companies it requires a mixture of the two. These policies should be clearly defined, transmitted to employees, and should be followed in all cases.

❖ Consider requiring hearing protection in all areas where the exposure is above the action level of 50 percent of the permitted dose. Due to variations in employee susceptibility to hearing loss, this approach would provide protection for a greater number of employees.

❖ Be consistent in the requirement to wear hearing protection. Visitors and employees who infrequently enter high noise level areas should be treated the same when entering these areas as those employees normally working there. When an employee sees a senior level manager, office worker, or visitor in a high noise level area who is not wearing hearing protection, he may perceive the use of hearing protection as a burden or "punishment." This perception, especially if it is a common occurrence, can jeopardize the entire program.

❖ Offer a *wide* variety of hearing protectors. This is not to say that each employee should be provided with a different type or brand, but offering a wide variety allows the employee to choose the one which meets his needs and provides maximum comfort. Some employees' ears may be irritated by certain types of hearing protection, and, therefore, they will not wear them. Employees having problems wearing the available hearing protection should be accommodated and provided with hearing protection with which they are comfortable.

❖ Ensure that hearing protectors are evaluated for the specific work area where they are used. Evaluate existing protectors provided to the employees, as well as protectors which are under consideration.

This section has dealt with the ways in which management can do more than just meeting the requirements of the OSHA noise standard. Although some of these guidelines may not apply in all cases, or may be impossible in other cases, the development and maintenance of an effective Hearing Conservation Program is not out of reach. With proper planning, sustained interest, and adequate resources, the essence of the proactive approach can be instilled into each and every Hearing Conservation Program to make it effective.

CHAPTER
25

ROPES, CHAINS, AND SLINGS

This chapter describes the various types of hoisting apparatus in the form of ropes, chains, and slings, and their proper selection, use, maintenance, and inspection of their component parts. Ropes, chains, and slings are commonly used between cranes, derricks, and hoists and the load so that the load may be lifted and moved to a desired location. Knowledge of the properties and limitations of the slings, the type and condition of material being lifted, weight and shape of the object being lifted, angle of the lifting sling to the load being lifted, and environment in which the lift is to be made are all important considerations to be evaluated before the transfer of material can take place in a safe manner.

ROPES AND SLINGS

Ropes and slings are usually divided into two main classes: fiber rope slings and wire rope slings. Fiber ropes are further divided into natural and synthetic fibers depending on their construction.

Natural Fiber

Natural fiber ropes and slings are usually made from manila, sisal, or henequen fibers. Most natural fiber ropes and sings used in industry today are made from manila fibers due to its superior breaking strength, consistency between grades, excellent wear properties in both fresh and salt water atmospheres, and its elasticity. The main reasons for selecting natural fiber rope slings are price and the ability to form or bend around angles of the object being lifted. This produces a more secure grip and reduces the chance of marring a finished or painted surface.

Disadvantages of using natural fiber ropes are susceptibility to cuts and abrasions when wrapped around sharp corners, dragged on the floor, or run over by equipment if left lying on the floor. The cutting problem encountered with sharp corners can be eliminated by the use of pads placed between the rope and cutting edge. Another disadvantage of these ropes is reduced capability or inability to be used to lift materials at elevated temperatures. Reductions by as much as 50 percent in safe lifting capacity of the rope can occur when surface temperatures of the object to be lifted are below $100^{\circ}F$ over ambient (room) temperature. If surface temperatures are greater, the rope may become charred or burned, which would require removal from service. The temperature range to which the rope will be exposed should be determined so the appropriate rope material can be chosen. In addition, these types of ropes should never be used in atmospheres where they may come in contact with acids and caustics (or their vapors or mists) since they will degrade the fibers. If any of the above are combined with hot or humid conditions, the service life of the rope or sling will be significantly reduced.

To reduce the chances of employee injury and property damage, the following general rules should be observed:

1. Do not attempt lifts which exceed the rated load capacity of the rope.

2. Fiber rope slings should have a diameter of curvature meeting at least minimum OSHA or manufacturer's specifications.

3. Natural fiber and synthetic fiber rope slings, except for wet frozen slings, may be used in a temperature range from minus $20^{\circ}F$ to plus $180^{\circ}F$ without decreasing the work load limit. For operations outside this temperature range and for wet frozen ropes, the sling manufacturer's recommendations should be followed.

4. Spliced fiber rope slings should not be used unless they have been spliced in accordance with the requirements of the manufacturer.

5. Natural and synthetic fiber rope slings should be immediately removed from service if any of the following conditions are present:

   —Abnormal wear

   —Powdered fibers between strands

   —Broken or cut fibers

   —Variations in the size or roundness of strands

   —Discoloration or rotting

   —Distortion of hardware in the sling.

6. Only fiber rope slings made from new rope should be used. Use of repaired or reconditioned fiber rope slings is prohibited by law.

7. Nylon web slings should not be used where fumes, vapors, sprays, mists, or liquids of acids or phenolics are present.

8. Web slings with aluminum fittings should not be used where fumes, sprays, mists, or liquids of caustics are present.

9. Synthetic web slings of polyester and nylon should not be used at temperatures in excess of 180°F. Polypropylene web slings should not be used at temperatures in excess of 200°F.

10. Synthetic web slings which are repaired should not be used unless repaired by a sling manufacturer or an equivalent entity.

11. Each repaired sling should be tested by the manufacturer or the equivalent entity to twice the rated capacity prior to return to service. The employer should retain a certificate of the test and have it available for examination.

12. Synthetic web slings should be immediately removed from service if any of the following conditions are present:

    —Acid or caustic burns

    —Melting or charring of any part of the sling surface

    —Snags, punctures, tears, or cuts

    —Broken or work stitches

    —Distortion of fittings.

13. Sling legs should not be kinked.

14. Slings used in a basket hitch should have the loads balanced to prevent slippage.

15. Slings should be securely attached to their loads.

16. Suspended loads should be kept clear of all obstructions.

17. All employees should be kept clear of loads about to be lifted and of suspended loads.

18. Hands and fingers should not be placed between the sling and its load while the sling is being tightened around the load.

19. Shock loading is dangerous and should be prohibited.

20. A sling should not be pulled from under a load when the load is resting on the sling.

Natural fiber rope slings should not be used to lift loads which exceed rated capacity. This can result in permanent damage to the device and a permanent reduction in its load limit. If a natural fiber rope is used to lift a load which exceeds 50 percent of its rated capacity, the manufacturer of the rope should be contacted so new safe load limits for that rope can be determined.

Information on the load limits for specific natural as well as synthetic fiber rope slings is provided by the supplier or manufacturer. The overall strength of a fiber rope sling is usually determined by its diameter and material of construction. Each sling should be marked or coded to show its rated capacity for each type of hitch and type of material. All employees who may have the responsibility to use lifting devices should be trained in how to identify, inspect, and use proper rope for the job. If a coding system is employed to help determine safe loading limits, charts showing the color to weight and angle of lift to weight correspondence should be located in appropriate areas throughout the facility (rope storage and maintenance locations).

Although there is no uniform guide specifying how or how often inspection of natural fiber rope is to occur or at what point it must be taken out of service, some common practices do exist.

1. Before being put into service, all new rope slings should be inspected along the entire length for damage or defects. If any inconsistencies are noted the device should be returned to the supplier or manufacturer for replacement.

2. Employees should be trained to visibly inspect all lifting devices daily or whenever there is suspicion of damage after a lift. These types of inspections should include the following:

   —Wear of the outer fibers. If the rope or sling shows more than 5 percent wear, take it our of service.

   —Abrasions, a sign of abuse from dragging the rope on the floor.

   —Discoloration, due to rotting or mildew from improper cleaning, drying, and storage.

   —Charring of the rope, due to misuse on hot surfaces.

   —Broken or cut fibers, can be a sign of misuse due to improper storage or failure to use pads between the device and sharp corners or edges.

   —Inconsistencies between strands or yarns in size, color, or texture.

   —Powder between strands can only be seen if the rope is slightly untwisted at a number of places along its length. This is a sign of abuse from overloading the rope and is also an indication of inadequate training and or supervision.

   —Perform a fingernail cut test on the fibers. If natural fiber rope has been excessively exposed to atmospheres containing acids or caustics the fibers will easily part when scratched.

3. Ropes and slings that are in service should receive a thorough cleaning and inspection once a month. Ropes showing inner fiber damage or other severe deterioration must be discarded.

Ropes and slings should be stored in cool dry places away from direct sunlight, either hung straight down or preferably over graded racks, permitting air to move freely about them to promote drying and reduce the likelihood of mildew or rotting of the fibers.

When new, a permanent identification number should also be attached to the rope sling and a file should be established to record all maintenance repairs and monthly inspection findings over the life of the lifting device. These records can also be retained after the device has been discarded if the user wants to establish life expectancy data for the various types of lifting devices used in different operations.

Synthetic Fiber

The other major class of fiber rope slings is those made from synthetic fibers. The fibers of these devices are usually made from nylon, polyester, polypropylene, polyethylene, or a combination of these to obtain the desired properties. Fiber rope slings of this type have many of the same qualities as natural fiber rope slings, but are in much wider use throughout the industry because they can be engineered to fit a particular operation. Some of the advantages of this type of rope include:

1. Increased strength over natural fiber rope

2. Increased elasticity over natural fiber rope

3. Better resistance to shock loading

4. Better resistance to abrasion than natural fiber rope

5. Resistant to bacterial attack which causes rot or mildew

6. More resistant to acids, caustics, alcohol based solvents, bleaching solutions, and their atmospheres

7. Fibers do not swell when they become wet

8. Properly made splices have little effect on overall rope strength

9. More strength resistant over time to the effects of sun and ultra-violet light than natural fiber rope

10. Some types of synthetic fibers are not hygroscopic (do not absorb moisture from the air)

11. Better resistance to heat than natural fibers.

Each synthetic material will only possess some of the properties listed above. The rope supplier or manufacturer should be consulted to help you determine which base material of rope construction will best fit your needs.

Diagram 25-1 Synthetic Web Sling Constructions

Basic Synthetic Web Sling
Constructions

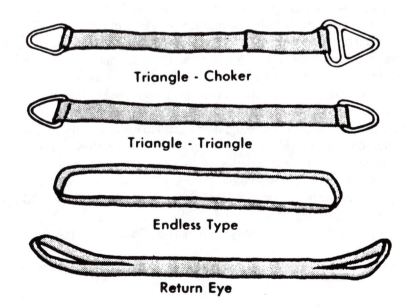

Triangle - Choker

Triangle - Triangle

Endless Type

Return Eye

Some of the disadvantages of synthetic fibers include:

1. Although they have a higher resistance to heat than natural fiber, if temperatures exceed certain limits, the rope may soften or even melt causing permanent reduction in lifting capacity.

2. Synthetic fibers cost more than natural ones and can be significantly more expensive for blended synthetic ropes.

3. Polyester fibers are susceptible to damage from alkalis.

4. Polyolefin fibers are susceptible to abrasion damage.

5. Polyolefin fibers will swell and soften when used in atmospheres containing hydrocarbons.

The inspection, care, and maintenance routine for synthetic ropes are similar to those used for natural rope slings. One additional area for inspection is the presence of melted fibers on the surface of the rope. This is an indication that the rope was used on a piece of equipment or in an atmosphere which exceeded its temperature limits. This type of defect will at a minimum reduce the safe lifting capacity of the rope and possibly damage it to the point that permanent removal from service is required. If these or any other defects are observed in or on the rope or core, contact the supplier or manufacturer for advice.

Wire Rope

The most widely used type of rope sling in industry is wire rope. By definition, wire rope is a twisted bundle of cold drawn steel wires. It is usually composed of wires, strands, and a core. The wires can be one of a number of grades depending on characteristics desired. These wires are cold drawn to a predetermined size and laid together in various arrangements having a definite pitch (or lay) to form a strand. The strands are then formed around a core, which may be made of sisal rope, synthetic fibers, metallic strand, or independent wire rope core. The size, number and arrangement of wires, the number of strands, the lay, and the type of core in the rope are determined largely by the service for which the rope is to be used. In general, the greater the number of wires per strands, the more flexible the rope.

A device called a wire rope clip is used for securing a loop at the end of a wire rope. The size of the clip needed will be determined by the diameter of the wire rope in which the loop is to be made. The clip should be drop forged and should use U-bolts provided with two finished hex nuts. The distance between clips should not be less than six times the diameter of the rope. The saddle of the clip should be in contact with the long end of the rope, and the U-bolt in contact with the short end. During installation, the clip furthest from the loop should be put on first, the clip nearest the loop next, and then install the in-between clips. After the rope has been broken in, and on subsequent inspections, all nuts should be tightened again to make sure they are holding.

The main reasons for the wide acceptance of wire over fiber rope are its greater strength, durability, and stable physical characteristics over a wide variety of environmental conditions. Another feature which make wire rope more attractive include its predictable stretch characteristics when placed under heavy stresses. Preformed wire rope is also less likely to unwind, set, kink, or generate protruding wires.

Some of the major safety precautions for wire rope listed by OSHA are:

1. Wire rope slings shall not be used with loads in excess of their rated capacities.

Diagram 25-2
U-Bolt for Wire Rope

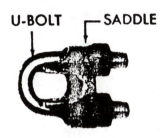

2. Cable laid and 6 × 19 and 6 × 37 slings shall have a minimum clear length of wire rope 10 times the component rope diameter between splices, sleeves, or end fittings.

3. Braided slings shall have a minimum clear length of wire rope 40 times the component rope diameter between the loops or end fittings.

4. Cable laid grommet, strand laid grommets, and endless slings shall have a minimum circumferential length of 96 times their body diameter.

5. Fiber core wire rope slings of all grades shall be permanently removed from service if they are exposed to temperatures in excess of 200°F.

6. When nonfiber core wire rope slings of any grade are used at temperatures above 400°F or below minus 60°F, recommendations of the sling manufacturer regarding use at that temperature shall be followed.

7. Welding of end attachments, except covers to thimbles, shall be performed prior to the assembly of the sling.

8. All welded end attachments shall not be used unless tested by the manufacturer or equivalent entity at twice their rated capacity prior to initial use. The manufacturer shall retain a certificate of the proof test, and make it available for examination.

9. Wire rope slings shall be visually inspected at the start of each shift, before each use, and on regular intervals based on conditions of use. If damaged or defective,

the sling shall be immediately removed from service. The following are some things to look for when inspecting a wire rope sling:

—Ten randomly distributed broken wires in one rope lay, or five broken wires in one strand in one rope lay.

—Wear or scraping of one-third the original diameter of outside individual wires.

—Kinking, crushing, bird caging, or any other damage resulting in the distortion of the wire rope structure.

—Evidence of heat damage.

—End attachments that are cracked, deformed, or worn.

—Hoods that have been opened more than 15 percent of the normal throat opening measured at the narrowest point or twisted more than 10 degrees from the plane of the unbent hook.

—Corrosion of the rope or end attachments.

10. Never lift a heavy load at less than a 45 degree angle.

11. Never place a cable between the floor and load.

12. Never use a sling against the sharp edge of a load.

13. Never use or modify a sling with knots, bolts, or other makeshift devices.

14. Always be sure that a sling is securely attached to or that it supports its load. This can be achieved initially by visual inspection and then by slowly lifting the object an inch or two off the floor for a period before proceeding with the lift.

All slings, fasteners, and attachments must be visually inspected daily by a competent person before being put into use. More thorough periodic inspections should be made on a regular basis. The time interval is based on frequency of wire rope or sling use, severity of service conditions, nature of the lifts being made, and experience gained on the service life of slings used in similar circumstances. Such inspections shall in no event be at intervals greater than once every twelve months.

To be in compliance with OSHA you must maintain a record of the most recent month in which each wire rope sling was thoroughly inspected. OSHA requires you to make these records available for examination.

Before attempting a lift, the size, shape, weight of the load and the number and location of points at which slings can be attached to the load must be determined. For every kind of load and lift point, there is a correct sling and hitch to use. The following is a brief summary of the common types of hitches together with their respective advantages and limitations. See Diagram 25-3.

Single Leg Hitch

A single leg hitch is made with one sling, one end usually attached to a hook and the other to the object being lifted. These types of lifts should be done vertically to optimize the load limit characteristics of the rope. The biggest danger of this type of hitch is that it does not provide the use optimum control of large or awkward loads in case of slippage or spin during the lift and move. Therefore this type of hitch is best used on small, regularly shaped objects which can be easily controlled by one person. A modification to this hitch can be employed by the use of a spreader bar. As the name implies, this is a bar which has a rope hanging vertically down from each end. This arrangement puts less stress on each of the ropes by dividing the weight of the load and provides more control to the operator while reducing the chance of spin.

Basket Hitch

Another hitch which is commonly employed in industry is the basket hitch. To employ this type of attachment the rope is placed in a "U" shape with the load being at the base and both ends attached to a hook. This type of hitch is easy to attach and when properly applied, provides good load control. The stress on each leg of the sling is automatically equalized. This type of hitch is best used on straight lifts which are of such shape that sliding of the sling over the surface of the load will not occur and control will not be lost.

Choker Hitch

The third type of hitch most frequently encountered is the choker. Choker hitches are easy to attach and by their tightening action provide a good level of load control. This type of hitch is created by passing the loop at one end of the sling through the other. This arrangement forms a loop which when lifted tightens like a noose about the object. When loads are to be turned, a choker hitch, especially a double choker hitch, is the arrangement to employ. A properly made double choker hitch will "choke" through the eyes with the sling center bearing on the crane hook. In this way, the two legs of the wire rope sling are automatically equalized as the load is lifted. When a double choke hitch is made incorrectly, there is no automatic stress equalization and it is possible that one leg will take the full weight of the load. This type of hitch is most suited when transporting bundles of bars or pipes or for turning loads (use double choker).

Diagram 25-3
Types of Hitches

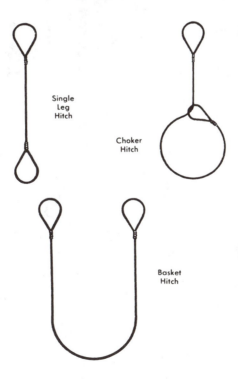

Single
Leg
Hitch

Choker
Hitch

Basket
Hitch

CHAINS AND CHAIN SLINGS

Steel and alloys such as stainless steel, monel, bronze, and other metals are commonly used for lifting slings made of chain. Regular hardware chain or other chain not specifically designed for use in slings should not be used for load lifting. The best alloy steels to use for slings are those that have been specially heat treated for hardness, tensile strength, and abrasion resistance. Chain slings offer many advantages over other devices including greater lifting capacities, greater resistance to mechanical damage and much greater heat resistance. OSHA requires that alloy steel chain slings have permanently affixed and durable identification stating size, grade, rate capacity, and reach.

Normally, the chains will utilize hooks, rings, or other couplings to allow for fashioning of the load lifting device. These attachments must have a rated capacity at least equal to that of the chain elements. Like other slings, all parts should be inspected for:

Diagram 25-4
Couplings for Chains

1. Evidence of corrosion or other wear

2. Evidence of faulty welds

3. Cracking of the links or other parts

4. Evidence of the chain links stretching or bending

5. Evidence of the hooks stretching beyond 15 percent of the normal throat openings or twisting beyond 10 degrees from the plane of the unbent hook.

Chains and slings showing evidence of damage must be immediately removed until repaired and recertified for use. Some general rules of safe use include the following:

1. Do not remove the OSHA required identification and load ratings from the chain slings

2. Use only those attachments which have a rated capacity at least as great as the chain

3. Do not straighten kinks or force links over hooks by using a hammer

4. Keep chains and slings free of grease and store them properly

5. Do not splice chains using bolts or other inappropriate devices.

Finally, flat synthetic fiber webs and metal mesh slings can be used for loads that could be damaged by other sling types. These belt-like devices require a similar inspection schedule and level of care as the synthetic ropes and metal chains and slings. See Table 25-1.

Table 25-1
Fiber and Synthetic Rope Inspection

FIBER & SYNTHETIC ROPE INSPECTION

ROPE ID NO. _____   DATE ROPE PUT IN SERVICE _____

ROPE MATERIAL _____   DATE ROPE LAST INSPECTED _____

SIGNS OF WEAR	SIGNS OF ABRASION	POWDERED FIBERS	BROKEN OR CUT FIBERS	DISPLACEMENT OF YARNS OR STRANDS	VARIATIONS IN SIZE OR ROUNDNESS OF STRANDS	DISCOLORATION	ROTTING

CHAPTER
26

CONSTRUCTION SAFETY

Most businesses, regardless of their size or the type of business activities, are likely to perform some type of construction work. This might be accomplished by using maintenance employees or by hiring a contractor that specializes in the type of construction work needed. In either case the business will be subject to the OSHA Construction Industry Standards, including Part 1926. Specific responsibilities of the business for outside contractors or in-house employees will depend upon the control the business establishes over the construction project.

1. Using in-house employees for construction work will require compliance with all of the construction industry standards.

2. Maintaining direct control of the construction project means the business will be viewed as the general contractor (GC) for the project and will be subject to all of the regulatory requirements of a general contractor.

3. Exercising no direct project control limits the business to complying with OSHA regulations specifying required communications between the contractor and the business.

What is the construction standard? Where do you find it? When does it apply to businesses that are already covered by the OSHA general industry or other standards?

What? The construction standard regulates safety and health for employees working for businesses within the construction industry and for employees not working in the construction industry but performing work tasks associated with the construction industry.

Where? The sub-parts of the construction regulation are shown in Table 26-1.

When? If your employees are required to perform work tasks that are considered construction related you are obligated to follow the applicable OSHA regulations.

The following examples illustrate when employees are covered by General Industry Standards (29 CFR 1910) versus the Construction Industry Standards (29 CFR 1926).

- ❖ Employees overhauling, repairing, or modifying machinery or equipment. (1910)
- ❖ Installing electrical service to machines and equipment. (1910)
- ❖ Employees excavating a trench to repair or install piping, electrical services, etc. (1926)
- ❖ Removal or construction of partitions and walls. (1926)

Table 26-1
Topics and Work Tasks Covered by 29 CFR 1926
Construction Industry Standards

- ❖ Subpart C–General Safety and Health Provisions
- ❖ Subpart D–Occupational Health and Environmental Controls
- ❖ Subpart E–Personal Protective and Life Saving Equipment
- ❖ Subpart F–Fire Protection and Prevention
- ❖ Subpart G–Signs, Signals, and Barricades
- ❖ Subpart H–Materials Handling, Storage, Use, and Disposal
- ❖ Subpart I–Tools - Hand and Power
- ❖ Subpart J–Welding and Cutting
- ❖ Subpart K–Electrical
- ❖ Subpart L–Scaffolding
- ❖ Subpart M–Floor and Wall Openings
- ❖ Subpart N–Cranes, Derricks, Hoists, Elevators, and Conveyors
- ❖ Subpart O–Motor Vehicles, Mechanized Equipment, and Marine Operations
- ❖ Subpart P–Excavations, Trenching, and Shoring
- ❖ Subpart Q–Concrete and Masonry Construction
- ❖ Subpart R–Steel Erection
- ❖ Subpart S–Underground Construction, Caissons, Cofferdams and Compressed Air
- ❖ Subpart T–Demolition
- ❖ Subpart U–Blasting and Use of Explosives
- ❖ Subpart V–Power, Transmission, and Distribution
- ❖ Subpart W–Rollover Protective Structures; Overhead Protection
- ❖ Subpart X–Stairways and Ladders

Note: subparts A and B cover employers under the Contract Work Hours and Safety Standards Act (Construction Safety Act 40 U.S.C. 333) where federal government contract dollars are involved.

USE OF CONTRACTORS

Employers who use contractors to perform work in their establishments are required to communicate safety and health information to the contractor. Conversely, the contractor must provide the facility manager with the contractor's safety and health information. The following regulations cover communications between contractors and facility managers:

❖ Hazard Communication Standard—Facility managers and contractors are obligated to inform each other of the hazardous materials that are on-site, or that will be brought on-site.

❖ Control of Hazardous Energy Sources (Lockout/Tagout Standard)—Facility managers and contractors are to inform each other of their respective lockout/tagout programs.

❖ Additional Communications—It is widely recognized by the safety profession that persons not familiar with their work environment are at a greater risk of having an accident. Therefore, it is suggested that a notice with important work site safety information be prepared and posted at all contractor job sites for the duration of the construction project. An example of a safety information posting is shown in Figure 26-1. It is also prudent to provide contractors with the following emergency information:

–how to report a fire

–how to request medical assistance

–emergency evacuation alarms & procedures

–specific hazards within the facility and/or the work site.

It is recommended that the individual responsible for hiring the contract labor be responsible for communicating safety and health information, completion of the safety forms, and records retention.

Although the OSHA standards do not stipulate that you must retain records of exchanges of information, it is strongly advised. It is to your advantage to retain these records in the event of an accident and as an aid to verify compliance with OSHA regulations. Figure 26-2 on page 347 provides an example of a standardized document for recordkeeping purposes.

MULTI-EMPLOYER WORK SITES

When construction projects require more than one contractor on your site, you become a multi-employer work site. In this situation you can be held accountable for conditions that expose your employees to hazards for which you have no direct control. OSHA has a citation policy that holds each employer on a multi-employer work site responsible for the safety of their employees independently, even though the condition may be the fault of another contractor. Important aspects of this policy include the following:

1. Regardless of contract arrangements between the prime contractor and sub-contractors, the prime contractor can not be relieved of the responsibility for compliance with the OSHA standard.

2. Citations for safety violations will be issued to:

—Employers responsible for the unsafe conditions.

Ajax Manufacturing
Emergency Procedures for Contractors

MANDATORY POSTING REQUIRED BY CONTRACTOR AT ALL WORK SITES

Date Posted:_____   Date Posting Removed_____

PHONE

TO REPORT A FIRE OR NON-MEDICAL   6911
EMERGENCIES

TO REPORT MEDICAL EMERGENCIES   6912

WHEN REPORTING ANY EMERGENCY, STATE:

1. THE NATURE OF THE EMERGENCY
2. THE TELEPHONE NUMBER YOU ARE CALLING FROM OR YOUR LOCATION IN THE
   FACILITY
3. YOUR NAME & COMPANY.

EMERGENCY EVACUATIONS

A CONTINUOUS SIREN WILL BE SOUNDED. WHEN THE SIREN IS SOUNDED YOU MUST IMME-
DIATELY SHUT OFF ALL WELDING GASES AT THE MAIN SHUT OFF VALVES, EXTINGUISH ALL
IGNITION SOURCES, AND EXIT THE PLANT THROUGH THE CONTRACTOR'S GATE.

AN AJAX SECURITY GUARD WILL BE AT THE GATE TO VERIFY THAT ALL CONTRACTOR PER-
SONNEL HAVE BEEN EVACUATED AND TO GIVE FURTHER INSTRUCTIONS.

DO NOT REENTER THE PLANT FOR ANY REASON UNTIL THE ALL CLEAR SIGNAL IS GIVEN.
THE ALL CLEAR SIGNAL IS A SERIES OF THREE SHORT SIREN BLASTS, PAUSE, THREE SHORT
SIREN BLASTS, PAUSE, THREE SHORT SIREN BLASTS.

IN THE EVENT OF A POWER FAILURE OR THE NEED FOR A SPECIFIC EVACUATION PROCE-
DURE, YOU WILL BE GIVEN INSTRUCTIONS BY AJAX EMERGENCY RESPONSE TEAM MEM-
BERS. EMERGENCY RESPONSE TEAM MEMBERS WILL BE WEARING RED HARD HATS WITH
"EMERGENCY RESPONSE TEAM" IN WHITE LETTERS ON THE FRONT AND THE BACK OF THE
HATS.

—Employers that allow their employees to be exposed to the unsafe conditions by
not taking steps to prevent the exposure.

—The employer responsible for the unsafe conditions, the employer allowing his
employees to be exposed to the unsafe conditions, and to the prime contractor (general
contractor).

3. You may avoid an OSHA violation by proving:

—You did not create the hazard.

—You did not have the authority or the ability to correct the hazard.

FIGURE 26-2
Ajax Manufacturing
Contractor Notification Form

This form is to document that the following contractor:

_____

has been apprised of Ajax Manufacturing's LOCKOUT/TAGOUT PROGRAM, APPROPRIATE
HAZARD COMMUNICATION INFORMATION, AJAX'S EMERGENCY NOTIFICATION PROCE-
DURES and other pertinent Ajax safety and health information listed in the space below. The identified
contractor has provided Ajax Manufacturing with corresponding information.

```
┌─────────────────────────────────────────────────────────────────────┐
│                                                                     │
│                                                                     │
│                                                                     │
│                                                                     │
│                                                                     │
│                                                                     │
└─────────────────────────────────────────────────────────────────────┘
```

The purpose of this document is to preserve a record for compliance to the Occupational Safety and
Health Administration (OSHA) standards 29 CFR 1910.147 and 29 CFR 1910.1200 and to Ajax
Manufacturing's safety and health program.

_____	_____	_____
Contractor Representative	Title	Date

_____	_____	_____
Ajax Project Manager	Department	Date

—You specifically requested the controlling employer and/or the hazard-creating
employer to correct the hazard to which your employees were exposed and made a
reasonable effort to persuade the responsible employer to correct the hazard.

—You have instructed and, where necessary, informed your employees how to avoid
or minimize the dangers associated with the hazard and, where feasible, have taken
alternative means of protecting your employees from the hazard, short of taking your
employees off the job.

If you plan to use your employees for construction work you will be required to have
a variety of authorized/designated persons, qualified individuals, and competent persons to
meet the OSHA regulatory requirements. Although one person can satisfy several requirements, additional employees may be required to comply with specific standards. The
following definitions for employee classifications are found in 1926.32 of the construction
standard:

❖ **Authorized/Designated Person**   A person approved or assigned by the employer
to perform a specific type of duty or duties or to be at a specific location or locations
at the job site.

❖ **Qualified Person**   One who by degree, certificate, or professional standing, or by
extensive knowledge, training and experience, has successfully demonstrated the
ability to solve or resolve problems relating to the subject matter, work, or project.

❖ **Competent Person** One who is capable of identifying existing and predictable hazards in the surroundings or working conditions which are unsanitary, hazardous, or dangerous to employees, **and** who has the authority to take prompt corrective measures to eliminate them. Table 26-2 identifies sections of the standard that require authorized/designated, qualified, and competent persons.

In conclusion, you may certainly use your employees for construction related projects provided you are prepared to meet all of the requirements of the OSHA construction standards.

Table 26-2
Construction Standard Requirements by Section for
Authorized or Designated, Qualified and Competent Persons

	Reference	*Section*	*Required Designation*
1.	1926.20	General Safety and Health Provisions	Competent
2.	1926.50	Medical Services and First Aid	Qualified
3.	1926.53	Ionizing Radiation	Competent
4.	1926.54	Nonionizing Radiation	Qualified
5.	1926.55	Gases, Vapors, Fumes, Dusts, and Mists	Competent or Qualified
6.	1926.58	Asbestos, Tremolite, Anthophyllite, and Actinolite	Competent
7.	1926.59	Hazard Communication	Competent
8.	1926.103	Respiratory Protection	Competent
9.	1926.155	Fire Protection and Prevention Definitions	Competent
10.	1926.251	Rigging Equipment for Material Handling	Competent
11.	1926.302	Power-Operated Hand Tools	Qualified
12.	1926.354	Welding, Cutting, & Heating...Coatings	Competent
13.	1926.404	Electrical - Wiring Designs and Protection	Competent
14.	1926.451	Scaffolding	Qualified or Competent
15.	1926.502	Guarding Low-Pitched Roof Perimeters	Competent
16.	1926.550	Cranes and Derricks	Qualified, Competent, or Designated
17.	1926.552	Material Hoists, Personnel Hoists, and Elevators	Qualified or Competent
18.	1926.556	Aerial Lifts	Authorized
19.	1926.601	Motor Vehicles	Competent
20.	1926.603	Pile Driving Equipment	Designated
21.	1926.650	Excavations, Trenching, and Shoring - General Protection Requirements	Competent
22.	1926.651	Specific Excavation Requirements	Qualified or Competent
23.	1926.700	Concrete and Masonry Construction - Scope, Application, and Definitions	Competent
23.	1926.703	Cast-In-Place Concrete	Competent
23.	1926.752	Bolting, Riveting, Fitting-Up and Plumbing-Up	Competent
24.	1926.800	Underground Construction	Qualified, Competent or Authorized
25.	1926.803	Compressed Air	Competent
26.	1926.850	Demolition-Preparatory Operations	Competent
27.	1926.859	Mechanical Demolition	Competent
28.	1926.900	Blasting and Use of Explosives-General Provisions	Qualified, Competent or Authorized

RESOURCES: WHERE TO GO FOR HELP AND INFORMATION

PROFESSIONAL ORGANIZATIONS AND SOCIETIES

Air and Waste Management Association
P.O. Box 2861
Pittsburgh, PA 15230
Telephone: (412) 232-3444

A not-for-profit professional association of air and waste management scientists and engineers. AWMA publishes a wide variety of technical references and sponsors a large number of training courses.

Alliance of American Insurers
1501 Woodfield Road
Suite 400
Schaumburg, IL 60195
Telephone: (708) 330-8500

A national organization of leading property and casualty insurance companies which publishes a variety of safety and health related materials.

American Association of Occupational Health Nurses
50 Lenox Pointe
Atlanta, GA 30324
Telephone: (404) 262-1162

A not-for-profit professional association of occupational health nurses which publishes various safety related publications.

American Industrial Hygiene Association
2700 Prosperity Avenue
Suite 250
Fairfax, VA 22031
Telephone: (703) 849-8888

A not-for-profit professional association of industrial hygienists that publishes a variety of safety related materials and offers a number of safety related training courses.

American Red Cross
2025 E Street, N.W.
Washington, D.C. 20006
Telephone: (202) 728-6629

A well known national organization that provides courses in first aid, cardiopulmonary resuscitation and other related courses through chapters located throughout the country.

American Society of Heating, Refrigerating, and Air Conditioning Engineers
1791 Tullie Circle, N.E.
Atlanta, GA 30329
Telephone: (404) 636-8400

A not-for-profit professional organization which publishes standards for heating, refrigeration, air conditioning, and general building ventilation for industrial and other systems.

American Society of Safety Engineers
1800 East Oakton Street
Des Plaines, IL 60018
Telephone: (708) 692-4121

A not-for-profit professional association of persons working or with an interest in safety. This organization also provides training and publications in the areas of safety and industrial hygiene.

National Association of Manufacturers
1331 Pennsylvania Avenue, N.W.
Suite 1500N
Washington, D.C. 20006
Telephone: (202) 637-3000

A not-for-profit national trade association for manufacturers. Provides safety related training courses and publishes materials on safety through the national and state associations.

National Environmental Health Association
720 South Colorado Blvd.
Suite 970
Denver, CO 80222
Telephone: (303) 756-9090

A not-for-profit professional association of persons working in a broad range of environmental areas including safety. They offer a number of publications and training courses related to safety.

National Fire Protection Agency
Batterymarch Park
Quincy, MA 02269
Telephone: 1-800-344-3555

The NFPA is a not-for-profit national association which develops and issues technical standards for fire protection and electrical safety (National Electrical Code).

National Safety Council
444 North Michigan Avenue
Chicago, IL 60611
Telephone: (312) 527-4800

A not-for-profit national organization (with local chapters in many states) that provides publications, consulting services, training, and other resources and activities in safety.

STANDARDS ORGANIZATIONS

American National Standards Institute
11 West 42nd Street
New York, NY 10036
Telephone: (212) 642-4900

A not-for-profit national organization that issues consensus standards (includes safety standards) for the United States.

American Society for Testing and Materials
1916 Race Street
Philadelphia, PA 19103
Telephone: (215) 299-5400

A not-for-profit consensus standards setting organization with emphasis on test methods for materials. ASTM publishes standards for safety and safety related areas.

FEDERAL ORGANIZATIONS

The Occupational Safety and Health Administration (OSHA) has offices in all 50 states and other U.S. possessions. Some states are "agreement" states which enforce their own standards which must be as strict as the federal standards for the sectors of commerce and industry they chose to cover (may be all or partial). States or possessions administering their own programs in part or fully are:

Alaska	New York
Arizona	North Carolina
California	Oregon
Connecticut	Puerto Rico
Hawaii	South Carolina
Indiana	Tennessee
Iowa	Utah
Kentucky	Vermont
Maryland	Virginia
Michigan	Virgin Islands
Minnesota	Washington
Nevada	Wyoming
New Mexico	

Each local and area office of OSHA whether state or federally run will offer a wide variety of publications and information free of charge. Additionally, technical publications and assistance are available through the OSHA regional offices.

Region I

Connecticut, Maine, Massachusetts, New Hampshire, Rhode Island, and Vermont

133 Portland St.
First Floor
Boston, MA 02114
(617) 565-7164

Region II

New Jersey, New York, and Puerto Rico
201 Varick Street
Room 670
New York, NY 10014
(212) 337-2378

Region III

Delaware, District of Columbia, Maryland, Pennsylvania, Virginia, and West Virginia
Gateway Building
Suite 2100
3535 Market Street
Philadelphia, PA 19104
(215) 596-1201

Region IV

Alabama, Florida, Georgia, Kentucky, Mississippi, North Carolina, South Carolina, and Tennessee
1375 Peachtree Street
N.E., Suite 587
Atlanta, GA 30367
(404) 347-3573

Region V

Illinois, Indiana, Michigan, Minnesota, Ohio, and Wisconsin
230 South Dearborn Street
32nd Floor - Room 3244
Chicago, IL 60604
(312) 353-2220

Region VI

Arkansas, Louisiana, New Mexico, Oklahoma, and Texas
525 Griffin Street
Room 602
Dallas, TX 75202
(214) 767-4731

Region VII

Iowa, Kansas, Missouri, and Nebraska
911 Walnut Street
Room 406
Kansas City, MO 64106
(816) 426-5861

Region VIII

Colorado, Montana, North Dakota, South Dakota, Utah and Wyoming
Federal Building
Room 1576
1961 Stout Street
Denver, CO 80294
(303) 844-3061

Region IX

Arizona, American Samoa, California, Guam, Hawaii, Nevada, and Trust Territory of the Pacific Islands
71 Stevenson Street
Suite 415
San Francisco, CA 94105
(415) 744-6670

Region X

Alaska, Idaho, Oregon, and Washington
111 Third Avenue
Suite 715
Seattle, WA 98101-3212
(206) 442-5930

Each state also has a separate on-site consultation program. These programs provide free (paid through state-wide workers' compensation premiums, taxes, or other means) assistance to employers without fear of citation unless a serious or life threatening situation is noted and not corrected. The main contact for these services by state are as follows:

Directory of Approved Sources of OSHA-Funded Consultation

Alabama

7(c)(1) Onsite Consultation Program
Martha Parham West
P.O. Box 70388
Tuscaloosa, Alabama 35487
(205) 348-3033

Alaska

Division of Occupational Safety and Health
Alaska Department of Labor
3301 Eagle Street, Suite 303
Pouch 7-022
Anchorage, Alaska 99510
(907) 264-2599

Arizona

Consultation and Training

Division of Occupational Safety and Health
Industrial Commission of Arizona
P.O. Box 19070
Phoenix, AZ 85005
(602) 255-5795

Arkansas

OSHA Consultation
Arkansas Department of Labor
10421 West Markham
Little Rock, Arkansas 72205
(501) 682-4522

California

CAL/OSHA Consultation Service
525 Golden Gate Avenue
2nd Floor
P.O. Box 603
San Francisco, California 94102
(415) 557-2870

Colorado

Occupational Safety and Health Section
Institute of Rural Environmental Health
Colorado State University
110 Veterinary Science Building
Fort Collins, Colorado 80523
(303) 491-6151

Connecticut

Division of Occupational Safety and Health
Connecticut Department of Labor
200 Folly Brook Blvd.
Wethersfield, Connecticut 06109
(203) 566-4550

Delaware

Occupational Safety and Health
Division of Industrial Affairs
Delaware Department of Labor
820 North French Street, 6th Floor
Wilmington, Delaware 19801
(302) 571-3908

District of Columbia

Office of Occupational Safety and Health
District of Columbia Department of Employment Services
950 Upshur Street, N.W.

Washington, D.C. 20011
(202) 576-6339

Florida

7(c)(1) Onsite Consultation Program
Bureau of Industrial Safety and Health
Department of Labor and Employment Security
Forrest Building, Suite 349
2728 Center View Drive
Tallahassee, Florida 32399-0663
(904) 488-3044

Georgia

7(c)(1) Onsite Consultation Program
Georgia Institute of Technology
O'Keefe Building - Room 23
Atlanta, Georgia 30332
(404) 894-8274

Guam

OSHA Onsite Consultation
Government of Guam
Int'l Trade Center
3rd Floor
P.O. Box 9970
Tamuning, Guam 96911
(671) 646-9244

Hawaii

Division of Occupational Safety and Health
830 Punchbowl Street
Honolulu, Hawaii 96813
(808) 548-4155

Idaho

Safety and Health Consultation Program
Boise State University
Department of Commerce and Environmental Health
1910 University Drive, MG-110
Boise, Idaho 83725
(208) 385-3283

Illinois

Illinois Onsite Consultation
Industrial Services Division
Department of Commerce and Community Affairs
State of Illinois Center
100 West Randolph Street

Suite 3-400
Chicago, Illinois 60601
(312) 814-2339

Indiana

Division of Labor
Bureau of Safety, Education and Training
1013 State Office Building
Indianapolis, Indiana 46204-2287
(317) 232-2688

Iowa

7(c)(1) Consultation Program
Iowa Bureau of Labor
1000 East Grand Avenue
Des Moines, Iowa 50319
(515) 281-5352

Kansas

Kansas 7(c)(1) Consultation Program
Kansas Department of Human Resources
512 West 6th Street
Topeka, Kansas 66603
(913) 296-4386

Kentucky

Consultation and Training
Kentucky OSH Program
Kentucky Labor Cabinet
U.S. Highway 127, South, Bay 4
Frankfort, Kentucky 40601
(502) 564-6895

Louisiana

7(c)(1) Consultation Program
Office of Workers' Compensation
Louisiana Department of Labor
1001 North 23rd Street
Baton Rouge, Louisiana 70804-9094
(504) 342-9601

Maine

Division of Industrial Safety
Maine Department of Labor
State Home Station 82
Hallowell Annex
Augusta, Maine 04333
(207) 289-6460

Maryland

7(c)(1) Consultation Services
Division of Labor and Industry
501 Saint Paul Place
Baltimore, Maryland 21202
(301) 333-4218

Massachusetts

7(c)(1) Consultation Program
Division of Industrial Safety
Massachusetts Department of Labor and Industries
100 Cambridge Street
Boston, Massachusetts 02202
(617) 727-3463

Michigan

Michigan Department of Public Health
Division of Occupational Health
3423 N. Logan Street
P.O. Box 30195
Lansing, Michigan 48909
(517) 335-8250

Bureau of Safety and Regulation
Michigan Department of Labor
7150 Harris Drive
P.O. Box 30015
Lansing, Michigan 48909
(517) 332-1814

Minnesota

Consultation Division
Department of Labor and Industry
443 Lafayette Road
St. Paul, Minnesota 55155
(612) 297-2393

Mississippi

7(c)(1) Onsite Consultation Program
Division of Occupational Safety and Health
Mississippi State Board of Health
305 West Lorenz Blvd.
Jackson, Mississippi 39219-1700
(601) 987-3981

Missouri

Onsite Consultation Program

Division of Labor Standards
Department of Labor and Industrial Relations
3315 West Truman Blvd.
Jefferson City, Missouri 65109
(314) 751-3403

Montana

Montana Bureau of Safety and Health
Division of Workers' Compensation
5 South Last Chance Gulch
Helena, Montana 59601
(406) 444-6401

Nebraska

Division of Safety, Labor and Safety Standards
Nebraska Department of Labor
State Office Building
301 Centennial Mall, South
Lincoln, Nebraska 68509-5024
(402) 471-4717

Nevada

Training and Education Section
Division of Occupational Safety and Health
4600 Kietzke Lane, Building D, Suite 139
Reno, Nevada 89502
(702) 789-0546

New Hampshire

Onsite Consultation Program
New Hampshire Department of Labor
19 Pillsbury Street
Concord, New Hampshire 03301
(603) 271-3170

New Jersey

Division of Workplace Standards
New Jersey Department of Labor
CN953
Trenton, New Jersey 08625-0953
(609) 984-3507

New Mexico

OSHA Consultation
Occupational Health and Safety Bureau
1190 St. Francis Drive, Room N-2200
Santa Fe, New Mexico 87504-0968
(505) 827-2885

New York

Division of Safety and Health
New York State Department of Labor
One Main Street, Room 809
Brooklyn, New York 11201

North Carolina

North Carolina Consultative Services
North Carolina Department of Labor
Shore Memorial Building
214 West Jones Street
Raleigh, North Carolina 27603
(191) 733-3949

North Dakota

Division of Environmental Engineering
North Dakota State Department of Health
1200 Missouri Avenue, Room 304
Bismarck, North Dakota 58502-5520
(701) 224-2348

Ohio

Division of Onsite Consultation
Department of Industrial Relations
2323 West 5th Avenue
P.O. Box 825
Columbus, Ohio 43216
(614) 644-2631

Oklahoma

OSHA Division
Oklahoma Department of Labor
30 North West 7th Street
Oklahoma City, Oklahoma 73102
(405) 239-6823

Oregon

7(c)(1) Consultation Program
Department of Insurance and Finance/APD
Labor and Industries Building
Salem, Oregon 97310
(503) 378-3272

Pennsylvania

Indiana University of Pennsylvania
Safety Sciences Department
205 Uhler Hall

Indiana, Pennsylvania 15705
(Toll-free in state)
(800) 382-1241

Puerto Rico

Occupational Safety and Health Office
Puerto Rico Department of Labor and Human Resources
505 Munoz Rivera Avenue, 21st Floor
Hato Rey, Puerto Rico 00918
(809) 754-2134/2171

Rhode Island

Division of Occupational Health
Rhode Island Department of Health
206 Cannon Building
75 Davis Street
Providence, Rhode Island 02908
(401) 277-2438

South Carolina

7(c)(1) Onsite Consultation Program
Consultation and Monitoring, SC DOL
3600 Forest Drive
P.O. Box 11329
Columbia, South Carolina 29211
(803) 734-9599

South Dakota

S.T.A.T.E. Engineering Extension
Onsite Technical Division
South Dakota State University
P.O. Box 2218
Brookings, South Dakota 57007
(605) 688-4101

Tennessee

OSHA Consultative Services
Tennessee Department of Labor
501 Union Building, 6th Floor
Nashville, Tennessee 37219
(615) 741-7036

Texas

Occupational Safety and Health Division
Texas Department of Health
1100 West 49th Street
Austin, Texas 78756
(512) 458-7254

Utah

Utah Safety and Health
Consultation Service
160 East 300 South, 3rd Floor
Salt Lake City, Utah 84151-0870
(801) 530-6868

Vermont

Division of Occupational Safety and Health
Vermont Department of Labor and Industry
118 State Street
Montpelier, Vermont 05602
(802) 828-2765

Virginia

Virginia Department of Labor and Industry
Voluntary Safety and Health Compliance
2201 West Broad Street
Richmond, Virginia 23220
(804) 367-9980

Virgin Islands

Division of Occupational Safety and Health
Virgin Islands Department of Labor
Lagoon Street
Frederiksted, Virgin Island 00840
(809) 772-1315

Washington

Voluntary Services
Washington Department of Labor and Industries
1011 Plum Street, M/S HC-462
Olympia, Washington 98504
(206) 586-0961

West Virginia

West Virginia Department of Labor
State Capitol, Building 3, Room 319
1800 E. Washington Street
Charleston, West Virginia 25305
(304) 348-7890

Wisconsin

Section of Occupational Health
Wisconsin Department of Health and Human Services
1414 E. Washington Avenue
Room 112

Madison, Wisconsin 53703
(608) 266-8579

Wisconsin Department of Industry, Labor and Human Relations
Bureau of Safety Inspection
1570 East Moreland Blvd.
Waukesha, Wisconsin 53186
(414) 521-5063

Wyoming

Occupational Health and Safety
State of Wyoming
122 West 25th, Herschler Blvd.
Cheyenne, Wyoming 82002
(307) 777-7786

The OSHA National Office

The national office of OSHA also provides technical assistance and publications. Each contact address and phone number are shown below:

Technical/General Information:

> OSHA
> U.S. Department of Labor
> 200 Constitution Avenue, N.W.
> Washington, D.C. 20210
> Telephone: (202) 523-7075

Publications:

> OSHA Publications Office
> Room N-3101
> Third Street & Constitution Avenue, N.W.
> Washington, D.C. 20210
> Telephone: (202) 523-9667

NIOSH

The National Institute for Occupational Safety and Health (NIOSH) publish a large number of materials which are available without cost through the following address:

> NIOSH
> Publications Dissemination
> 4676 Columbia Parkway
> Cincinnati, OH 45226
> Telephone: (513) 533-8287

NIOSH also provides free technical assistance and health hazard evaluations. NIOSH is not a regulatory agency and does not issue fines or citations. The toll-free number for technical assistance and health hazard evaluations is 1-800-356-4674.

INDEX